Finite-element plasticity and metalforı

Finite-element plasticity and metalforming analysis

G.W. ROWE

Professor of Mechanical Engineering
University of Birmingham

C.E.N. STURGESS

Jaguar Professor of Automotive Engineering
University of Birmingham

P. HARTLEY

Senior Lecturer, Department of Manufacturing Engineering
University of Birmingham

I. PILLINGER

Senior Computer Officer, Department of Mechanical Engineering
University of Birmingham

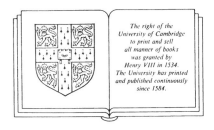

The right of the
University of Cambridge
to print and sell
all manner of books
was granted by
Henry VIII in 1534.
The University has printed
and published continuously
since 1584.

CAMBRIDGE UNIVERSITY PRESS
Cambridge
New York   Port Chester
Melbourne   Sydney

CAMBRIDGE UNIVERSITY PRESS
Cambridge, New York, Melbourne, Madrid, Cape Town, Singapore, São Paulo

Cambridge University Press
The Edinburgh Building, Cambridge CB2 2RU, UK

Published in the United States of America by Cambridge University Press, New York

www.cambridge.org
Information on this title: www.cambridge.org/9780521383622

© Cambridge University Press 1991

First published 1991
This digitally printed first paperback version 2005

A catalogue record for this publication is available from the British Library

ISBN-13 978-0-521-38362-2 hardback
ISBN-10 0-521-38362-5 hardback

ISBN-13 978-0-521-01731-2 paperback
ISBN-10 0-521-01731-9 paperback

Contents

Preface

More and more sectors of the metalforming industry are beginning to recognise the benefits that finite-element analysis of metal-deformation processes can bring in reducing the lead time and development costs associated with the manufacture of new components. Finite-element analyses of non-linear problems, such as metal deformation, require powerful computing facilities and large amounts of computer running time but advances in computer technology and the falling price of hardware are bringing these techniques within the reach of even the most modest R & D department.

Finite-element programs specially designed for the analysis of metalforming processes (such as the program EPFEP3*, which is used for the majority of examples in this monograph) are now commercially available. As a result, there is an urgent need for a text that explains the principles underlying finite-element metalforming analysis to the people who are starting to use these techniques industrially, and to those undertaking the undergraduate and postgraduate engineering courses in the subject that industry is beginning to demand. One of the aims of this monograph is to fulfil that need.

The first chapters provide an introduction to the application of finite-element analysis to metalforming problems, starting from the basic ideas of the finite-element method, and developing these ideas firstly to the study of linear elastic deformation and then to the examination of non-linear elastic–plastic processes involving small amounts of deformation. No previous knowledge of finite elements is required, although the reader is assumed to be familiar with the use of matrices.

Chapter 4 describes a program, written in BASIC for a PC-type desktop computer, that uses the simple, small-deformation formulation. Although restricted to two-dimensional metal flow, low levels of deformation and a small number of elements, this program demonstrates the use of finite-element analysis in metalforming and is particularly useful for tuition purposes. A full listing is given in Appendix 11. Chapter 4 also discusses implementations of the small-deformation theory on more powerful microcomputers that have larger memories and access to a FORTRAN compiler. These programs are capable of more detailed analysis, and can be used for the study of many simple two-dimensional processes.

The application of finite-element techniques to metalforming is still the subject of much research, and the second aim of this monograph is to present the current state of knowledge to those new to the field. Chapter 5 therefore examines in detail the theoretical aspects of a finite-element model of large-deformation plasticity, and Chapter 6 explains how this theory can be implemented in a practical program for the study of complex metalforming problems. Chapter 7 describes a wide variety of practical applications of the finite-element programs.

Although, where appropriate, we mention some of the other finite-element techniques that have been applied to plasticity problems, this monograph is written from the standpoint of an elastic–plastic approach to metal deformation. We are firmly of the opinion that the various simplifications that are sometimes made are a false economy and a serious hindrance to the further development of the finite-element study of metal flow.

The last chapter of this monograph identifies the main areas where we believe that future developments will take place, or at least those areas in which work needs to be done. Unfortunately, the gaps in the present state of knowledge of metalforming processes and the deficiencies in the available models of metal deformation are all too obvious. We hope that this monograph will provide a stimulus to others to attempt to remedy these failings.

G.W.R.
C.E.N.S.
P.H.
I.P.

Acknowledgements

Many people have been associated over the years with the Finite-element Plasticity Group at the University of Birmingham and have contributed, directly or indirectly, to the development of large-deformation elastic–plastic finite-element techniques. We are particularly indebted to Profs. C. Liu and K. Wang, Drs S.E. Clift, A.A.M. Hussin, A.A.K. Al-Sened, A.J. Eames and J. Salimi and Mr K. Kawamura, whose research work has influenced the contents of this monograph. We should also like to thank Mr S.K. Chanda, Profs. S. Ikeda and K. Kato, as well as the above persons, for providing the numerous examples of finite-element metalforming applications that are to be found in the text.

In addition, both undergraduate and postgraduate students have contributed through a variety of short projects. Many of the overseas research students and visiting research fellows have been able to work with us at Birmingham through the generosity of their respective governments and, in some cases, with additional help from the British Council. We are grateful to these bodies for their support.

Much of our research work has been carried out with the financial support of the UK Science and Engineering Research Council, and more recently the ACME Directorate, and with computer facilities provided by the Centre for Computing and Computer Science at the University of Birmingham and the University of Manchester Regional Computer Centre.

Our contacts with industry have been particularly useful to us and we are grateful for the advice and help we have obtained, especially from Alcan Plate, Austin-Rover, Automotive Products, GKN and High Duty Alloys.

We are very grateful to Mr B. Van Bael for checking the manuscript and particularly for wading through the mathematics. It is to his credit if the text emerges free of error. If any mistakes remain, they are entirely our fault, not his.

We were very saddened by the death of Professor Rowe during the preparation of this monograph. Much of the work described in the following pages was begun at his instigation and carried out under his guidance. Indeed, that the monograph came into existence at all is very much due to his energy and his enthusiasm. We owe a great debt to him and feel privileged to have known him and to have worked with him over many years. In completing the book, we have been very conscious of the high standards he set in everything he wrote. We hope that the result does not compare too unfavourably with those standards. We should like to dedicate this work to him.

C.E.N.S
P.H.
I.P.

Nomenclature

Use of subscripts and superscripts

lower-case subscript in italic type:
 usually indicates a Cartesian component of a quantity, e.g. x_i**.** If preceded by a comma, it indicates a derivative with respect to a particular Cartesian co-ordinate, e.g. $u_{i,j} = \partial u_i / \partial x_j$.

lower-case subscript in bold type:
 indicates the quantity associated with a particular element, e.g. $L_{\mathbf{j}}$.

upper-case subscript in italic type:
 indicates the quantity associated with a particular node of an element or body, e.g. N_I.

Greek subscript:
 indicates the quantity associated with a particular degree of freedom of the whole workpiece. In the simple one-dimensional examples considered in the early part of this monograph, a Greek subscript therefore denotes the value of a quantity for a particular node of the workpiece, e.g. d_α.

lower-case superscript in italic type:
 used in the notation for contravariant and mixed tensors, but only in Appendix 8 (vectors and tensors).

superscript in normal type:
 is a label, rather than a numerical index, denoting a particular type of subset of a quantity, e.g. $\triangle T^{\text{0est}}$. Some frequently-used superscripts are:
 T transpose of vector or matrix
 p plastic portion
 $'$ (prime) for tensors – deviatoric component, e.g. σ'_{ij}
  for positional quantities – transformed position, e.g. x'_i

for scalars – derivative, e.g. Y'

Other examples are given in the list of symbols below.

lower-case superscript in parentheses:

indicates the value of a quantity for a particular iteration, e.g. $\triangle d^{(2)}$. The parentheses are used to avoid confusion with an exponent.

Greek superscript:

indicates the quantity associated with a particular face of the element under consideration, e.g. A^α.

Specific values of bold subscripts are printed using bold numerals but when numbers are substituted for italic subscripts normal type is used. No distinction is made between numbers substituted for lower-case subscripts and those substituted for Greek subscripts. The meaning is made clear by the context. The context will also make clear whether, for example, X^2 means the square of the value of X, or the second contravariant component of the vector $\mathbf{X}$.

The different types of subscript and superscript may be used together, e.g. $f_{i\alpha}$ or $q^{(i)}_{ij}$.

Frequently-used qualifying symbols

A dot above a quantity denotes the time derivative, e.g. $\dot{d}_{lm}$.

A bar above a tensor denotes the generalised value, e.g. $\bar{\sigma}$.

Representation of vectors and matrices

A vector is represented by bold type, a matrix by a symbol enclosed in square brackets. Both may also be notated by specifying the symbol (without bold type or brackets) with algebraic subscripts (and possibly superscripts). This may represent a particular element of the vector or matrix, or may be intended to stand for the whole array, with the indices implicitly varying over all their possible values, e.g. $[K] \equiv K_{\alpha\beta}$.

When the elements of a vector are written out explicitly, they are enclosed in parentheses. If they are written as a row vector, the elements are separated by commas. The elements of a matrix are simply enclosed in square brackets, without any commas.

List of symbols

A	area
A^{1234}	area of quadrilateral 1234
A_i	area of element i
A^α	area of face α of element
a	coefficient of general polynomial shape function. Also coefficient

	of 1-D linear displacement function
a_i	coefficient of linear function for displacement in x_i direction
a_{ij}	coefficient of expression determining ith reference co-ordinate of a point as a quadratic function of local curvilinear co-ordinates X^j
da	infinitesimal area at P at time t
da'	infinitesimal area at P' at time $t+dt$
B	coefficient of yield-stress function
$[B]$	matrix that may be used to express the strain at a point within an element in terms of the displacement of its nodes
$[B_i]$	strain/nodal displacement matrix for element i
B_{ijlm}	coefficient of LCR strain increment/nodal displacement increment matrix $(= \frac{1}{2}\,(\bar{B}_{ijlm} + \bar{B}_{jilm}))$
$\bar{B}_{ijlm}$	coefficient of defining matrix for LCR strain increment/nodal displacement increment matrix $(= r_{mi}N_{I,j} - <r_{ik}>(\delta_{jm}N_{I,k} - \delta_{km}N_{I,j})/4)$
b	coefficient of general polynomial shape function. Also coefficient of 1-D linear displacement function
b_i	coefficient of general linear shape function
b_{ij}	coefficient of linear function for displacement in x_i direction. Also coefficient of expression determining ith reference co-ordinate of a point as a quadratic function of local curvilinear co-ordinate X^j
$C^{j+},\ C^{j-}$	coefficients of expression determining rate of flow of heat at element centroid as a function of the difference in temperature between the centroid and the point on the jth local curvilinear axis with local co-ordinate $+1$ or -1
c	coefficient of general polynomial shape function. Also thermal capacity per unit volume
c_i	coefficient of linear combination of functions that form $\hat{\phi}$
$[D]$	6×6 stress/strain constitutive matrix such that $\boldsymbol{\sigma} = [D]\boldsymbol{\epsilon}$
$[D^e]$	elastic part of elastic–plastic constitutive matrix
$[D^p]$	plastic part of elastic–plastic constitutive matrix
$[D_i]$	constitutive matrix for element i
D_{ij}	entry in row i, column j of $[D]$
D_{ijkl}	coefficient of constitutive tensor relating Jaumann rate of stress $\overset{*}{\tau}_{ij}$ to component $\dot{\epsilon}_{kl}$ of stain-rate tensor
d	displacement. Also thickness of friction layer
$\mathbf{d}$	global nodal displacement vector
$\mathbf{d}^*$	approximate solution of equations derived by incomplete Choleski conjugate-gradient method $(= [U^*]\mathbf{d})$
d_α	component α of the nodal displacement vector. For 1-D problems, this is the displacement of node α

d_α^*	component α of nodal displacement vector after modification during Gaussian elimination procedure
$\mathbf{d}_I$	vector of components of displacement of node I
$\mathbf{d}_i$	vector of components of displacement of nodes of element i
$\mathbf{d}_I'$	vector of components of displacement of node I in rotated axis system
d_{Ii}	displacement of node I of element in x_i direction
$\triangle d$	representative displacement increment
$\triangle d^a$	first estimate of displacement increment during secant-modulus technique, evaluated by solving stiffness equations obtained for initial strain ϵ^0 and stress σ^0
$\triangle d^c$	second estimate of displacement increment during secant-modulus technique, evaluated by solving stiffness equations obtained for mid-increment strain ϵ^m and stress σ^m
$\triangle d^{(i)}$	ith correction to the incremental displacement in initial-stiffness iteration
$\triangle d_{Ii}$	change in d_{Ii} during an increment of deformation
$\delta\mathbf{d}$	error in estimate of $\mathbf{d}$
E	Young's Modulus of elasticity
E_i	Young's Modulus for the material of element i
e	small extension of an element
F	functional of state functions in variational method. Also number of faces of an element in contact with a die
f	force
$\mathbf{f}$	global nodal force vector. Also force acting at a point P of a body
$\mathbf{f}'$	force acting at a point P' of a body at time $t+dt$
$\mathbf{f}^*$	right-hand side of equations derived by incomplete Choleski conjugate-gradient method $(= [L^*]^{-1}\mathbf{f})$
$\hat{\mathbf{f}}$	force acting at time $t+dt$ on an infinitesimal plane with area da' and a normal that has *reference* components equal to N_i'
$\mathbf{f}^+$	global force vector corresponding to a global displacement vector of $\mathbf{d}+\delta\mathbf{d}$
f_α	component α of the nodal force vector. For 1-D problems, this is the resultant force at node α
f_α^*	component α of the nodal force vector after modification during the Gaussian elimination procedure
$\hat{f}^i$	contravariant reference component of $\hat{\mathbf{f}}$
$\mathbf{f}_I$	vector of components of force applied to node I
$\mathbf{f}_i$	vector of components of force applied to nodes of element i
$\mathbf{f}_I'$	vector of components of force applied to node I in rotated axis system

f'^i	contravariant reference component of $\mathbf{f}'$
f_{Ii}	component of force in x_i direction applied to node I of element
$f_{i\alpha}$	force applied to element i at node α
$\triangle f$	representative force increment
$\triangle f^{(i)}$	resultant force calculated in ith iteration of initial-stiffness technique
$\triangle f^{(i)*}$	force in equilibrium with $\triangle d^{(i)}$ in initial-stiffness iteration
$\triangle f_{Ii}$	change in f_{Ii} during an increment of deformation
G	Rigidity (shear) Modulus of elasticity $(= E/2(1+\nu)\ )$. Also functional of state functions used in variational method
$\mathbf{G}_i$	basis of 3-D vector space
$\mathbf{G}^i$	basis of 3-D vector space (dual of $\mathbf{G}_i$)
g_i	stiffness coefficient of element i $(= A_i E_i / L_i)$
$g_i(\mathbf{x})$	one of the simple functions of position used to define an approximation to the function of state in variational method
$\mathbf{g}_i$	basis of 3-D vector space
$\mathbf{g}^i$	basis of 3-D vector space (dual of $\mathbf{g}_i$)
H	initial height of billet
h	final height of billet
$h(\mathbf{d})$	scalar function of n variables that has a minimum value for the vector solution of the stiffness equations
h_{ij}	matrix used in the evaluation of rotational values $(= q_{ik} q_{kj})$
I	potential energy; more generally, functional that needs to be minimised in order to determine the correct solution for the function of state in a physical system
I_i	potential energy of element i
J	Jacobian of transformation matrix $(= \det (x'^{i,j})\ )$
K	stiffness
$[K]$	global stiffness matrix
$[K^*]$	matrix derived by incomplete Choleski conjugate-gradient method $(= [L^*]^{-1} [K][U^*]^{-1})$
$[K_i]$	stiffness matrix of element i
$K_{\alpha\beta}$	entry in global stiffness matrix
$K^*_{\alpha\beta}$	entry in global stiffness matrix after modification during Gaussian elimination procedure
$[K_{IJ}]$	2×2 (for 2-D) or 3×3 (for 3-D) submatrix of coefficients of global stiffness matrix relating components of force at node I to components of displacement at node J
K_{ImJn}	coefficient of element stiffness matrix relating component n of incremental displacement of node J to component m of incremental force applied to node I

K^ϵ_{ImJn}	coefficient of element deformation stiffness matrix
K^σ_{ImJn}	coefficient of element stress-increment correction matrix
K^ϕ_{ImJn}	coefficient of element dilatation matrix
k	shear yield stress. Also thermal conductivity
k^{j+}, k^{j-}	coefficients of heat transfer between die and face of element with centre on positive or negative jth local curvilinear axis
L	length
$[L]$	lower-triangular matrix derived from $[K]$ during Choleski decomposition
$[L^*]$	approximation to $[L]$ having a pattern of sparseness similar to that of $[K]$
L_i	length of element i
l^{j+}, l^{j-}	distance between centroid of element and face with centre on positive or negative jth local curvilinear axis
m	friction factor. Also exponent in yield-stress function
m^α	friction factor associated with die in contact with face α of element
$\triangle m$	proportionality factor in modified Prandtl–Reuss equations
N	number of nodes in an FE discretisation. Also number of steps in thermal calculation for an increment of deformation
$N(\mathbf{x})$	general shape function of position
N'_i	covariant component of $\mathbf{n}'$ in the convected co-ordinate system
$N_I(\mathbf{x})$	shape function of position relating to the contribution from node I
$N_{I,i}$	gradient of $N_I (x_j) (= \partial N_I (x_j) /\partial x_i)$
n	number of degrees of freedom in an FE discretisation
$\mathbf{n}$	normal to infinitesimal surface at P at time t. Also normal to boundary surface at P at start of increment
$\mathbf{n}'$	normal to infinitesimal surface at P' at time $t+dt$. Also normal to boundary surface at P' at end of increment
n_i	covariant components of $\mathbf{n}$ in reference co-ordinate system
n'_i	covariant components of $\mathbf{n}'$ in reference co-ordinate system
O	origin of reference co-ordinate system
O'	origin of convected co-ordinate system
P	location of infinitesimal region of body at time t. Also coefficient used in calculating number of thermal steps
P'	location of infinitesimal region of body at time $t+dt$
p	coefficient of general polynomial shape function. Also coefficient of quadratic function of $\triangle m (= 3\hat{\sigma}'_{ij}\hat{\sigma}'_{ij}/2)$
$p(R)$	function of residual R in weighted residual method
p^1, p^2, p^3	planes of constraint of nodes
$p^{(i)}_{ij}$	result of dividing h_{ij} by ith estimate of q_{ij}
q	coefficient of quadratic function of $\triangle m (= -3\hat{\sigma}'_{ij} (\sigma^{0'}_{ij}+\triangle\sigma^{e'}_{ij}))$

q^{ij}	symmetric part of deformation matrix $x'^{i,j}$
q_{ij}	symmetric part of deformation matrix $x'_{i,j}$
$q_{ij}^{(i)}$	ith estimate of q_{ij} using Newton's method of solving equation $q_{ik}q_{kj} = h_{ij}$
δQ^c	increase in energy of element in one step of thermal calculation due to conduction of heat into and out of element
δQ^d	increase in energy of element in one step of thermal calculation due to work of deformation
δQ^f	increase in energy of element in one step of thermal calculation due to frictional heating at boundaries
R	residual functional of state functions in weighted residual method
$[R]$	orthonormal (rotational) matrix relating global to rotated components of a vector
r	order of polynomial interpolation function. Also coefficient of quadratic function of $\triangle m$ $(= 3(\sigma_{ij}^{0\prime} + \triangle \sigma_{ij}^{e\prime})^2 /2)$
$\mathbf{r}$	right-hand side vector of equations obtained by Gauss–Jordan elimination
r_i	coefficient of expression determining temperature of a point as a quadratic function of local curvilinear co-ordinates X^j
r^{ij}	rotational (orthogonal) part of deformation matrix $x'^{i,j}$
r_{ij}	rotational (orthogonal) part of deformation matrix $x'_{i,j}$
S	coefficient in elastic–plastic constitutive relationship $(=3/2\bar{\sigma}^2 [1+(Y'/3G)])$
S^{j+}, S^{j-}	coefficients generalising element heat-flow expression to external faces
s	thickness of an element
$\mathbf{s}$	right-hand side vector of equations obtained by Gaussian elimination
s_i	coefficient of expression determining temperature of a point as a quadratic function of local curvilinear co-ordinates X^j
s^{ij}	contravariant component of nominal or Piola–Kirchhoff I stress
s_{ij}	covariant component of nominal or Piola–Kirchhoff I stress
T	absolute temperature. Also tensile stress acting in single-element example
T^f	temperature at end of increment of deformation
T^i	temperature at start of increment of deformation
T^0, T^1	temperature at beginning and end of time step $\triangle t$. T^0 is also temperature at the centroid of an element
T^{j+}, T^{j-}	temperature at point on jth local curvilinear axis with local co-ordinate $+1$ or -1
$[T]$	diagonal matrix derived from $[K]$ by Gauss–Jordan elimination

$T^{0(i)}$	estimate of temperature of centroid at end of ith thermal step
T^{ij}	contravariant component of a general tensor in co-ordinate system X^i
$\triangle T^0$	change in temperature during an increment of deformation
$\triangle T^{0\text{est}}$	estimated change in temperature during an increment of deformation
t	time
t^{ij}	contravariant component of a general tensor in co-ordinate system x^i
t^0	initial time
δT	change in temperature during one step of thermal calculation
dt	infinitesimal time increment
$\triangle t$	time interval for increment of deformation
U	strain energy
$[U]$	upper-triangular matrix derived from $[K]$ by Gaussian elimination or Choleski decomposition
$[U^*]$	approximation to $[U]$ having a pattern of sparseness similar to that of $[K]$
U_i	strain energy of element i
$u(x)$	linear displacement of point from its initial position x in 1-D example
$\mathbf{u}$	vector of displacement of a point or sometimes a typical vector
u^i	contravariant component of displacement of a point
u_i	covariant component of displacement of a point; for Cartesian co-ordinate system, displacement of a point in x_i direction
u_i	displacement of a point of element i
$u_{i,j}$	gradient of component i of displacement in direction x_j
$\dot{u}^{i,j}$	rate of deformation matrix in terms of contravariant components $(= d(x'^{i,j})/dt = \dot{x}'^{i,j})$
$\dot{u}_{i,j}$	rate of deformation matrix in terms of covariant components $(= d(x'_{i,j})/dt = \dot{x}'_{i,j})$
$\triangle u_{i,j}$	deformation gradient; change in $u_{i,j}$ during an increment of deformation $(= \triangle x'_{i,j})$
V	volume
V^*	volume in local element co-ordinate system
V^i	contravariant component of $\mathbf{v}$ in co-ordinate system X^i
V_i	covariant component of $\mathbf{v}$ in co-ordinate system X^i
V_i	volume of element i
V^{ij}	jth convected component of $\mathbf{v}^i$
V'^{ij}	jth convected component of $\mathbf{v}'^i$
$\mathbf{v}$	typical vector

v^i	contravariant component of $\mathbf{v}$ in co-ordinate system x^i
v_i	covariant component of $\mathbf{v}$ in co-ordinate system x^i
$\mathbf{v}^i$	arbitrary infinitesimal vector at P at time t
$\mathbf{v}'^i$	arbitrary infinitesimal vector $\mathbf{v}^i$ after deformation of P to P'
v^{ij}	jth reference component of $\mathbf{v}^i$
v'^{ij}	jth reference component of $\mathbf{v}'^i$
W	work done by external forces
$W_\mathbf{i}$	work done by external forces on element i
w	multiplication factor in Gaussian elimination procedure
X	coefficient used in calculation of number of thermal steps
X^i	co-ordinate i of a point P in rotated, convected or transformed axis system (specifically, contravariant component with respect to basis $\mathbf{G}_i$)
X_i	co-ordinate i of a point P in rotated, convected or transformed axis system (specifically, covariant component with respect to basis $\mathbf{G}^i$)
X'^i	co-ordinate i of a point P' in rotated, convected or transformed axis system (specifically, contravariant component with respect to basis $\mathbf{G}_i$)
x	linear co-ordinate
x^1, x^2	lower and upper limits of integration in variational expression
$\mathbf{x}$	position vector of a point
x^i	co-ordinate i of a point P in initial reference (usually Cartesian) axis system (specifically, contravariant component with respect to basis $\mathbf{g}_i$)
x_i	co-ordinate i of a point P in initial reference (usually Cartesian) axis system (specifically, covariant component with respect to basis $\mathbf{g}^i$)
x_i^0	ith reference co-ordinate of centroid of an element
x_i^{j+}, x_i^{j-}	ith reference co-ordinate of point on jth local curvilinear axis with local co-ordinate $+1$ or -1
x'^i	co-ordinate i of point P' in initial reference axis system (specifically, contravariant component with respect to basis $\mathbf{g}_i$)
x'_i	co-ordinate i of point P' in initial reference axis system (specifically, covariant component with respect to basis $\mathbf{g}^i$)
x_{Ii}	co-ordinate i of node I of an element
$x'^{i,j}$	contravariant co-ordinate transformation matrix defining deformation of infinitesimal region at point P at time t into infinitesimal region at point P' at time $t+\mathrm{d}t$ $(= \partial x'^i / \partial x^j)$
$x'_{i,j}$	covariant co-ordinate transformation matrix defining deformation of infinitesimal region at point P at time t into infinitesimal region

	at point P' at time $t+dt$ $(= \partial x'_i / \partial x_j)$
$x^{*i,j}$	transformation matrix defining a rigid-body rotation; rotational part of $x'^{i,j}$ $(= r^{ij})$
$\triangle x'_{i,j}$	deformation gradient; change in $x'_{i,j}$ during an increment of deformation $(= \triangle u_{i,j})$
Y	yield stress in simple tension, a function of strain, strain rate and temperature
Y'	rate of change of Y with respect to plastic strain
Z	coefficient used in calculation of number of thermal steps
α	angle. Also proportion of work of deformation converted into heat
β	angle
γ_{ij}	engineering shear strain in $x_i x_j$ plane $(= 2\epsilon_{ij}, i \neq j)$
$\triangle$	change in the value of a quantity during an increment of deformation
δ	increment of a quantity
$\delta( )$	arbitrary variation of function enclosed in parentheses
δ_{ij}	Kronecker delta $(=1, i=j; =0, i \neq j)$. Also δ^i_j, δ^{ij} etc
ϵ	strain
ϵ^{AB}	normal strain in AB direction
ϵ^0	representative strain at start of increment
ϵ^c	representative strain at end of increment calculated by secant-modulus technique $(= \epsilon^0 + \triangle \epsilon^c)$
ϵ^m	representative strain during an increment calculated by secant-modulus technique $(= \epsilon^0 + \frac{1}{2}\triangle \epsilon^a)$
$\bar{\epsilon}^P$	accumulated generalised plastic strain $(= \int d\bar{\epsilon}^P)$
$\bar{\epsilon}^{0p}$	plastic strain at start of plastic part of increment of deformation
$\dot{\bar{\epsilon}}^P$	rate of change of plastic strain with respect to time
$\boldsymbol{\epsilon}$	vector of 3 normal components of strain and 3 engineering shear components of strain
ϵ_i	principal component of strain
$\epsilon_{\mathbf{i}}$	strain in element i
ϵ_{ij}	normal component of strain in x_i direction if $i=j$; shear component of strain in $x_i x_j$ plane if $i \neq j$
$\dot{\epsilon}^{ij}$	contravariant component of strain-rate tensor $(= \frac{1}{2}(\dot{u}^{i,j} + \dot{u}^{j,i}))$
$\dot{\epsilon}_{ij}$	covariant component of strain-rate tensor, equivalent to $\dot{\epsilon}^{ij}$ for Cartesian (orthogonal) co-ordinate system $(= \frac{1}{2}(\dot{u}_{i,j} + \dot{u}_{j,i}))$
ϵ'_{ij}	deviatoric component of strain $(= \epsilon_{ij} - \delta_{ij}\epsilon_{kk}/3)$
$d\bar{\epsilon}^P$	increment in generalised plastic strain $(= (2d\epsilon^{p'}_{ij}d\epsilon^{p'}_{ij}/3)^{1/2})$
$d\epsilon^e_{ij}$	elastic part of incremental strain component
$d\epsilon^p_{ij}$	plastic part of incremental strain component

$\triangle\epsilon^a$	strain increment corresponding to a displacement of $\triangle d^a$, the first estimate of displacement, in secant-modulus technique
$\triangle\epsilon^c$	strain increment corresponding to a corrected displacement of $\triangle d^c$, in secant-modulus technique
$\triangle\bar{\epsilon}^p$	change in plastic strain during an increment of deformation
$\triangle\epsilon^{(i)}$	strain increment calculated from $\triangle d^{(i)}$ in initial-stiffness iteration
$\triangle\epsilon_{ij}$	component of LCR increment of strain (specifically, change in strain during plastic part of an increment)
η	coefficient of viscosity
κ	Bulk Modulus of elasticity ($= E/3(1-2\nu)$)
$d\lambda$	proportionality factor in Prandtl–Reuss elastic–plastic flow equations
$\triangle\lambda$	proportionality factor in incremental form of Prandtl–Reuss elastic–plastic flow equations
ν	Poisson's Ratio
σ	Cauchy or True stress
σ^{AB}	normal stress acting in AB direction
σ^0	representative stress at start of increment
σ^c	representative stress at end of increment corresponding to a strain of ϵ^c
σ^h	hydrostatic component of stress ($= \sigma_{kk}/3$)
σ^m	representative stress during an increment corresponding to a strain of ϵ^m
$\bar{\sigma}$	generalised stress ($= (3\sigma'_{ij}\sigma'_{ij}/2)^{1/2}$)
$\boldsymbol{\sigma}$	vector of 6 unique components of stress (since $\sigma_{ij} = \sigma_{ji}$)
σ_i	principal component of stress
σ_{i}	stress in element i
σ_{ij}	normal component of stress in x_i direction if $i=j$; shear stress resulting from force in x_j direction acting on surface with normal in x_i direction if $i\neq j$
σ'_{ij}	deviatoric component of stress ($= \sigma_{ij} - \delta_{ij}\sigma^h$)
$\sigma^{0'}_{ij}$	deviatoric stress at beginning of plastic part of deformation increment
$\hat{\sigma}'_{ij}$	deviatoric stress half-way through hypothetical elastic step of deformation increment ($= \sigma^{0'}_{ij} + \tfrac{1}{2}\triangle\sigma^{e'}_{ij}$)
$\tilde{\sigma}'_{ij}$	deviatoric stress half-way through plastic part of step ($= \sigma^{0'}_{ij} + \tfrac{1}{2}\triangle\sigma'_{ij}$)
$\triangle\sigma^{(i)}$	stress increment calculated from $\triangle\epsilon^{(i)}$ in initial-stiffness iteration
$\triangle\sigma'_{ij}$	change in deviatoric stress during plastic part of deformation increment ($= \triangle\sigma^{e'}_{ij} - 2G\triangle\lambda\tilde{\sigma}'_{ij}$)
$\triangle\sigma^{e'}_{ij}$	hypothetical elastic change in stress; change in stress that would

	result if plastic part of deformation increment took place elastically $(= 2G\triangle\epsilon_{ij})$
τ	maximum permitted error in calculated incremental temperature change
τ^{ij}	contravariant component of Kirchhoff or Piola–Kirchhoff II stress
$\dot{\tau}^{ij}$	time derivative of τ^{ij}
$\tau^{\circ ij}$	rotationally-invariant Kirchhoff stress; contravariant component of Kirchhoff stress at a point in a co-ordinate system that rotates with the deforming infinitesimal region at that point
$\dot{\tau}^{\circ ij}$	contravariant component of Jaumann rate of Kirchhoff stress
$\dot{\tau}^{\circ}_{ij}$	covariant component of Jaumann rate of Kirchhoff stress
$\phi(\mathbf{x})$	function of state in a physical system to be determined by variational method
ϕ', ϕ''	first and second derivatives of ϕ with respect to position
$\hat{\phi}(\mathbf{x})$	approximation to ϕ in the form of linear combination of simple functions
$\triangle\phi(x_i)$	dilatation increment function of position
$\triangle\phi_i$	coefficient of linear dilatation function
$<>$	skew-symmetric part of enclosed matrix : $<[A]> = ([A]-[A]^{\mathrm{T}})/2$
$\mid\mid$	magnitude of enclosed vector
$\{\ \}$	set of objects
$\cdot$	scalar product
$*$	vector product
$\det(\ )$	determinant of enclosed matrix

1 General introduction to the finite-element method

1.1 INTRODUCTION

Metalforming is an ancient art and was the subject of closely-guarded secrets in antiquity. In many respects the old craft traditions have been retained until the present time, even incorporating empirical rules and practices in automated production lines. Such techniques have been successful when applied with skill, and when finely adjusted for specific purposes. Unfortunately, serious problems arise in commissioning a new production line or when a change is made from one well-known material to another whose characteristics are less familiar. The current trend towards adaptive computer control and flexible manufacturing systems calls for more precise definition and understanding of the processes, while at the same time offering the possibility of much better control of product dimensions and quality. This is especially important in the advanced aerospace and other industries, where the advantages of new materials have to be fully exploited as quickly as possible.

Alloys that have been specifically designed to withstand high temperatures provide one example where the tools are stressed to their limits and the alloys themselves may be deformable only within narrow ranges of temperature and strain-rate. Practical tests to determine the best tool shapes and forming conditions can be very expensive and wasteful of scarce tool and workpiece material. A single turbine blade may cost several thousand pounds, so any reduction in prototype manufacturing costs is clearly valuable. Even the common forgings used in large numbers for automobiles may be subject to cracking during production, or be weakened by high residual stresses, if the preforms are incorrectly designed.

It is therefore important to provide guidance from theoretical models which can easily be modified. Until quite recently the only available theories of metalforming operations were based on simple equilibrium, ignoring internal distortion, or else made gross assumptions about the properties of the materials [1.1]. Both slip-line field theory and upper-bound theory, in the latter category, have been advanced by the introduction of computer methods, but there is no doubt that the greatest improvement in detail and accuracy has come from the application of finite-element (FE) analysis. Originally the FE method of stress analysis was restricted to linear elastic deformation, in which form it has many uses in structural design. Many books have been written on this subject [1.2, 1.3, 1.4] but the main concern of this monograph will be with plastic deformation. It is now possible to take into account the practical non-linearities in shape and material properties, and to produce accurate predictions of stress, strain, strain-rate and temperature distributions throughout a plastically-deforming workpiece. The analyses can even be used to determine whether the alloy is likely to fail by tensile cracking or catastrophic shear during processing. Information about the surface pressures, tractions and overall load is also made available and can be used as input conditions for elastic FE analysis of tool designs. The results of such comprehensive analysis provide a sound basis for development of control strategy in metalforming.

A complete processing route, including the effects of interstage annealing, can thus be examined theoretically to determine optimum preform shapes from the point of view of product quality or economic manufacture. The information provided is fully detailed and can be incorporated in Computer-Aided Design/Computer-Aided Manufacturing (CAD/CAM) sequences for tool production, or as a basis for analyses of fatigue resistance or fracture liability during normal use of the product.

This monograph develops the analytical procedures from an elementary starting point, needing no prior knowledge of FE methods. The basis of the programs is fully explained, without requiring the reader actually to write programs personally unless so desired, and examples of their use in practical metalforming situations are given.

Most FE analysis up to the present date has required very large main-frame computers, but the rapid advances in microcomputers during the last decade are changing the pattern, especially in civil engineering [1.5]. Chapter 4 shows that, at the expense of greater computing time but not necessarily of overall turn-round time, FE analyses for both elastic and plastic problems can be carried out on the latest large-memory micros. There is a very great saving in capital cost, and a further industrial advantage of immediate in-house availability of the results. As in all computer packages however, there is danger in using the programs without a background understanding of the principles involved and the limitations imposed by the assumptions and structure of the analysis.

It is intended that this monograph will provide a basis for confidence in using FE plasticity analysis in industrial metalforming, as well as being a self-contained text-book for specialist graduate or undergraduate courses in forging, extrusion, rolling and related processes. It also provides a background for more general non-linear FE analysis, as applied, for example, to polymeric materials.

1.2   EARLIER THEORETICAL METHODS FOR METALFORMING ANALYSIS

The earliest theoretical models were based on the concept that the energy expended in forging could be attributed to the internal deformation, including both shape change and internal distortion, and the external resistance at the tool faces [1.6]. Empirical formulae were developed containing terms dependent upon the flow stress of the alloy and upon the coefficient of friction. These allowed fair predictions of overall forces to be made.

Further detail of pressure distributions can be obtained by analysing the local stress equilibrium. This also predicts the forces, and the results are helpful when designing tools and choosing tool materials. Although the method is intended for simple shapes, useful approximations can often be made for real problems, especially in rolling [1.7].

Slip-line theory gives much more detailed information about the flow patterns, as well as the stress distributions and overall forces. This technique has been applied to forging, forward and backward extrusion, drawing and other processes, but it makes very restrictive assumptions about the material properties [1.8]. Only the average yield stress is included as a variable, and elastic deformation is ignored. The rigorous theory is, moreover, confined to plane-strain deformation. Originally the method was very slow, requiring manual calculation and drawings, but it can now be used on a computer [1.9].

The only serious competitors to the FE Method (FEM) among the currently available theories are Finite-Difference Methods, Boundary-Element Methods and the Upper-Bound (UB) technique [1.10]. In its computerised form, the latter offers a rapid and relatively simple way of calculating the major force requirements [1.11]. It does however require experience in choosing the appropriate basic units of the shear field. Computer packages are now available for well-established configurations or assemblies of proven elements, applicable to plane strain or axial symmetry. More elaborate versions are becoming available [1.12], some of which include temperature and strain-rate effects, but all suffer from the necessity to make assumptions about the internal deformation patterns. They are consequently unlikely to be reliable for prediction of strain and strain-rate distributions in unknown configurations, and it is difficult to obtain higher accuracy. UB methods ignore elastic strains and stresses.

Finite-Difference methods have proved useful in solving thermal problems [1.13] and have some proponents for structural analysis [1.14]. These will be

discussed briefly in Chapter 2. Boundary-Element methods have been suggested
for small plastic deformation [1.15].

1.3   BASIC FINITE-ELEMENT APPROACH

The use of finite elements for stress analysis of aircraft structures was proposed
in 1956, using pin-jointed bars and triangular plates as the elements [1.16], though
the fundamental concept was used by ancient geometricians in approximating
to a circle by a polygon. The accuracy can, in theory, be made as high as is
desired or as time permits, and upper and lower bounds can be established. The
circle, for example, must lie between the inscribed and circumscribed polygons
(figure 1.1).

In modern usage, a limitation is imposed only by the size of the computer
memory and the computational time required.

As applied to structural mechanics, the fundamental concept of FE formulations is the stiffness K of each element. This relates the force applied to the
displacement or deformation produced. Thus, for example, in a simple tensile
test a force f produces a stress σ in a bar of cross-section A, and an increase δL
in the length L. Since the strain $\epsilon = \delta L / L$ is directly related to the stress by the
Young's Modulus E in an elastic body:

$$f = A\sigma = AE\epsilon = AE\frac{\delta L}{L} = K \cdot \delta L \qquad (1.1)$$

The stiffness can also be deduced from an energy equation using the principle
of minimum potential energy or some related variational principle. Because
there is only one variable in the simple tension example, this is equivalent to

Fig. 1.1 A circle and the two polygons forming upper and lower bounds.

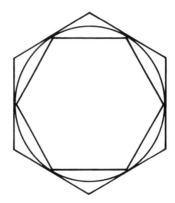

requiring that W, the work done by the external force, and the strain energy U in a specimen of volume V are related by:

$$\frac{\partial U}{\partial(\delta L)} = \frac{\partial W}{\partial(\delta L)} = \frac{\partial}{\partial(\delta L)}\left(f \cdot \delta L\right) = f \tag{1.2}$$

For an elastic body in simple tension:

$$U = \frac{ALE}{2}\epsilon^2 = \frac{AE}{2L}(\delta L)^2 \tag{1.3}$$

so:

$$f = \frac{\partial}{\partial(\delta L)}\left(\frac{AE}{2L}(\delta L)^2\right) = AE\frac{\delta L}{L} = K \cdot \delta L \tag{1.4}$$

as before. Note that:

$$U = \tfrac{1}{2}K(\delta L)^2 \tag{1.5}$$

Analogous quadratic expressions for strain energy may be derived for all deformation problems. In two- and three-dimensional (2-D and 3-D) elastic problems the Poisson's Ratio v is also involved, as seen later (Section 2.5.1), and in plastic deformation the relationship is not a linear one.

1.4   GENERAL PROCEDURE FOR STRUCTURAL FINITE-ELEMENT ANALYSIS

The structure, which may be a workpiece or other continuum, is first divided into an assemblage of subdivisions or elements, all interconnected at joints or nodes. For example, a beam might be represented by a number of triangles, as in figure 1.2, though in practice many other element configurations are possible [1.2]. See also Section 2.1.7.

Fig. 1.2 A beam represented by finite elements.

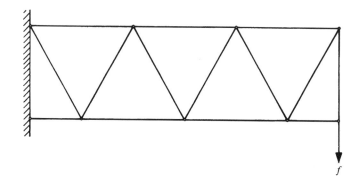

This representation can be considered as a framework, or more conveniently for present purposes and the subsequent discussion of plasticity, as a set of rigid triangular plates. In a further elaboration it could be assumed that each plate was itself capable of deforming uniformly or in some other specified way. The selected assumption defining the field variable throughout the continuum is described as an *interpolation function*.

The equilibrium equations for each of the elements are written down individually, containing the nodal values as the unknown quantities, and the set of simultaneous equations is then solved in matrix form. Once these nodal quantities are known, the interpolation function will provide the complete solution.

It is important that an orderly step-by-step procedure should be followed. For a static structure, the sequence of operations is given below. (Note that in the simple one-dimensional examples given in this chapter in which there is only one degree of freedom per node, a symbol with a Greek subscript represents a value at a node; the notation for use in 2-D and 3-D cases will be given later. A symbol with a lower-case subscript in bold type always represents an element quantity.)

(i)   Discretisation of the structure
(ii)   Selection of the interpolation model
(iii) Derivation of the stiffness matrix $[K_i]$ for the equations relating the load or force vectors $\mathbf{f_i}$ to the displacement vectors $\mathbf{d_i}$ for each element:

$$\mathbf{f_i} = [K_i]\mathbf{d_i} \tag{1.6}$$

(iv)   Assembly of all the element equations to give equilibrium equations for the whole body:

$$\mathbf{f} = [K]\mathbf{d} \tag{1.7}$$

where $\mathbf{d}$ is the global nodal displacement vector and $\mathbf{f}$ is the global load vector

(v)   Modification of the equations as necessary to incorporate boundary conditions, sometimes carried out before the assembly (iv)

(vi)   Solution of the set of matrix equations. In a linear elastic problem the displacement vector can be found in one step from the applied force vector. Plastic deformation is non-linear and requires incremental or iterative solutions

(vii) Computation of the stresses and strains in each element from the nodal displacements, using the interpolation function as necessary

This sequence of procedures can be exemplified by reference to a simple tensile test bar.

1.5 SIMPLE APPLICATION OF ELASTIC FE ANALYSIS: A TENSILE TEST BAR

1.5.1 Discretisation

As the first example, ignore the detailed shape and assume two steps (figure 1.3).

There are three elements, of which two require solutions. The displacements d_1, d_2, d_3, d_4 are unknown.

1.5.2 Interpolation

Assume a linear variation of axial displacement within each element:

$$u(x) = a + bx \tag{1.8}$$

where a and b are constants and x is the distance from the left-hand end of the element.

Consider element 1, of length L_1:

$$d_1 = a + b \cdot 0 = a$$
$$d_2 = a + b \cdot L_1; \; b = (d_2 - d_1)/L_1 \tag{1.9}$$

So:

$$u_1 = d_1 + (d_2 - d_1) \, x/L_1 \tag{1.10}$$

1.5.3 Stiffness matrices of the elements

To determine the stiffness of an element, the Hamiltonian principle, commonly known as the principle of minimum energy, is used. For a static system this states that the equilibrium configuration will be one in which the potential energy has a stationary value, usually a minimum. This principle can be applied to the body as a whole, or, as here, to the elements individually.

As applied to elastic deformation, the potential energy is the difference

Fig. 1.3 A tensile bar, regarded as three elements.

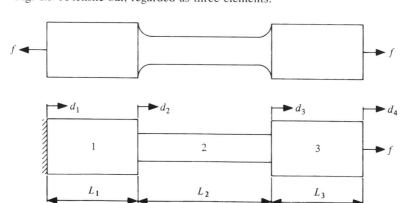

tween the total work expended by the relevant external forces and the strain energy stored within the specimen. For pure elastic strain, the latter is recovered in entirety on release of the forces, though in some materials there is appreciable elastic hysteresis loss.

Thus for element i, the potential energy:

$$I_i = \text{Strain energy of element} - \text{Work done by external forces on element}$$
$$= U_i - W_i \tag{1.11}$$

where, for example, for element 1:

$$U_1 = \int \frac{\sigma_1}{2}\epsilon_1 dV_1 = A_1 \int_0^{L_1} \frac{\sigma_1 \epsilon_1}{2} dx = \frac{A_1 E_1}{2} \int_0^{L_1} \epsilon_1^2 dx \tag{1.12}$$

But:

$$\epsilon_1 = \frac{\partial u_1}{\partial x} = \frac{d_2 - d_1}{L_1} \tag{1.13}$$

$$U_1 = \frac{A_1 E_1}{2} \int_0^{L_1} \frac{d_2^2 + d_1^2 - 2d_1 d_2}{L_1^2} dx \tag{1.14}$$

Thus:

$$U_1 = \frac{A_1 E_1}{2L_1}(d_1^2 + d_2^2 - 2d_1 d_2) \tag{1.15}$$

This can also be expressed in matrix form using standard matrix algebra with the displacement vectors for element 1:

$$\mathbf{d}_1 = \begin{pmatrix} d_1 \\ d_2 \end{pmatrix} \quad ; \mathbf{d}_1^T = (d_1, d_2) \tag{1.16}$$

Now:

$$\mathbf{d}_1^T \begin{bmatrix} 1 & -1 \\ -1 & 1 \end{bmatrix} = (d_1, d_2) \begin{bmatrix} 1 & -1 \\ -1 & 1 \end{bmatrix} \tag{1.17}$$
$$= (d_1 - d_2, -d_1 + d_2)$$

which can be directly multiplied into the vector $\mathbf{d}_1$ to give:

$$(d_1 - d_2, -d_1 + d_2) \begin{pmatrix} d_1 \\ d_2 \end{pmatrix} = d_1^2 - d_2 d_1 - d_1 d_2 + d_2^2 \tag{1.18}$$

Thus from equation 1.15:

$$U_1 = \frac{A_1 E_1}{2L_1} \mathbf{d}_1^T \begin{bmatrix} 1 & -1 \\ -1 & 1 \end{bmatrix} \mathbf{d}_1 \qquad (1.19)$$

$$U_1 = \tfrac{1}{2} \mathbf{d}_1^T [K_1] \mathbf{d}_1 \qquad (1.20)$$

so, by analogy with equation 1.5, the stiffness matrix $[K_i]$ for the ith element is:

$$[K_i] = \frac{A_i E_i}{L_i} \begin{bmatrix} 1 & -1 \\ -1 & 1 \end{bmatrix}$$

$$= g_i \begin{bmatrix} 1 & -1 \\ -1 & 1 \end{bmatrix} \qquad (1.21)$$

where the stiffness coefficient for element i:

$$g_i = \frac{A_i E_i}{L_i} \qquad (1.22)$$

Returning now to equation 1.11, the work done by the forces acting on the element i is:

$$W_i = \mathbf{d}_i^T \mathbf{f}_i \qquad (1.23)$$

where, for example, $\mathbf{f}_1 = (f_{11}, f_{12})^T$ and $f_{i\alpha}$ is the force acting upon element i at node α.

According to the principle of minimum potential energy, for element i:

$$\frac{\partial I_i}{\partial d_\alpha} = \frac{\partial}{\partial d_\alpha}(U_i - W_i) = 0, \; \alpha = 1,2,3,4 \qquad (1.24)$$

so the element stiffness equations are:

$$0 = [K_1]\mathbf{d}_1 - \mathbf{f}_1 \qquad (1.25a)$$

$$0 = [K_2]\mathbf{d}_2 - \mathbf{f}_2 \qquad (1.25b)$$

$$0 = [K_3]\mathbf{d}_3 - \mathbf{f}_3 \qquad (1.25c)$$

Using equation 1.21, these equations may be written in terms of the global displacement vector $\mathbf{d} = (d_1, d_2, d_3, d_4)^T$:

$$\begin{bmatrix} g_1 & -g_1 & 0 & 0 \\ -g_1 & g_1 & 0 & 0 \\ 0 & 0 & 0 & 0 \\ 0 & 0 & 0 & 0 \end{bmatrix} \begin{pmatrix} d_1 \\ d_2 \\ d_3 \\ d_4 \end{pmatrix} = \begin{pmatrix} f_{11} \\ f_{12} \\ 0 \\ 0 \end{pmatrix} \qquad (1.26a)$$

$$\begin{bmatrix} 0 & 0 & 0 & 0 \\ 0 & g_2 & -g_2 & 0 \\ 0 & -g_2 & g_2 & 0 \\ 0 & 0 & 0 & 0 \end{bmatrix} \begin{Bmatrix} d_1 \\ d_2 \\ d_3 \\ d_4 \end{Bmatrix} = \begin{Bmatrix} 0 \\ f_{22} \\ f_{23} \\ 0 \end{Bmatrix} \qquad (1.26b)$$

$$\begin{bmatrix} 0 & 0 & 0 & 0 \\ 0 & 0 & 0 & 0 \\ 0 & 0 & g_3 & -g_3 \\ 0 & 0 & -g_3 & g_3 \end{bmatrix} \begin{Bmatrix} d_1 \\ d_2 \\ d_3 \\ d_4 \end{Bmatrix} = \begin{Bmatrix} 0 \\ 0 \\ f_{33} \\ f_{34} \end{Bmatrix} \qquad (1.26c)$$

1.5.4   Assembly of element stiffness matrices

Combining equations 1.26:

$$\begin{bmatrix} g_1 & -g_1 & 0 & 0 \\ -g_1 & g_1+g_2 & -g_2 & 0 \\ 0 & -g_2 & g_2+g_3 & -g_3 \\ 0 & 0 & -g_3 & g_3 \end{bmatrix} \begin{Bmatrix} d_1 \\ d_2 \\ d_3 \\ d_4 \end{Bmatrix} = \begin{Bmatrix} f_{11} \\ f_{12}+f_{22} \\ f_{23}+f_{33} \\ f_{34} \end{Bmatrix} \qquad (1.27)$$

But, for example, $f_{12} + f_{22}$ is the *total* force acting at node 2, so the right-hand side of equation 1.27 is simply **f**, the global force vector $(f_1, f_2, f_3, f_4)^{\text{T}}$. Hence equation 1.27 may be expressed in terms of the global stiffness matrix $[K]$:

$$[K]\mathbf{d} = \mathbf{f} \qquad (1.28)$$

1.5.5   Boundary conditions

In this example the boundary conditions are very simple and have already been specified in figure 1.3. The element 1 is fixed at the left-hand side and the load f is applied to the end of element 3.

1.5.6   Numerical solution for the displacements

Let:

$$A_1 = A_3 = 200 \text{ mm}^2; \quad A_2 = 100 \text{ mm}^2$$

$$L_1 = L_3 = 50 \text{ mm}; \quad L_2 = 100 \text{ mm}$$

$$E_1 = E_2 = E_3 = 2 \times 10^5 \text{ N/mm}^2$$

$$f = 100 \text{ N}$$

Then:

$$g_1 = \frac{200 \times 2 \times 10^5}{50}$$

$$g_2 = \frac{100 \times 2 \times 10^5}{100} \tag{1.29}$$

$$g_3 = \frac{200 \times 2 \times 10^5}{50}$$

$$g_1 = g_3 = 8 \times 10^5; \; g_2 = 2 \times 10^5$$

Hence from equation 1.27:

$$[K] = 2 \times 10^5 \begin{bmatrix} 4 & -4 & 0 & 0 \\ -4 & 4+1 & -1 & 0 \\ 0 & -1 & 1+4 & -4 \\ 0 & 0 & -4 & 4 \end{bmatrix} \tag{1.30}$$

and so:

$$\mathbf{f} = \begin{pmatrix} f_1 \\ f_2 \\ f_3 \\ f_4 \end{pmatrix} = 2 \times 10^5 \begin{bmatrix} 4 & -4 & 0 & 0 \\ -4 & 5 & -1 & 0 \\ 0 & -1 & 5 & -4 \\ 0 & 0 & -4 & 4 \end{bmatrix} \begin{pmatrix} d_1 \\ d_2 \\ d_3 \\ d_4 \end{pmatrix} \tag{1.31}$$

The boundary conditions must now be included.

Assuming that the bar is anchored at the left-hand end, $d_1 = 0$. The force f_1 is a reaction force and thus does not need to be considered in the equation; the row and column (1) of $[K]$ can thus be eliminated. Furthermore, f_2 and f_3 are zero so:

$$\begin{pmatrix} 0 \\ 0 \\ 100 \end{pmatrix} = 2 \times 10^5 \begin{bmatrix} 5 & -1 & 0 \\ -1 & 5 & -4 \\ 0 & -4 & 4 \end{bmatrix} \begin{pmatrix} d_2 \\ d_3 \\ d_4 \end{pmatrix} \tag{1.32}$$

Writing the equations in full and solving:

$$5d_2 - d_3 + 0 = 0 \qquad \rightarrow d_3 = 5d_2$$

$$-d_2 + 5d_3 - 4d_4 = 0 \qquad \rightarrow d_4 = 6d_2$$

$$0 - 4d_3 + 4d_4 = 50 \times 10^{-5} \rightarrow d_2 = 1.25 \times 10^{-4} \tag{1.33}$$

$$\rightarrow d_3 = 6.25 \times 10^{-4}$$

$$\rightarrow d^4 = 7.5 \times 10^{-4}$$

1.5.7   Strains and stresses in the elements

$$\epsilon_1 = \frac{\partial u_1}{\partial x} = \frac{d_2-d_1}{L_1} = \frac{1.25 \times 10^{-4}}{50} = 2.5 \times 10^{-6}$$

$$\epsilon_2 = \frac{\partial u_2}{\partial x} = \frac{d_3-d_2}{L_2} = \frac{5 \times 10^{-4}}{100} = 5 \times 10^{-6} \tag{1.34}$$

$$\epsilon_3 = \frac{\partial u_3}{\partial x} = \frac{d_4-d_3}{L_3} = \frac{1.25 \times 10^{-4}}{50} = 2.5 \times 10^{-6}$$

$$\sigma_1 = E_1\epsilon_1 = 2 \times 10^5 \times 2.5 \times 10^{-6} = 0.5\,\text{N/mm}^2$$

$$\sigma_2 = E_2\epsilon_2 = 2 \times 10^5 \times 5 \times 10^{-6} \quad = 1.0\,\text{N/mm}^2 \tag{1.35}$$

$$\sigma_3 = \sigma_1$$

These are, of course, trivial solutions that can be obtained by inspection, so this lengthy procedure would certainly not be used for so simple a problem. It is however an example that illustrates the way in which problems are solved by the FE method.

In the next chapter, we shall consider in greater detail the application of the FE method to small-strain elastic deformation and in particular the structure of various FE matrices, solution methods and other numerical techniques.

References

[1.1]   Alexander, J.M., Brewer, R.C. and Rowe, G.W. *Manufacturing Properties of Materials*, Ellis Horwood (1987).

[1.2]   Zienkiewicz, O.C. *The Finite-element Method*, McGraw-Hill, 3rd Edn, (1977).

[1.3]   Irons, B. and Ahmad, S. *Techniques of Finite-elements*, Ellis Horwood (1980).

[1.4]   Livesley, R. *Finite-elements : An Introduction for Engineers*, Cambridge University Press (1984).

[1.5]   Various, *International Conference on Education, Practice and Promotion of Computational Methods in Engineering using Small Computers*, Int. Assoc. Computational Mechanics, Macao (1985).

[1.6]   Nadai, A. *Plasticity, a Mechanics of the Plastic State of Matter*, McGraw-Hill (1931).

[1.7]   Hoffmann, O. and Sachs, G. *Introduction to the Theory of Plasticity for Engineers*, McGraw-Hill (1953).

[1.8]   Johnson, W. and Mellor, P.B. *Engineering Plasticity*, von Nostrand, Reinhold (1973).

[1.9]   Rowe, G.W. *Principles of Industrial Metalworking Processes*, Arnold (1977).

[1.10] Johnson, W. and Kudo, H. *The Mechanics of Metal Extrusion*, Manchester University Press (1962).

[1.11] Avitzur, B. *Metalforming Processes and Analysis, McGraw-Hill* (1968).

[1.12] McDermott, R.P. and Bramley, A.N. Forging analysis – a new approach. *Proc. 2nd Nth. American Metalworking Res. Conf.*, SME, pp. 35–47 (1974).

[1.13] Marti, J., Kalsi, G. and Atkins, A.G. A numerical and experimental study of deep elastoplastic indentation. *Proc 1st Int. Conf. on Numerical Methods in Industrial Forming Processes*, ed. J.F.T. Pittman, R.D. Wood, J.M. Alexander and O.C. Zienkiewicz, Pineridge Press, pp. 279–87 (1982).

[1.14] de Arantes e Oliviera, E.R. On the accuracy of finite difference solutions to the biharmonic equation. *International Conference on Education, Practice and Promotion of Computational Methods in Engineering using Small Computers*, Int. Assoc. Computational Mechanics, Macao **IA**, pp. 121–9 (1985).

[1.15] Brebbia, C.A. *Boundary Element Methods in Engineering*, Halsted Press (1978).

[1.16] Turner, M.J., Clough, R.W., Martin, H.C. and Topp, L.J. Stiffness and deflection analysis of complex structures. *J. Aero. Sci.* **23**, 805–24 (1956).

2 Basic formulation for elastic deformation

2.1 TYPES OF ELEMENTS

Discretisation is a fundamental feature of all FE analysis of a continuum [2.1]. In principle many types of element could be used but in practice a small number of types cover almost all metalforming requirements. These are shown in figure 2.1.

Fig. 2.1 A selection of various linear and higher-order elements used in finite-element elastic and plastic analysis.

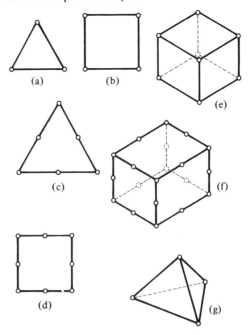

2.1.1   Linear elements

We have implicitly used this very simple type in the tensile bar example (figure 2.2).

In general terms, the value of a force applied to a particular node of the isolated element will affect the displacement of all the nodes to an extent determined by the stiffness K:

$$\mathbf{f} = [K]\mathbf{d} \qquad (2.1)$$

or explicitly:

$$\begin{pmatrix} f_1 \\ f_2 \end{pmatrix} = \begin{bmatrix} K_{11} & K_{12} \\ K_{21} & K_{22} \end{bmatrix} \begin{pmatrix} d_1 \\ d_2 \end{pmatrix} \qquad (2.2)$$

in which, for the sake of simplicity, it is assumed that there is only one element in the body.

The *stiffness influence coefficient* $K_{\alpha\beta}$ is defined as the force needed at node α to produce a unit displacement at node β, while all other nodes are restrained.

If for example, node 1 is restrained and a unit displacement is assumed at node 2, as in figure 2.3, the necessary force induced at node 2 will be found from:

$$f = A\sigma = AE\epsilon = AE\,\frac{d_2}{L} = \frac{AE}{L}\cdot 1 \qquad (2.3)$$

The stiffness influence coefficient K_{22} is then equal to AE/L. Similarly the stiffness influence coefficient at node 1 due to the force applied at 2 is the reaction force in the opposite direction, so $K_{21} = -AE/L$.

Fig. 2.2 A simple tensile specimen (a), represented by a linear model (b), showing the forces acting on the central element (c).

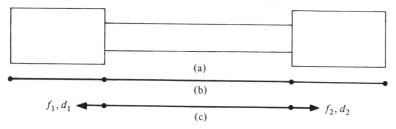

Fig. 2.3 The forces and displacements on a single element.

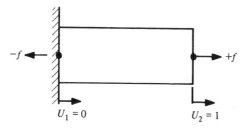

The forces at each node due to a unit displacement at node 1 can also be found, so the full stiffness matrix of the element becomes:

$$\begin{bmatrix} K_{11} \, K_{12} \\ K_{21} \, K_{22} \end{bmatrix} = \begin{bmatrix} +AE/L & -AE/L \\ -AE/L & +AE/L \end{bmatrix} = \frac{AE}{L} \begin{bmatrix} +1 & -1 \\ -1 & +1 \end{bmatrix} \qquad (2.4)$$

Like many stiffness matrices, this is symmetric. It has already been deduced in a different way in Chapter 1, equation 1.21, where it was used to solve the first tensile specimen problem.

2.1.2 Plane-strain triangular elements

Triangular elements (figure 2.1a) are widely used for plane strain and axisymmetric problems.

Assuming the strain to be uniform in an element:

$$\epsilon = \begin{pmatrix} \epsilon_{11} \\ \epsilon_{22} \\ \gamma_{12} \end{pmatrix} = \begin{pmatrix} \dfrac{\partial u_1}{\partial x_1} \\ \dfrac{\partial u_2}{\partial x_2} \\ \dfrac{\partial u_1}{\partial x_2} + \dfrac{\partial u_2}{\partial x_1} \end{pmatrix} \qquad (2.5)$$

where u_1 and u_2 are the components of the displacements in the x_1 and x_2 directions, ϵ_{11}, ϵ_{22} are direct strains and γ_{12} is an engineering shear strain.

2.1.3 Linear quadrilateral elements

Very often the geometric shape can conveniently be filled with rectangular elements (figure 2.1b). These are less stiff than the triangles in plane strain; in an extreme example, under constant-volume conditions the apex of a triangle can move only parallel to its base when the base nodes are fixed.

2.1.4 Higher-order (quadratic) elements

Because of this problem of stiffness it is often desirable to use higher-order elements (figure 2.1c,d).

Greater flexibility is obtained, at the expense of larger matrices and longer solution times, by introducing mid-side nodes. The edge of an element is then, in general, a parabola instead of a straight line.

Quadrilateral elements can also be provided with mid-side nodes, producing eight-node elements. Again flexibility is improved at the cost of memory size and computing time. The advantage is usually less for quadrilaterals than for triangles.

In some circumstances even the use of two intermediate nodes on each side may be desirable to model complex distortion accurately. Such third-order or cubic elements are seldom required.

2.1.5   Three-dimensional elements

The most popular element types for 3-D plastic deformation are rectangular bricks (figure 2.1e,f). These may have eight nodes, one at each corner, and can be more flexible with additional mid-side nodes. The size of matrix is a critical feature of 3-D analysis. Even a coarse mesh of 10 elements on each side requires 1000 elements and the number of nodes must also be controlled.

Tetrahedral elements are also widely used (fig. 2.1g). These can often be fitted around corners. Different types of element may be incorporated in a single network, for example to model complex shapes more easily.

2.1.6   Size of elements

Generally speaking, the accuracy of FE solutions improves as the number of elements is increased. The number is normally limited by the size of the computer memory and the time for solution of the simultaneous equations. As in the example of figure 1.2, the general pattern of discretisation is quickly established with relatively few elements. Beyond a certain limit, the computing cost of refining the mesh rises rapidly without corresponding gain in useful information. For plane-strain or axi-symmetric plasticity problems, it is seldom desirable to use more than 1000 elements. In 3-D though, this gives a rather coarse mesh and even a ten-fold increase in the number of elements only halves the linear mesh dimensions, so definition of detail presents serious problems in 3-D.

To save computer space and time, it is usual to combine large-element meshes in regions of secondary interest with finer meshes in critical zones. For example figure 2.4 shows a mesh prepared for analysis of indentation. The region around the corner of the indenting punch must be well defined as it is known that the deformation is symmetrical about the centre line and occurs mainly in a narrow band with steep lateral strain gradients, especially near this corner.

Fig. 2.4 Fine and coarse-mesh elements used for indentation.

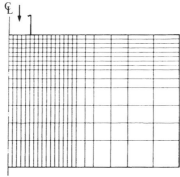

Small mesh size also usually improves the convergence of an iterative solution.

2.1.7   Shape and configuration of elements: aspect ratio

It is convenient to use simple triangular and quadrilateral elements, usually chosen with an aspect ratio of about unity. This provides satisfactorily uniform deformation and allows simple interpolation functions to be used. Nevertheless, if the pattern of deformation permits, it is possible to economise on computer storage and time by using long, thin elements instead of small squares in certain regions. This can be seen in figure 2.4 and another example is given in figure 2.5 for open-die extrusion.

In many structural analyses, the most efficient mesh may be much more complex, as shown in figure 2.6 [2.1]. Such meshes can be generated by one of the commercial optimising algorithms. This is very useful but is not strictly necessary for many metalforming analyses.

2.2   CONTINUITY AND EQUILIBRIUM

The basic FE equations for plastic or elastic deformation are simple, relying on equilibrium of forces at every node and the compatibility of strains between elements.

In a linear problem, for example in elastic deformation of metals, the displacement vector can be found directly from the force vector, or the reverse, using the stiffness relationship for each element given in equation 2.1.

Fig. 2.5 A non-uniform mesh used for extrusion analysis.

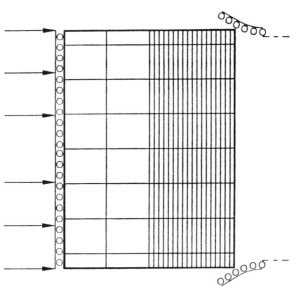

As explained in Chapter 1, Section 1.5, a stiffness relationship can be established for plastic deformation using the principle of minimum potential energy.

Because the FE solution is obtained only at the nodes, there will be discontinuous changes from one node to the next. It is therefore necessary to introduce interpolation functions to describe the behaviour of the solution across an element while still satisfying the equilibrium conditions [2.2].

2.3   INTERPOLATION FUNCTIONS

2.3.1   Polynomials

These functions should be as simple as possible, whilst being compatible with the required accuracy and detail of the solution. It is usual to select a polynomial function, which can often be simply a linear relationship:

$$N(x) = a + bx + cx^2 + \cdots + px^r \qquad (2.6)$$

The common linear approximation for 2-D is:

$$N(x_1, x_2) = a + b_1 x_1 + b_2 x_2 \qquad (2.7)$$

It is usual to apply the same interpolation function to both the geometric description of the element and the displacements. The elements are then known as isoparametric, but sometimes one interpolation function may be of higher order than the other. We shall not consider these here.

2.3.2   Convergence

The interpolation function must satisfy the requirements of convergence or the finite-element solution will not converge on the true solution as the element size is reduced.

Fig. 2.6 A mesh for analysing a block containing a large hole.

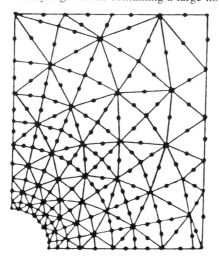

The first requirement is that the interpolation function must be continuous. Polynomial functions satisfy this condition.

Secondly, uniform states of the field variable and its derivatives must be represented when the element size reduces to zero. In particular, zero strain and constant strain must be compatible with the assumed displacement model.

Thirdly the variable and its derivatives (up to one order less than the highest found in the functional $I(N)$) must be continuous at element boundaries and interfaces. In solid mechanics this means that there can be no discontinuities or overlaps between elements, and the slopes of beams, etc, must be continuous.

The elements and functions normally used in plasticity satisfy all these conditions. In other words they are 'complete'.

2.4   DISPLACEMENT VECTOR u AND MATRIX [B]

The first requirement for a solution, once the element mesh has been established and the interpolation function has been selected, is to define the displacements of the nodes and from these to deduce the strains in each element.

The basic technique can be explained with reference to simple triangular elements.

The displacement vector **u** of any point (x_1, x_2) within the element has components u_1, u_2 parallel to the two axes.

Assume these to be linear functions of the co-ordinates:

$$u_i = a_i + b_{i1}x_1 + b_{i2}x_2 \, , \, i = 1,2 \tag{2.8}$$

This displacement must be equal to the nodal displacement when evaluated at the nodal positions. The equations thus obtained are presented and solved in Appendix 1.

A typical expression of one of the coefficients is given in equation A1.10:

$$b_{12} = \frac{1}{2A} \left((x_{31}-x_{21})d_{11} + (x_{11}-x_{31})d_{21} + (x_{21}-x_{11})d_{31} \right) \tag{2.9}$$

where A is the area of the triangle 123 in figure 2.7, x_{Ii} is the ith co-ordinate of node I and d_{Ii} is the displacement of node I in the x_i direction.

Substituting for the coefficients in equation 2.8:

$$u_i = \frac{1}{2A} \left(x_{21}x_{32} - x_{22}x_{31} + (x_{22}-x_{32})x_1 + (x_{31}-x_{21})x_2 \right) d_{1i}$$

$$+ \frac{1}{2A} \left(x_{12}x_{31} - x_{11}x_{32} + (x_{32}-x_{12})x_1 + (x_{11}-x_{31})x_2 \right) d_{2i} \tag{2.10}$$

$$+ \frac{1}{2A} \left(x_{11}x_{21} - x_{12}x_{21} + (x_{12}-x_{22})x_1 + (x_{21}-x_{11})x_2 \right) d_{3i}$$

or:

$$u_i = \begin{pmatrix} N_1, N_2, N_3 \end{pmatrix} \cdot \begin{pmatrix} d_{1i} \\ d_{2i} \\ d_{3i} \end{pmatrix} \qquad \text{(2.10 continued)}$$

where N_I is the interpolation function of position relating to node I.

As shown in Appendix 1, the strain vector $\boldsymbol{\epsilon}$ is given by:

$$\boldsymbol{\epsilon} = \begin{pmatrix} \epsilon_{11} \\ \epsilon_{22} \\ \gamma_{12} \end{pmatrix}$$

$$= \frac{1}{2A} \begin{bmatrix} x_{22}-x_{32} & 0 & x_{32}-x_{12} & 0 & x_{12}-x_{22} & 0 \\ 0 & x_{31}-x_{21} & 0 & x_{11}-x_{31} & 0 & x_{21}-x_{11} \\ x_{31}-x_{21} & x_{22}-x_{32} & x_{11}-x_{31} & x_{32}-x_{12} & x_{21}-x_{11} & x_{12}-x_{22} \end{bmatrix} \begin{pmatrix} d_{11} \\ d_{12} \\ d_{21} \\ d_{22} \\ d_{31} \\ d_{32} \end{pmatrix}$$

or

$$\boldsymbol{\epsilon} = [B]\mathbf{d} \qquad (2.11)$$

The matrix $[B]$ is generally a function of position, though for the constant-strain triangle of this example it is independent of position within the element. The $[B]$ matrix is of great importance in all finite-element analysis.

Fig. 2.7 Deformation of a simple triangular element.

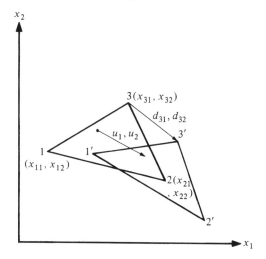

2.5   THE [D] MATRIX OF ELASTIC CONSTANTS

2.5.1   General expression

We consider first the [D] matrix in terms of linear elastic constants. This will be developed for non-linear plastic deformation in Chapter 3.

In one dimension (1-D) Young's Modulus E is the ratio of stress σ to strain ϵ:

$$E = \frac{\sigma}{\epsilon}; \quad \epsilon_{11} = \frac{\sigma_{11}}{E} \tag{2.12}$$

In 2-D the Poisson's Ratio ν must be included:

$$\epsilon_{11} = \frac{\sigma_{11}}{E} - \nu \frac{\sigma_{22}}{E} \tag{2.13}$$

and similarly in 3-D:

$$\epsilon_{11} = \frac{\sigma_{11}}{E} - \nu \frac{\sigma_{22}}{E} - \nu \frac{\sigma_{33}}{E} \tag{2.14}$$

The shear strain in 2-D can be derived by considering a cube subjected to a shear stress (Appendix 7). For example:

$$\gamma_{12} = \frac{2(1+\nu)}{E} \sigma_{12} = \frac{\sigma_{12}}{G} \tag{2.15}$$

where G is the elastic Modulus of Rigidity.

Collecting these expressions into a matrix equation, the isotropic elastic modulus is given by the equations:

$$
\begin{pmatrix} \epsilon_{11} \\ \epsilon_{22} \\ \epsilon_{33} \\ \gamma_{12} \\ \gamma_{23} \\ \gamma_{13} \end{pmatrix} = \frac{1}{E}
\begin{bmatrix}
1 & -\nu & -\nu & 0 & 0 & 0 \\
-\nu & 1 & -\nu & 0 & 0 & 0 \\
-\nu & -\nu & 1 & 0 & 0 & 0 \\
0 & 0 & 0 & 2(1+\nu) & 0 & 0 \\
0 & 0 & 0 & 0 & 2(1+\nu) & 0 \\
0 & 0 & 0 & 0 & 0 & 2(1+\nu)
\end{bmatrix}
\begin{pmatrix} \sigma_{11} \\ \sigma_{22} \\ \sigma_{33} \\ \sigma_{12} \\ \sigma_{23} \\ \sigma_{13} \end{pmatrix} \tag{2.16}
$$

In practice, this relationship is required in the form:

$$\boldsymbol{\sigma} = [D]\boldsymbol{\epsilon} \tag{2.17}$$

In Appendix 2 it is shown that: (2.18)

$$
\begin{pmatrix} \sigma_{11} \\ \sigma_{22} \\ \sigma_{33} \\ \sigma_{12} \\ \sigma_{23} \\ \sigma_{13} \end{pmatrix} = \frac{E}{1+\nu}
\begin{bmatrix}
\dfrac{1-\nu}{1-2\nu} & \dfrac{\nu}{1-2\nu} & \dfrac{\nu}{1-2\nu} & 0 & 0 & 0 \\[2mm]
\dfrac{\nu}{1-2\nu} & \dfrac{1-\nu}{1-2\nu} & \dfrac{\nu}{1-2\nu} & 0 & 0 & 0 \\[2mm]
\dfrac{\nu}{1-2\nu} & \dfrac{\nu}{1-2\nu} & \dfrac{1-\nu}{1-2\nu} & 0 & 0 & 0 \\[2mm]
0 & 0 & 0 & \dfrac{1}{2} & 0 & 0 \\[2mm]
0 & 0 & 0 & 0 & \dfrac{1}{2} & 0 \\[2mm]
0 & 0 & 0 & 0 & 0 & \dfrac{1}{2}
\end{bmatrix}
\begin{pmatrix} \epsilon_{11} \\ \epsilon_{22} \\ \epsilon_{33} \\ \gamma_{12} \\ \gamma_{23} \\ \gamma_{13} \end{pmatrix}
$$

There are three important special conditions often met in 2-D analysis that permit the [D] matrix to be simplified. These are the conditions of plane stress, plane strain and axial symmetry.

2.5.2   Plane elastic stress

Under plane-stress conditions:

$$\sigma_{33} = \sigma_{13} = \sigma_{23} = 0 \tag{2.19}$$

Furthermore, since the two shear stresses above are zero:

$$\gamma_{13} = \gamma_{23} = 0 \tag{2.20}$$

and equation 2.18 can be re-written immediately:

$$
\begin{pmatrix} \sigma_{11} \\ \sigma_{22} \\ \sigma_{12} \end{pmatrix} = \frac{E}{1+\nu}
\begin{bmatrix}
\dfrac{1-\nu}{1-2\nu} & \dfrac{\nu}{1-2\nu} & \dfrac{\nu}{1-2\nu} & 0 \\[2mm]
\dfrac{\nu}{1-2\nu} & \dfrac{1-\nu}{1-2\nu} & \dfrac{\nu}{1-2\nu} & 0 \\[2mm]
0 & 0 & 0 & \dfrac{1}{2}
\end{bmatrix}
\begin{pmatrix} \epsilon_{11} \\ \epsilon_{22} \\ \epsilon_{33} \\ \gamma_{12} \end{pmatrix}
\tag{2.21}
$$

But:

$$\sigma_{33} = \frac{E}{(1+\nu)\,(1-2\nu)}[\nu\epsilon_{11} + \nu\epsilon_{22} + (1-\nu)\epsilon_{33}] \tag{2.22}$$

so:

$$\epsilon_{33} = -\frac{\nu}{1-\nu}(\epsilon_{11} + \epsilon_{22}) \tag{2.23}$$

Substituting for ϵ_{33} into equation 2.21:

$$\begin{pmatrix} \sigma_{11} \\ \sigma_{22} \\ \sigma_{12} \end{pmatrix} = \frac{E}{1+\nu} \begin{bmatrix} \dfrac{1-\nu}{1-2\nu} - \dfrac{\nu^2}{(1-\nu)(1-2\nu)} & \dfrac{\nu}{1-2\nu} - \dfrac{\nu^2}{(1-\nu)(1-2\nu)} & 0 \\ \dfrac{\nu}{1-2\nu} - \dfrac{\nu^2}{(1-\nu)(1-2\nu)} & \dfrac{1-\nu}{1-2\nu} - \dfrac{\nu^2}{(1-\nu)(1-2\nu)} & 0 \\ 0 & 0 & \dfrac{1}{2} \end{bmatrix} \begin{pmatrix} \epsilon_{11} \\ \epsilon_{22} \\ \gamma_{12} \end{pmatrix}$$

$$= \frac{E}{1+\nu} \begin{bmatrix} \dfrac{1}{1-\nu} & \dfrac{\nu}{1-\nu} & 0 \\ \dfrac{\nu}{1-\nu} & \dfrac{1}{1-\nu} & 0 \\ 0 & 0 & \dfrac{1}{2} \end{bmatrix} \begin{pmatrix} \epsilon_{11} \\ \epsilon_{22} \\ \gamma_{12} \end{pmatrix} \tag{2.24}$$

2.5.3 Plane elastic strain

Under plane-strain conditions:

$$\epsilon_{33} = \gamma_{13} = \gamma_{23} = 0 \tag{2.25}$$

Furthermore, since the two shear strains above are zero:

$$\sigma_{13} = \sigma_{23} = 0 \tag{2.26}$$

and since the strain ϵ_{33} is zero, the corresponding normal stress does not enter into the strain energy expression and equation 2.18 can be re-written as:

$$\begin{pmatrix} \sigma_{11} \\ \sigma_{22} \\ \sigma_{12} \end{pmatrix} = \frac{E}{1+\nu} \begin{bmatrix} \dfrac{1-\nu}{1-2\nu} & \dfrac{\nu}{1-2\nu} & 0 \\ \dfrac{\nu}{1-2\nu} & \dfrac{1-\nu}{1-2\nu} & 0 \\ 0 & 0 & \dfrac{1}{2} \end{bmatrix} \begin{pmatrix} \epsilon_{11} \\ \epsilon_{22} \\ \gamma_{12} \end{pmatrix} \tag{2.27}$$

2.5.4 Axial symmetry

If x_1 is identified with the radial direction, x_2 with the circumferential direction and x_3 with the axial direction, the conditions of axial symmetry require that:

$$\gamma_{12} = \gamma_{23} = 0 \tag{2.28}$$

Hence:

$$\sigma_{12} = \sigma_{23} = 0 \tag{2.29}$$

and equation 2.18 becomes:

$$
\begin{pmatrix} \sigma_{11} \\ \sigma_{22} \\ \sigma_{33} \\ \sigma_{13} \end{pmatrix}
= \frac{E}{1+\nu}
\begin{bmatrix}
\dfrac{1-\nu}{1-2\nu} & \dfrac{\nu}{1-2\nu} & \dfrac{\nu}{1-2\nu} & 0 \\[2mm]
\dfrac{\nu}{1-2\nu} & \dfrac{1-\nu}{1-2\nu} & \dfrac{\nu}{1-2\nu} & 0 \\[2mm]
\dfrac{\nu}{1-2\nu} & \dfrac{\nu}{1-2\nu} & \dfrac{1-\nu}{1-2\nu} & 0 \\[2mm]
0 & 0 & 0 & \dfrac{1}{2}
\end{bmatrix}
\begin{pmatrix} \epsilon_{11} \\ \epsilon_{22} \\ \epsilon_{33} \\ \gamma_{13} \end{pmatrix}
\tag{2.30}
$$

2.6   FORMULATION OF THE ELEMENT STIFFNESS MATRIX $[K_i]$

As in equation 1.11, the potential energy I is found from the difference between the work done by external forces and the strain energy stored. For example, in plane stress ($\sigma_{33} = \sigma_{13} = \sigma_{23} = 0$):

$$
I = U - W
$$

$$
= \frac{1}{2} \int (\sigma_{11}\epsilon_{11} + \sigma_{22}\epsilon_{22} + \sigma_{12}\gamma_{12})\mathrm{d}V \tag{2.31}
$$

$$
- f_{11}d_{11} - f_{12}d_{12} - f_{21}d_{21} - f_{22}d_{22} - f_{31}d_{31} - f_{32}d_{32}
$$

in which f_{Ii} denotes the component of force in the x_i direction acting at node I.

In matrix form the expression for element i is:

$$
I_i = \frac{1}{2} \int \boldsymbol{\sigma}_i^{\mathrm{T}} \boldsymbol{\epsilon}_i \, \mathrm{d}V_i - \mathbf{d}_i^{\mathrm{T}} \mathbf{f}_i
$$

$$
= \frac{1}{2} \int \boldsymbol{\epsilon}_i^{\mathrm{T}} [D_i] \boldsymbol{\epsilon}_i \, \mathrm{d}V_i - \mathbf{d}_i^{\mathrm{T}} \mathbf{f}_i \tag{2.32}
$$

$$
= \frac{1}{2} \mathbf{d}_i^{\mathrm{T}} \left(\int [B_i]^{\mathrm{T}} [D_i] \, [B_i] \mathrm{d}V_i \right) \mathbf{d}_i - \mathbf{d}_i^{\mathrm{T}} \mathbf{f}_i
$$

Differentiating with respect to each of the components of displacement and equating the derivatives to zero leads to a set of conditions for the minimum of the potential energy:

$$
\mathbf{0} = \int [B_i]^{\mathrm{T}} [D_i] \, [B_i] \mathrm{d}V_i \cdot \mathbf{d}_i - \mathbf{f}_i \tag{2.33}
$$

or:

$$
\int [B_i]^{\mathrm{T}} [D_i] \, [B_i] \mathrm{d}V_i \cdot \mathbf{d}_i = \mathbf{f}_i \tag{2.34}
$$

Thus the elastic stiffness relationship for element i may be written as:

$$[K_i]\mathbf{d}_i = \mathbf{f}_i \qquad (2.35)$$

where:

$$[K_i] = \int [B_i]^T[D_i] [B_i]\mathrm{d}V_i \qquad (2.36)$$

The derivation of this stiffness matrix for the case of a constant-strain triangle deforming under plane-stress conditions is given in detail in Appendix 4. For a constant-strain triangle, the integrand in equation 2.36 is the same throughout the element and so:

$$[K_i] = s{\cdot}A_i [B_i]^T[D_i] [B_i] \qquad (2.37)$$

where s is the assumed thickness of all the elements.

2.7   FORMULATION OF THE GLOBAL STIFFNESS MATRIX $[K]$

2.7.1   Assembly of element matrices

The vector $\mathbf{f}_i$ in equation 2.35 contains all the components of nodal force acting upon element i and, for an isolated element, is simply the vector of applied nodal loads. However, in general, the element will not be isolated, and in this case the forces acting upon element i will be the sum of the applied loads and the reactions from any adjoining elements. We do not know, and often do not need to know, the values of these reactions, but this does not matter because it is found that they cancel (action and reaction being equal and opposite) when the value of force acting at a given node is summed over all the elements containing that node. In this process, the individual element stiffness matrices are assembled into a *global stiffness matrix* $[K]$, the coefficients of which determine the influence of a component of displacement of one node in the mesh upon a component of force applied at another node. These coefficients are simply the sums of the corresponding coefficients in all the element stiffness matrices. An example of the assembly process was given in Chapter 1 (Section 1.5.4).

The global stiffness equations may be written as:

$$[K]\mathbf{d} = \mathbf{f} \qquad (2.38)$$

where the global displacement vector $\mathbf{d}$ is the vector of the components of displacement of all the nodes in the mesh, and the global force or load vector $\mathbf{f}$ contains the components of force acting at all the nodes.

In an elastic deformation problem $[K]$ is usually linear (though large strains in rubbers for example are recoverable and therefore elastic, but are not linearly related to stresses). There is considerable interest in the solution of large-deformation elastic problems, but these do not concern us here.

2.7.2   Properties of $[K]$

The global stiffness matrix $[K]$ has several properties that have important consequences for the computer implementation of the FEM. Although we are at
present assuming that the deformation is purely elastic, these properties are also
exhibited by the stiffness matrix for small-deformation plasticity derived in the
next chapter, and by the finite-deformation elastic–plastic stiffness matrix to be
introduced in Chapter 5.

Symmetry

Since the individual matrices $[D_i]$ are symmetric, it can readily be seen, by
transposing equation 2.36, that the element stiffness matrices $[K_i]$ also have this
property. Consequently $[K]$ is also symmetric because the element stiffness matrices are assembled into the global matrix in a symmetric manner. Since $[K]$ is
symmetric, only about half of the coefficients need be calculated and stored.

As will be seen in the next chapter, the constitutive matrices used in elastic–plastic analyses are also symmetric and so lead to symmetric stiffness matrices.

Positive semi-definiteness

$[K]$ is positive semi-definite since, for all non-zero vectors $\mathbf{v}$, the quadratic form
$\mathbf{v}^T[K]\mathbf{v}$ (a scalar quantity) is greater than or equal to zero. This property can be
deduced from the principle of minimum potential energy that forms the basis
for the derivation of the element stiffness matrices $[K_i]$.

It can easily be shown that any semi-definite matrix is singular, and so does
not have an inverse. The stiffness matrix in equation 2.4 is an example of this.
The singularity of $[K]$ is a consequence of one of the requirements for the
completeness of the FE model. In simple terms, the displacement vector obtained
by adding any rigid-body translation or rotation to a solution of equation 2.38
will also be a solution of this equation.

Since the basic stiffness equations permit an infinite number of solutions for
the displacement vector, it is necessary to specify values for a minimum number
of independent components of nodal displacement (five for 3-D problems). In
metalforming analysis, these will be determined in the normal course of things
by the nature of the process model, such as the existence of dies or planes of
symmetry. The specification of values for certain components of displacement is
equivalent to converting the set of stiffness equations into a lower-order set for
which the matrix of coefficients is positive definite (that is, has quadratic forms
that are strictly greater than zero.)

It should be remembered however, that any computer representation of the
stiffness matrix will always be an approximation. An FE program may therefore
succeed in obtaining a solution displacement vector even when insufficient num-

bers of displacement components are prescribed. The value of a solution under
such circumstances is questionable.

Bandedness

Since the global stiffness matrix is assembled on an element by element basis,
the value of force at a particular node can depend only upon the displacement
of nodes that belong to the same element as the one under consideration. Thus
it is found that each row of $[K]$ contains many zero coefficients. If the nodal
entries in the global vectors **d** and **f** are ordered correctly, it is possible to confine
the non-zero coefficients of $[K]$ to a band either side of the leading diagonal.
The zero coefficients outside the band do not have to be stored, and they do
not have to take part in any subsequent calculation.

The extent to which the width of the band can be reduced is obviously crucial

Fig. 2.8 Non-zero coefficients in the global stiffness matrices for typical 3-D
meshes with the diagonal bands bounded by the dashed lines: (a) 4×4×4
nodes, (b) 6×6×6 nodes, (c) 4×6×26 nodes, (d) square at top left-hand
corner of (c) enlarged to show sparse nature of band. Each square symbol
in these figures represents the 3×3 sub-matrix corresponding to the three
degrees of freedom at a node.

(a)

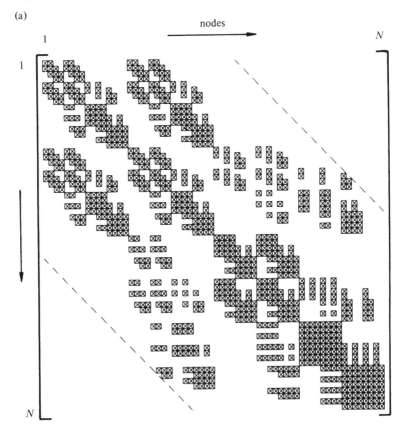

in determining the maximum size of problem that can be solved by an FE program, and in reducing the time of computation.

Figure 2.8 shows the patterns of non-zero coefficients in the global stiffness matrices for some typical 3-D meshes. In these examples, the order of the nodes in the global vectors is not optimal. These diagrams give an indication of the reductions in computer storage and computational time that are possible as a result of a suitable re-ordering of the nodal vectors.

Sparseness of diagonal band

Even when [K] has been arranged to have the narrowest possible width of band, it is often found that the band is sparse, that is, contains many zero coefficients.

Whereas the band-width is influenced by topology of the mesh (the types of element and their arrangement) and the ordering of the nodes, the number of non-zero entries in a row of [K] is independent of the nodal ordering.

The sparse nature of the diagonal band can be seen clearly in figure 2.8d. Not all techniques for solving the stiffness equations are able to exploit this sparseness.

Fig. 2.8 Continued

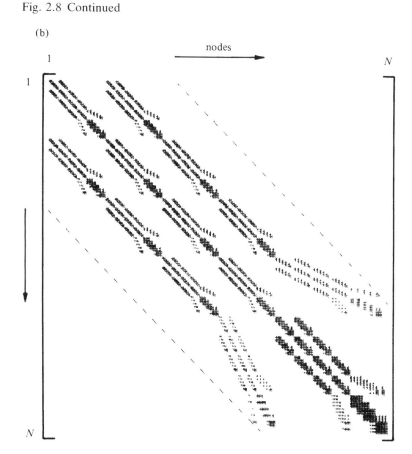

(b)

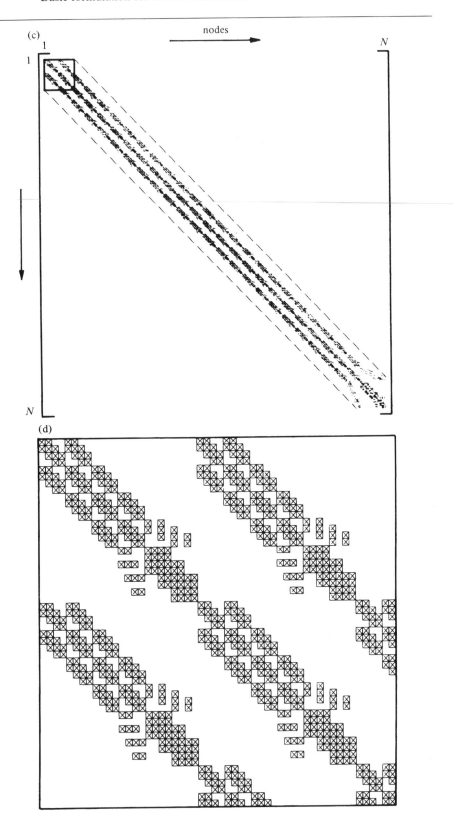

2.8 GENERAL SOLUTION METHODS FOR THE MATRIX EQUATIONS

As mentioned in the previous section, the solution of equation 2.38 must incorporate a certain number of prescribed components of displacement. We will consider first of all general solution methods for non-singular sets of matrix equations in which all the components of displacement are unknown, and examine ways of incorporating values of displacement into these solutions later.

Matrix equations such as 2.38 may be solved by direct or by indirect means. Direct methods obtain a solution by calculating the inverse relationship to equation 2.38; indirect methods iteratively modify a starting guess for **d** until a vector is obtained that satisfies the matrix equations.

2.8.1 Direct solution methods

Gauss–Jordan elimination

This is the method usually adopted when solving a small set of simultaneous equations by hand. By a sequence of row operations – in which multiples of one equation are added, coefficient by coefficient (including the right-hand side), to another – the matrix equations 2.38 are converted into an equivalent set that has the same solution vector:

$$[T]\mathbf{d} = \mathbf{r} \tag{2.39}$$

The new matrix $[T]$ has zeros everywhere except on its leading diagonal. The components of the displacement vector may then be obtained immediately by dividing each of the components of the right-hand side vector **r** by the corresponding diagonal term of $[T]$.

Gauss–Jordan elimination is not often used in FE work, partly because the number of computational operations involved is about 50% more than those required for a partial elimination scheme [2.3], but, more importantly, because the whole elimination has to be repeated if a solution is required for a new force vector. This makes the method unsuitable when multiple load cases and iterative techniques are considered.

Gaussian elimination and back-substitution

As in the previous method, row operations are performed on the stiffness matrix and force vector in order to obtain a set of matrix equations that have the same solution vector as the original ones:

$$[U]\mathbf{d} = \mathbf{s} \tag{2.40}$$

This time, $[U]$ is an upper-triangular matrix, that is, one having non-zero coefficients on and above the leading diagonal and zeros below the diagonal.

The back-substitution part of this method uses equations 2.40 to calculate the components of **d** in reverse order:

$$d_\alpha = \frac{1}{U_{\alpha\alpha}} \left(s_\alpha - \sum_{\beta=\alpha+1}^{n} U_{\alpha\beta} d_\beta \right) , \alpha = n, 1 \qquad (2.41)$$

in which n is the number of degrees of freedom in the FE discretisation and the summation is taken to be zero when $\alpha = n$.

It can be seen that when the time comes to evaluate d_α, all the values of the displacement components from $d_{\alpha+1}$ to d_n contained in the summation are already known.

The method of Gaussian elimination and back-substitution is described in detail in Appendix 5. It requires fewer computational operations than a complete Gauss–Jordan elimination, but for general non-symmetric matrix equations a complete re-calculation is still required to solve for each new right-hand side vector. It shall be seen that this is not the case for the symmetric matrices encountered in FE metalforming analysis.

Choleski LU decomposition

Any square, non-singular matrix can be expressed as a product of a lower-triangular matrix $[L]$ and an upper-triangular matrix $[U]$. Thus we can write equations 2.38 as:

$$[L][U]\mathbf{d} = \mathbf{f} \qquad (2.42)$$

There are infinitely many LU decompositions of a given matrix. However, if n entries are specified, the decomposition becomes unique. In most Choleski decomposition schemes, the n diagonal coefficients of $[L]$ are all chosen to be unity.

Making the substitution:

$$\mathbf{s} = [U]\mathbf{d} \qquad (2.43)$$

we obtain:

$$[L]\mathbf{s} = \mathbf{f} \qquad (2.44)$$

Since $[L]$ is in lower-triangular form with unit diagonal coefficients, equations 2.44 may be solved by forward-substitution:

$$(2.45)$$

$$s_\alpha = f_\alpha - \sum_{\beta=1}^{\alpha-1} L_{\alpha\beta} s_\beta , \alpha = 1, n$$

and then the values of the components of $\mathbf{s}$ may be used to solve equations 2.43 by back-substitution as in the previous method.

The advantage of this method is that, once the stiffness matrix has been decomposed in this way, solutions can be obtained for new force vectors with little extra computation (of the order of $n^2/2$ multiplications and additions for the forward substitution, compared with aproximately $n^3/3$ multiplications and additions for the elimination phase of the previous method, back-substitution being common to both).

The LU decomposition may be performed using the Crout algorithm [2.3]. However, since the stiffness matrix is symmetric, it can be shown that [L] is simply the transpose of [U], but with the entries in each column divided by the diagonal coefficient:

$$L_{\alpha\beta} = \frac{U_{\beta\alpha}}{U_{\beta\beta}}, \beta \leqslant \alpha \leqslant n \qquad (2.46)$$

In addition, the upper-triangular matrix [U] in the LU decomposition with unit diagonal [L] is precisely the same as the upper-triangular matrix obtained during the Gaussian elimination procedure (providing the rows of the latter matrix are *not* already normalised by dividing through by the diagonal coefficients). Thus for the symmetric stiffness matrices encountered in metalforming analysis, Gaussian elimination is equivalent to a Crout decomposition, although there may be slightly less round-off error accumulated in the latter process [2.3].

Whichever technique is used, only the coefficients of [U] need to be explicitly calculated and stored, since those of [L] can easily be obtained from these as required, whether for the first solution, or any subsequent re-solution.

It is a simple matter to modify the above solution methods so that only those coefficients lying within the non-zero diagonal band of [K] are involved in the computation. Unfortunately, during both the Gaussian elimination and Crout decomposition procedures, the zero coefficients within the band tend to be filled up with non-zero ones, so these methods cannot take advantage of the sparseness of the band itself.

Direct methods can however be modified so that only a small part of the band of the stiffness matrix need be in computer memory at any one time, the rest of the matrix being saved on secondary disc storage (Section 6.6.3). This permits a much larger stiffness matrix to be accommodated than would be the case if all the non-zero coefficients had to be in computer memory at the same time, though at the expense of increased program running time.

2.8.2 Indirect solution methods

Whereas the direct methods discussed in the previous section are almost certain to produce a solution of some sort after a predetermined amount of computation, it is not possible to determine beforehand how many iterations will be needed in order to converge to a solution using indirect techniques. For some types of problem indirect methods can be very efficient indeed; for others the computational effort may be many orders of magnitude greater than that expended during a direct solution. Moreover, since the stiffness matrix is repeatedly used in a large number of iterative steps, the storage of the stiffness matrix on disc would entail a prohibitively-large number of time-consuming disc accesses. It is therefore essential that all the non-zero coefficients of this matrix be stored in the main memory of the computer. This effectively limits the size of problem that can be examined.

For these reasons, indirect solution methods will not be considered in great detail here. However, since the effectiveness of these methods depends very much on the nature of the matrix equations being solved, and since larger and cheaper computer memory chips are continually being introduced, the reader is encouraged to experiment with the use of these techniques for particular problems. Also, it should be noted that iterative solutions are ideal candidates for parallel processing techniques, so the arrival of parallel processing hardware, such as Transputer boards, to be used in conjunction with computer workstations may cause a complete re-appraisal of iterative methods of solution.

The problem of finding the solution of equations 2.38 is equivalent to the problem of finding the minimum of the function h of n variables where:

$$h(\mathbf{d}) = \frac{1}{2}\mathbf{d}^{\mathrm{T}}[K]\mathbf{d} - \mathbf{d}^{\mathrm{T}}\mathbf{f} \tag{2.47}$$

Since $[K]$ is positive semi-definite, this function does have a minimum, and since $[K]$ is symmetric, the condition for a minimum of the function:

$$\frac{\partial h}{\partial \mathbf{d}} = [K]\mathbf{d} - \mathbf{f} = 0 \tag{2.48}$$

is precisely the condition that the vector $\mathbf{d}$ at which the minimum occurs is also the solution of equations 2.38.

Multi-dimensional minimisation techniques search for the minimum of a function by a succession of one-dimensional minimisations along lines in n-dimensional space. The group of techniques known as *conjugate-gradient* methods, choose the directions for the line minimisations in order to optimise the searching procedure [2.3]. Even so, a large number of steps may be required unless $[K]$ is approximately proportional to the identity matrix.

This fact is exploited by a hybrid technique particularly suited to the solution of sparse matrix equations, the *incomplete Choleski conjugate-gradient* method [2.4]. In this procedure, an approximate Choleski decomposition of $[K]$ is performed during which a coefficient of the upper-triangular matrix $[U^*]$ is only calculated if the corresponding coefficient of $[K]$ is non-zero, otherwise it is assumed to be zero. This considerably reduces the amount of computation and storage involved in the decomposition. The matrix $[L^*][U^*]$, where $[L^*]$ is calculated analogously to equation 2.46, is only approximately equal to $[K]$ and so cannot be used directly to solve for the vector $\mathbf{d}$. However, the equations:

$$[K^*]\mathbf{d}^* = \mathbf{f}^* \tag{2.49}$$

where:

$$[K^*] = [L^*]^{-1}[K][U^*]^{-1} \tag{2.50}$$

and:

$$\mathbf{f}^* = [L^*]^{-1} \mathbf{f} \qquad (2.51)$$

have the solution:

$$\mathbf{d}^* = [U^*]\mathbf{d} \qquad (2.52)$$

Since $[K^*]$ is an approximation to the identity matrix, conjugate-gradient methods quickly converge to the solution of equation 2.49. $[U^*]$ and $[L^*]$ are triangular matrices, so the inverses in equations 2.50 and 2.51 may be computed with little effort, and the required solution vector $\mathbf{d}$ may easily be obtained by inverting equation 2.52.

2.8.3 Iteration to improve a direct solution

Even if iteration is not used to produce the solution itself, it can be used to good effect to counter the accumulation of round-off error in a direct solution.

Suppose equations 2.38 have been solved to produce an estimate of the solution vector which, due to round-off error, differs by an unknown vector $\delta\mathbf{d}$ from the correct solution. Substituting the approximate solution $\mathbf{d} + \delta\mathbf{d}$ into equations 2.38 gives an approximate value $\mathbf{f}^+$ for the force vector:

$$[K](\mathbf{d} + \delta\mathbf{d}) = \mathbf{f}^+ \qquad (2.53)$$

Substituting from equation 2.38:

$$\mathbf{f} + [K]\delta\mathbf{d} = \mathbf{f}^+ \qquad (2.54)$$

and rearranging:

$$[K]\delta\mathbf{d} = \mathbf{f}^+ - \mathbf{f} \qquad (2.55)$$

Since the right-hand side vector of equations 2.55 is known, these equations can be solved to obtain the correction that needs to be applied to the original estimate of the displacement vector. This solution will require very little extra computation, providing $[U]$, the upper-triangular decomposition of $[K]$, has been saved.

Naturally, due to round-off effects, the value of $\delta\mathbf{d}$ thus obtained will itself be slightly in error, and the process of iterative correction may be repeated. However, it is quite likely that a single correction will give sufficient improvement in the accuracy of the displacement vector, particularly if the vector on the right-hand side of equations 2.55 is calculated using double-precision arithmetic.

2.9   BOUNDARY CONDITIONS

In the FE analysis of structural problems, it is usually required to find a set of nodal deflections that correspond to a specified set of nodal or distributed loads given that the displacement of supported nodes is zero in particular directions.

However, in metalforming analysis, the deforming loads are rarely known beforehand, whereas the components of displacement of certain nodes will be determined by the boundary conditions of the problem under consideration.

For example, if a node is in contact with a die during a forging operation, the displacement of that node perpendicular to the die surface will be the same as that of the die itself. (The displacement parallel to the die surface may not be known, depending as it does upon a number of factors, including the frictional conditions of the interface between the die and the workpiece.)

Although the *applied* nodal forces are not generally known, the resultant force will be known at many nodes, specifically those inside the body and those on the free surfaces of the workpiece. At these nodes, the resultant force is zero and it is the displacement that must be determined.

At the start of the analysis then, most of the components of **f** are known to be zero, but some components are unknown. The components of **d** corresponding to the latter will be specified by the boundary conditions of the problem. The solution of the stiffness equations must therefore take these known components of displacement into account and calculate the corresponding reactions, as well as all the other components of displacement.

There are several ways of accomplishing this. One method, which involves multiplying certain coefficients of $[K]$ by very large numbers is described in reference [2.5] and is incorporated into the BASIC program listed in Appendix 11. Appendix 6 describes another method, one which modifies the Gaussian elimination and back-substitution procedure. This appendix also considers the more complicated situation in which the specified components of displacement are not parallel to any of the global axes.

2.10   VARIATIONAL METHODS

Before going on to consider small-deformation elastic–plastic FE analysis, it is worthwhile mentioning some of the other numerical techniques that can be used to study plastic deformation problems.

2.10.1 Variational method of solution in continuum theory

Most plasticity analysis has used a variational approach, developed from conventional continuum methods [2.6].

In these, differential equations are set up, governing the behaviour of infinitesimal domains. An integral I is thus defined, which can sometimes be determined directly. If not, the variational principle is invoked, stating that the correct solution is the one that minimises the value of I.

Since several independent variables may be involved, I is expressed as a functional, that is, as a function of several other functions, for example:

$$I = \int_{x^1}^{x^2} F(x, \phi, \phi', \phi'') dx \qquad (2.56)$$

where ϕ' and ϕ'' are the first and second derivatives of the function $\phi(x)$. The problem is then to find the function $\phi(x)$ which gives a minimum (stationary) value of I.

Usually the function I has a clear physical meaning, for example it may be the potential energy, which is a function of both displacement and position. In some instances, it is possible to determine both maximum and minimum values of two functionals I, and thus to produce upper and lower bounds to the solution of a more complex problem, but this is unlikely in plasticity analysis.

2.10.2   Approximate solutions by the Rayleigh–Ritz method

The problem can be stated formally as the requirement to minimise the functional $I(\phi)$ in a volume V, subject to set boundary conditions on the boundary area S:

$$I = \int F(x, \phi, \phi') dV \qquad (2.57)$$

According to the Rayleigh–Ritz method [2.7], an approximate solution is assumed for the field variable:

$$\hat{\phi}(\mathbf{x}) = \sum_{i=1}^{n} c_i g_i(\mathbf{x}) \qquad (2.58)$$

where $g_i(\mathbf{x})$ are known functions defined over the volume and surface. The parameters c_i have to be determined, to satisfy the condition:

$$\frac{\partial I(\hat{\phi})}{\partial c_i} = 0, i = 1, n \qquad (2.59)$$

which leads to n simultaneous equations.

The accuracy of the solution depends on the choice of the 'trial' functions g_i. These must be part of a set that produces a convergence to the correct solution, and they must be continuous up to the necessary degree of differentiation in the functional, etc. Usually these functions are in the form of a polynominal or a trigonometric expression.

The Rayleigh–Ritz method resembles the general finite-element approach.

2.10.3   Weighted residuals

The method of weighted residuals [2.8], used to obtain approximate solutions to linear and non-linear differential equations in continuum problems, does not require a knowledge of the functional. The statement of the problem takes the form:

$$F(\phi) = G(\phi) \text{ in a volume } V \qquad (2.60)$$

subject to a set of boundary conditions on the surface S. The field variable itself is again approximated, as in equation 2.58.

The next step is to define a residual or error functional R, such that:

$$R(\hat{\phi}) = G(\hat{\phi}) - F(\hat{\phi}) \tag{2.61}$$

and R has to be minimised for the best solution.

A weighted function of this residual, $w \cdot p(R)$ is assumed to satisfy the acceptable residual condition. If the function is chosen so that $p(R) = 0$ when $R = 0$, then the corresponding function $\hat{\phi}(x)$ will be equal to the exact function $\phi(x)$.

The trial function $\hat{\phi}(x)$ is chosen to satisfy the boundary conditions and the acceptably small residual is given by the integration over the whole volume:

$$\int w \cdot p(R) \mathrm{d}V = 0 \tag{2.62}$$

There are several ways of using this approach, of which the Galerkin method is most widely used [2.9]. Another uses the 'least squares', by weighting the squares of residuals.

2.10.4 Galerkin's method

In this method the weighted functions are chosen to be the known functions g_i and the n separate integrals are each equated to zero:

$$\int g_i(R) \mathrm{d}V = 0, \ i = 1,n \tag{2.63}$$

There are thus n simultaneous equations to be solved for the n unknown coefficients c_i.

2.11   THE FINITE-DIFFERENCE METHOD

Mention should also be made of a completely different method of solving non-steady-state problems. The basis of this is to approximate a first derivative by a small finite step. For example if the quantity T, which might be temperature, varies with time:

$$\frac{\mathrm{d}T}{\mathrm{d}t} \simeq \frac{T^1 - T^0}{\Delta t} \tag{2.64}$$

where:

$$T^1 = T(t + \tfrac{1}{2}\Delta t) \ ; \ T^0 = T(t - \tfrac{1}{2}\Delta t) \tag{2.65}$$

and Δt is a small time interval.

This is known as the finite-difference method [2.10].

A known initial condition is used to find the solution at time $(t^0 + \Delta t)$, and the solution proceeds in small sequential steps.

This method will not be used for isothermal or adiabatic plasticity problems but it is useful for heat flow analysis and can be used in the temperature part of a coupled thermo-mechanical solution.

References

[2.1]   Zienkiewicz, O.C. *The Finite-element Method*, 3rd Ed., McGraw-Hill (1977).

[2.2]   Rao, S.S. *The Finite Element Method in Engineering*, Pergamon (1982).

[2.3]   Press, W.H., Flannery, B.P., Teukolsky, S.A. and Vetterling, W.T. *Numerical Recipes*, Cambridge University Press (1986).

[2.4]   Kershaw, D.S. The Incomplete Choleski-Conjugate Gradient method for the iterative solution of systems of linear equations. *J. Comp. Phys.* **26**, 43–65 (1978).

[2.5]   Cheung, Y.K. and Yeo, M.F. *A Practical Introduction to Finite Element Analysis*, Pitman (1979).

[2.6]   Strutt, J.W. (Lord Rayleigh) On the theory of resonance. *Trans. Roy. Soc.* **A161**, 77–118 (1870).

[2.7]   Ritz, W. Uber eine neue Methode zur Lösung gewissen Variations. *J. Reine Angew. Math.* **135**, 1–61 (1909).

[2.8]   Mikhlin, S.C. *Variational Methods in Mathematical Physics*, Macmillan (1964).

[2.9]   Galerkin, B.G. Series solution of some problems of elastic equilibrium of rods and plates. *Vestn. Inzh. Tech.* **29**, 897–908 (in Russian) (1915).

[2.10] Southwell, R.V. *Relaxation Methods in Theoretical Physics*, Clarendon Press (1946).

3   Small-deformation elastic–plastic analysis

3.1   INTRODUCTION

The methods described in Chapters 1 and 2 are generally applicable though the discussion so far has been related largely to linear elastic problems. In these problems the [D] matrix does not change during deformation and the changes in the [B] matrix due to the changes in nodal co-ordinates are small.

In plastic deformation, and especially in metalforming, the strains may be very large. A change of 30–40% in diameter of a drawn wire or height of a forging, for example, can often be required. In extrusion the strains may be much greater still, reaching logarithmic strains of 5 or more. This introduces very severe local distortion and we shall consider the problems involved later, in Chapters 5 and 6.

Even when the deformation involves only a small amount of plastic strain, the yield stress will change in a non-linear manner. The metal deforms elastically at first with essentially constant values of Young's Modulus and Poisson's Ratio, but as soon as the stress reaches the yield value, plastic deformation occurs with a much lower effective modulus. The yield stress itself increases with strain due to work hardening, and stress is no longer proportional to strain, but is related to the strain increment and, at least for hot working, also to the strain rate.

In this chapter, we shall consider only isothermal plastic deformation at temperatures such as the normal ambient, well below the range of hot working. To avoid serious mesh distortion problems we shall also assume that the deformation is moderate, say less than 10%.

3.2   ELEMENTS OF PLASTICITY THEORY

3.2.1   Yielding

Figure 3.1 shows two representative stress/strain curves, respectively for commercially pure aluminium, which is widely used as a model material, and for an aircraft alloy 7075.

These results are obtained by compression testing, since tensile tests are limited in total extension. For FE analysis they can be regarded as flow stress curves, showing how the flow stress Y increases with plastic strain. They are commonly represented by simple constitutive equations, for example:

$$Y = B \cdot (\bar{\epsilon}^{\mathrm{P}})^m \tag{3.1}$$

where $\bar{\epsilon}^{\mathrm{P}}$ is the accumulated plastic strain to be defined later.

More accurate analysis requires better fitting of the curves, using polynomials, or more complex expressions. For example, the yield stress of commercially-pure aluminium may be expressed as a logarithmic and exponential function of plastic strain for small strains, and as a linear function for larger values:

$$
\begin{aligned}
Y &= 162.4 + 21 \cdot \ln(\bar{\epsilon}^{\mathrm{P}} + 0.00465) + 9 e^{4.5(\bar{\epsilon}^{\mathrm{P}} - 0.6909)}, & \bar{\epsilon}^{\mathrm{P}} &< 0.7 \\
Y &= 164.8 + 72(\bar{\epsilon}^{\mathrm{P}} - 0.7), & \bar{\epsilon}^{\mathrm{P}} &> 0.7
\end{aligned}
\tag{3.2}
$$

As well as knowing the numerical value of the flow stress under these highly controlled and simplified conditions, it is essential to know how the material will react under a combination of stresses. For this purpose the well-known von Mises yield criterion is used. In terms of principal stresses (defined as the direct stresses on planes where the shearing stress is zero), this criterion is usually written:

$$(\sigma_1 - \sigma_2)^2 + (\sigma_2 - \sigma_3)^2 + (\sigma_3 - \sigma_1)^2 = 2Y^2 = 6k^2 \tag{3.3}$$

in which k is the yield stress in pure shear. The von Mises criterion is often interpreted as being determined by the shear strain energy.

Fig. 3.1 Representative stress/strain curves for (a) commercially-pure aluminium and (b) aluminium alloy 7075.

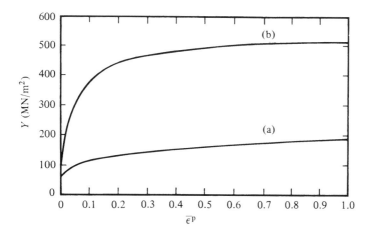

3.2.2   Deviatoric and generalised stresses

In plastic deformation, which is essentially a process of shear, we are concerned less with principal stress and more with the shear stresses, or in general terms with the *deviatoric* stresses σ'_{ij}. These are defined in contradistinction to the hydrostatic stress or mean stress σ^h, which, being uniform in all directions, causes only an elastic volume change, with no change in shape:

$$\sigma^h = \frac{1}{3}(\sigma_{11} + \sigma_{22} + \sigma_{33})$$

(3.4)

$$= \frac{1}{3}\sigma_{ii}$$

where the suffix notation implies summation of all the terms.

The deviatoric stress is then the difference between the total stress and the hydrostatic stress:

$$\sigma'_{ij} = \sigma_{ij} - \delta_{ij}\sigma^h$$

(3.5)

The Kronecker delta δ_{ij} is defined to be equal to one if $i=j$, and equal to zero if $i \neq j$. Thus a deviatoric shear component is the same as the total shear component.

It is convenient to express the yield criterion in terms of deviatoric stress components:

$$3(\sigma'^2_{11} + \sigma'^2_{12} + \sigma'^2_{13} + \sigma'^2_{21} + \sigma'^2_{22} + \sigma'^2_{23} + \sigma'^2_{31} + \sigma'^2_{32} + \sigma'^2_{33}) = 2Y^2 \quad (3.6)$$

or more concisely:

$$\frac{3}{2}\sigma'_{ij}\sigma'_{ij} = Y^2$$

(3.7)

The numerical coefficients in these expressions are required because Y was chosen to be the axial stress in a plastically-deforming uni-axial tension specimen. Under these conditions:

$$\sigma^h = \sigma_{11}/3; \quad \sigma'_{11} = 2\sigma_{11}/3; \quad \sigma'_{22} = \sigma'_{33} = -\sigma_{11}/3$$

(3.8)

and all the shear components are zero.

The function of deviatoric stress components on the left-hand side of equation 3.7 occurs frequently in plasticity theory and it is useful to define the square root of this to be the *generalised* stress $\bar{\sigma}$:

$$\bar{\sigma} = \left(\frac{3}{2}\sigma'_{ij}\sigma'_{ij}\right)^{1/2}$$

(3.9)

3.2.3   Constancy of volume in plastic deformation

An important consequence of the division of the stress into deviatoric and hydrostatic components is that it separates the distortional and dilatational elements of the strain.

Hydrostatic stress is uniform and has no effect on the conditions for yielding, though it may cause a normally brittle material to deform plastically by suppressing the tensile failure, which is a quite different matter.

Because hydrostatic stress does not produce shear, there is no plastic change of shape; the effect of hydrostatic stress is to produce elastic change in volume which is fully recoverable on release of the load.

A corollary of this is that the plastic deformation due to shear does not produce a change in volume. This is a well-established experimental fact, and is indeed one of the important constraints that must be imposed upon a plastic or elastic–plastic FE solution.

If a deformation is assumed to be purely plastic, with no elastic strain, then in terms of the principal strains ϵ_i:

$$dV = 0; \quad d\epsilon_1 + d\epsilon_2 + d\epsilon_3 = 0 \tag{3.10}$$

however large the individual strain components may be. This is the simplification made by rigid-plastic and visco-plastic FE formulations. As we have seen, in reality plastic deformation will be accompanied by elastic changes in volume. In the more rigorous elastic–plastic formulation that is the main subject of this monograph, the strain increments in equation 3.10 are interpreted as being the *plastic* parts of the total strain increments.

3.2.4. Decomposition of incremental strain

The assumption is made that the plastic and elastic components of incremental strain can be separated. This is believed to be strictly valid for infinitesimal strains and appears to represent real behaviour accurately also for finite strains. Thus:

$$d\epsilon_{ij} = d\epsilon_{ij}^e + d\epsilon_{ij}^p \tag{3.11}$$

3.2.5 Generalised plastic strain

In the same way that a generalised stress was defined, a generalised plastic strain increment can be written in the form:

$$d\bar{\epsilon}^p = \left(\frac{2}{3} d\epsilon_{ij}^p \, d\epsilon_{ij}^p \right)^{1/2} \tag{3.12}$$

in which it should be noted that $d\epsilon_{ij}^p = d\epsilon_{ij}^{p'}$, since the volume component of plastic strain increment is zero.

Once again, the coefficient in equation 3.12 is chosen so that $d\bar{\epsilon}^p = d\epsilon_{11}^p$ for simple plastic tension, since in this situation, the condition for zero plastic volume change gives that:

$$d\epsilon_{22}^p = d\epsilon_{33}^p = -\frac{1}{2} d\epsilon_{11}^p \tag{3.13}$$

and the shear components of incremental strain are again zero.

The accumulated plastic strain $\bar{\epsilon}^p$ is defined to be the sum of all the infinitesimal increments of plastic strain from the start of the deformation:

$$\bar{\epsilon}^p = \int d\bar{\epsilon}^p \qquad (3.14)$$

3.2.6 The relationship of stress to strain increment: the Prandtl–Reuss equations

In viscous flow of a Newtonian fluid, the shear stress is proportional to the shear strain rate:

$$\sigma'_{ij} = \eta \frac{d\epsilon'_{ij}}{dt} \qquad (3.15)$$

This equation, or a modified version in which the coefficient of viscosity η is not constant, can be used in the FE analysis of metal or polymer deformation at high temperature. Many metalforming problems have been handled by this method [3.1,3.2].

An alternative approach, adopted in this chapter, is to ignore in the first instance the influence of strain rate and to concentrate on the hardening effects of strain, as in cold metalforming.

According to Reuss [3.3], there is a direct proportionality between the deviatoric stress and the plastic strain-increment:

$$d\epsilon^{p\prime}_{ij} = \sigma'_{ij} d\lambda \qquad (3.16)$$

To this must be added the elastic strain:

$$d\epsilon^{e\prime}_{ij} = \frac{1}{2G} d\sigma'_{ij} \qquad (3.17)$$

where G is the Shear or Rigidity Modulus.

The condition of volume constancy in plastic deformation is:

$$d\epsilon^p_{ii} = 0 \qquad (3.18)$$

and for hydrostatic tension or compression (Appendix 7):

$$d\epsilon^e_{ii} = \frac{3(1-2\nu)}{E} d\sigma^h \qquad (3.19)$$

Collecting these components together:

$$d\epsilon'_{ij} = d\epsilon^{e\prime}_{ij} + d\epsilon^{p\prime}_{ij}$$

$$= \frac{1}{2G} d\sigma'_{ij} + \sigma'_{ij} d\lambda \qquad (3.20a)$$

$$d\epsilon_{ii} = d\epsilon^e_{ii} + d\epsilon^p_{ii}$$

$$= \frac{1-2\nu}{E} d\sigma_{ii} \qquad (3.20b)$$

These are the Prandtl–Reuss equations [3.4].

3.2.7   Elastic–plastic constitutive relationship

We require for an elastic–plastic FE formulation a relationship between the incremental strain and the incremental stress in a plastically-deforming body similar to equation 2.17.

The starting point is the Prandtl–Reuss equations 3.20. In Appendix 3 it is shown how these can be used to derive a relationship of the form:

$$d\sigma_{ij} = 2G \left(d\epsilon_{ij} + \delta_{ij} \left(\frac{\nu}{1-2\nu} \right) d\epsilon_{mm} - \frac{\sigma'_{ij}\sigma'_{kl}d\epsilon_{kl}}{S} \right) \qquad (3.21)$$

where:

$$S = \frac{2}{3}\bar{\sigma}^2 \left(1 + \frac{Y'}{3G} \right) \qquad (3.22)$$

and Y' is the rate of change of yield stress Y with respect to plastic strain $\bar{\epsilon}^p$.

In terms of the stress increment vector:

$$d\sigma = (d\sigma_{11}, d\sigma_{22}, d\sigma_{33}, d\sigma_{12}, d\sigma_{23}, d\sigma_{13})^T \qquad (3.23)$$

and the strain increment vector:

$$d\epsilon = (d\epsilon_{11}, d\epsilon_{22}, d\epsilon_{33}, d\gamma_{12}, d\gamma_{23}, d\gamma_{13})^T \qquad (3.24)$$

this relationship may be expressed as:

$$d\sigma = [D]d\epsilon \qquad (3.25)$$

where:

$$[D] = [D^e] - [D^p] \qquad (3.26)$$

$[D^e]$ is the elastic stress/strain matrix presented in Chapter 2; $[D^p]$ is described in Appendix 3.

3.2.8   Strain hardening in FE solutions

For small but finite increments of deformation, we may generalise equation 3.25:

$$\triangle\sigma = [D]\triangle\epsilon \qquad (3.27)$$

For an infinitesimal strain increment, the value of the slope Y' in equation 3.22 is found from the tangent at the appropriate strain. The slope, however, generally decreases with strain, so that at the end of a finite increment of deformation the value will be lower.

There are several ways of improving the accuracy of the constitutive relationship for finite deformation steps. The simplest of these uses the mean value of the tangents at the beginning and end of the step, which is essentially the same as using the slope of the secant, as in figure 3.2. This is therefore referred to as the secant modulus method.

Other approaches will be discussed later.

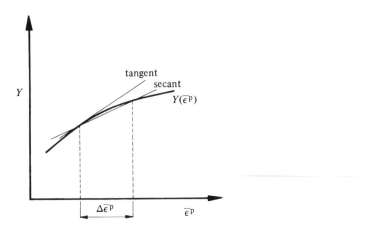

Fig. 3.2 The secant approximation to slope Y' for a finite deformation.

3.2.9   Force/displacement relationship: the stiffness matrix $[K_i]$

We assume a constant stiffness over each small increment of deformation and obtain a relationship between incremental displacement $\triangle \mathbf{d}_i$ and incremental force $\triangle \mathbf{f}_i$ for element i:

$$[K_i] \triangle \mathbf{d}_i = \triangle \mathbf{f}_i \qquad (3.28)$$

The element stiffness matrix takes the same form as that derived in Chapter 2 (equation 2.36):

$$[K_i] = \int [B_i]^T [D_i][B_i] \mathrm{d}V_i \qquad (3.29)$$

but now we use the elastic–plastic constitutive matrix associated with the current state of element i (q.v. equation 3.26).

3.3   EXAMPLE ANALYSIS USING THE SMALL-DEFORMATION FORMULATION

To illustrate the use of the small-deformation formulation, we return to the simple example of a tensile test considered in Chapter 1.

Figure 3.3 shows the FE mesh used in this example. As before, the mesh represents an axi-symmetric test specimen with gauge diameter 11.2 mm and gauge length 100 mm. Nodal points 7 and 8 are assumed to be fixed and prescribed axial displacements of 0.25 mm per increment are imposed upon nodes 1 and 2. The value of E is 2×10^5 N/mm^2 and ν is 0.33. The material yields initially at a stress of 80 N/mm^2 and the slope Y' is assumed to have constant value of 100 N/mm^2.

The deformation is modelled using the BASIC demonstration program listed in Appendix 11, and is continued for 10 increments – corresponding to 5% elongation. By this stage all the elements have started to deform plastically, and a neck at the mid-point (node 2) is clearly beginning to form (figure 3.4). The very coarse mesh results in the neck being far more diffuse than would be the case in practice, with large plastic deformation taking place at some distance away from the mid-point. This example is, however, a useful demonstration of the technique.

Fig. 3.3 (a) tensile-test specimen and (b) FE model.

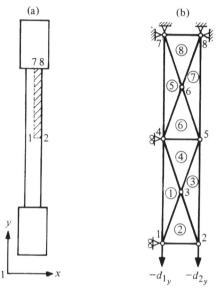

Fig. 3.4 Simulated deformation and distribution of generalised stress (N/mm²) for tensile test.

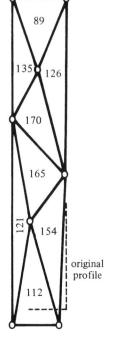

References

[3.1]   Rebelo, N. and Kobayashi, S. A coupled analysis of visco-plastic deformation and heat transfer. *Int. J. Mech. Sci.* **22**, 699–705, 707–18 (1980).

[3.2]   Pittman, J.F.T., Zienkiewicz, O.C., Wood, R.D. and Alexander, J.M. (Eds.) *Numerical Analysis of Forming Processes*, Wiley (1984).

[3.3]   Reuss, A., Berücksichtigung der elastichen Formänderung in der Plastizitätstheorie. *Zeits angew. Math. Mech.* **10**, 266–74 (1930).

[3.4]   Hill, R. *The Mathematical Theory of Plasticity*, Clarendon Press (1950).

4 Finite-element plasticity on microcomputers

4.1 MICROCOMPUTERS IN ENGINEERING

The previous chapters of this book have focussed attention on the basic FE theory for elastic and small-deformation elastic–plastic applications. The latter were illustrated by a very simple analysis of the tensile test.

In this chapter, we shall look at problems more closely associated with realistic metalforming processes. The first part of the chapter examines a PC-program written in BASIC that can be used for demonstrations or for student tutoring. A listing of this program is given in Appendix 11.

The second part of the chapter describes a FORTRAN program, implemented on a large-memory micro, that is designed for more serious metalforming studies.

No attention has been given so far to the machines on which FE analyses can be performed. Where mainframe computers or workstations are available, this is not important but in microcomputer applications, how the machine operates and the importance of selecting the right operating system, software and processors cannot be taken for granted. The decision as to what equipment is most appropriate must be made by the individual user, based on carefully assessed requirements.

A general introduction to microcomputers is given by Avison [4.1] in which many aspects of micro hardware and software are discussed, and comment on buying a micro is included. Hardware and software selection is also discussed by Samish [4.2] and Lane [4.3] respectively, while Bell [4.4] gives an overview of the role of computers in engineering. Some background knowledge is essential when trying to identify the merits or otherwise of various microcomputer systems. When the various single-user or multi-user operating systems are considered, together with different programming languages, central processors and also peripherals, the task of selecting the right system appears formidable. But with reference to the predetermined requirements a short list of suitable combinations can soon be produced from which the final selection can be made.

The use of micros to aid the solution to problems in many fields of engineering is rapidly expanding. These cover heat transfer, piping design, solids handling and transport modelling as well as the usual CAD and general stress analysis systems. O'Connell *et al.* [4.5] list 235 packages solely for mechanical engineering design. These commercial packages however, represent only the beginnings of a comprehensive software range for engineers.

Most of the software developed to date is for linear problems and once the appropriate algorithms have been established their implementation on a micro is not unduly demanding. The major problems, whether on a mainframe or a micro, are in the initial specification of the problem and in the interpretation of the mass of results that computers can quite easily produce. The solution of non-linear problems on micros introduces additional complexities. Either the random access memory (RAM) of the machine is too small or the running time is too long, or both. Careful attention must be given to how the program is structured to take full and efficient advantage of the micro. Killingbeck [4.6] describes many features of developing programs for scientific applications in physics. Brown [4.7] lists and describes the construction of various FE programs written in BASIC for solving mechanical engineering problems on a micro. In contrast to the linear analysis systems, very few non-linear programs have been written. Examples include time-dependent heat flow problems [4.8, 4.9] and the dynamic behaviour of structures [4.10].

The FE analysis of non-linear plasticity problems poses two major obstacles to its implementation on a micro. One is the large amount of data that needs to be handled and repetitively transferred to and from storage, and the second is the need to perform an analysis in numerous increments, each at least equivalent to a linear elasticity problem. The following sections describe how these obstacles may be overcome to solve realistic metalforming problems.

All the examples in the following sections of this chapter are based on small-deformation plasticity theory, as described in the previous chapter. The more correct, and more useful, large-deformation plasticity theory on which any critical analyses should be based is described in subsequent chapters.

4.2   NON-LINEAR PLASTICITY DEMONSTRATION PROGRAMS ON A MICROCOMPUTER USING BASIC

4.2.1   Introduction

Innumerable variations in microcomputers and operating systems together with the wide variety of peripherals, result in each particular combination having its own particular characteristics. The interchangeability of components and compatibility of microcomputers is still far from ideal. The introduction of non-linear programs to microcomputers is a demanding task on both developer and machine, and any system will require some modification when a program is transferred from one machine to another. Despite this, there are a number of techniques which can be used to aid the development of non-linear programs. The techniques themselves will be applicable to any micro system though of course their implementation will vary to suit the specific hardware and software requirements.

This section will describe techniques used for running a non-linear program on small (in terms of RAM capacity) machines. These programs are particularly suitable for demonstration or tuition purposes. Techniques for much larger micros for more realistic simulation will be described in Section 4.3.

4.2.2   Structure of the finite-element program in BASIC

The small micro used for the example here is an Olivetti M24, but similar programs have been run also on Sirius, IBM, Tektronix and BBC microcomputers, with minor changes in the file handling and graphics. A small-strain FE formulation, described in Chapter 3, is used and the program to examine 2-D metalforming problems is written in BASIC. This in itself does not pose any difficulties. The greatest problems arise in trying to structure the FE program into a form which could be run using the small amount of RAM available. A listing of the program developed is given in Appendix 11.

A simple flow chart for a mainframe program is shown in figure 4.1 and depicts the familiar sequence of calculating the $[B_i]$ and $[D_i]$ matrices, followed by the element stiffness matrices $[K_i]$. Once the global stiffness matrix $[K]$ is complete, the displacements, stresses and strains are found as usual. Each of these operations can easily be accommodated within a mainframe computer memory, but for the smaller machine this is not the case and frequent use of disc storage must be made. The new structure required to accomplish this is illustrated in the flow chart shown in figure 4.2.

The BASIC FE program consists of a series of separate files on a disc. One file contains all the instructions necessary to set up the computer model. This initial generation of the data may be accomplished with a small predetermined program or interactively. In either case the data must be transferred to a specified file on the disc. The initial program is then deleted from RAM (although retained on disc) and the next file loaded. This contains the routines to construct the $[B_i]$ and $[D_i]$ matrices for each element. This file must therefore recall the relevant mesh and property data from disc storage, evaluate the above matrices, then store this information back on the disc. This is continued until all the elements in the mesh have been considered, and the RAM is then cleared. The program is directed to the next file and, after retrieving the data previously set up for each element in turn, evaluates the element stiffness matrices [K$_i$], and finally the global stiffness [K]. Once again this information is transferred back to the disc and the RAM cleared. The final file is then loaded for retrieval of the stiffness matrix. The displacement of each nodal point is evaluated, and hence the strain and stress in each element may be determined. The material properties and mesh geometry are then updated ready for the next increment. Although this approach has been described as one of interchange between machine and disc, at no time is any of the program detail deleted from the disc.

4.2.3   Simple application and comparison to mainframe results

The familiar process of simple axi-symmetric upsetting is chosen in order to examine the operation of the BASIC program. The workpiece and FE mesh are shown in figure 4.3. The mesh is very crude but is sufficient to demonstrate the micro program.

Fig. 4.1 Simplified flow chart of FORTRAN FE program used on a main-frame computer.

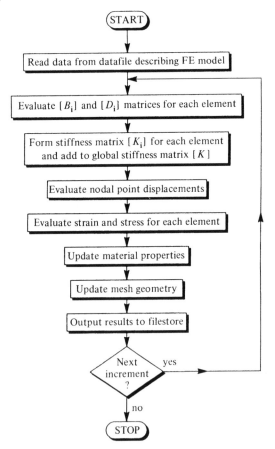

In this demonstration, only the two extreme boundary conditions of zero and sticking friction are considered. The process is modelled until a reduction of height of 20% is reached.

Fig. 4.2 Flow chart for BASIC FE program used on a microcomputer.

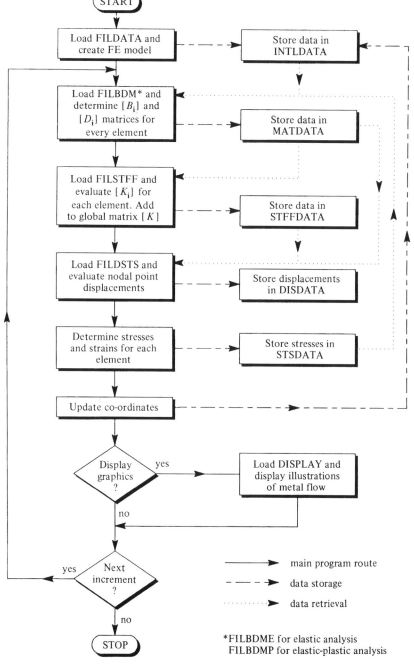

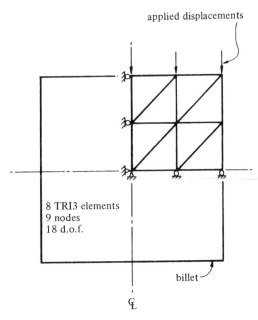

Fig. 4.3 Cylindrical billet and FE model.

In the first case, with theoretically no surface restraint, the workpiece should deform in a completely uniform manner. As a further check the uniform plastic strain that should develop in the workpiece can be evaluated easily from $\bar{\epsilon}^P = \ln(H/h)$, where H is the initial height and h is the final height of the billet. In this case, at 20% reduction, $\bar{\epsilon}^P = 0.22314$.

Figure 4.4a shows the FE predictions using the BASIC program. The model has clearly deformed homogeneously with a generalised plastic strain of 0.2209, an error of 1%.

Fig. 4.4 Distorted grids and distribution of generalised plastic strain for (a) zero friction and (b) high friction at 20% reduction in height. Results using Olivetti microcomputer.

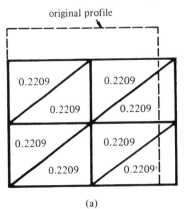

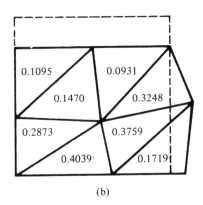

(a)   (b)

When the second limiting boundary condition is imposed, no relative movement of the die/billet interface occurs, and the deformation pattern is very different. The high interface friction results in a pattern of flow in which a central zone with little plastic deformation $\bar{\epsilon}^P = 0.1095$, is formed in the area in contact with the die. This accords with the known pattern of a conical dead zone, as suggested also by the plane-strain slip-line field analogy. The majority of plastic deformation, according to slip-line theory and experiment is restricted to diagonal regions located across the billet joining opposing corners on the section. This can be seen in broad terms in the simple model. Intermediate levels of deformation are found near the equatorial free surface. All these features, even in restricted detail due to the coarse mesh, can be identified on the FE predictions using the micro and illustrated in figure 4.4b.

Analyses with identical meshes, and another with a much finer mesh [4.11] have also been performed using a CDC7600 mainframe computer for which the programs were initially developed. In the case of zero friction the average plastic strain for the mainframe coarse mesh analysis is 0.222, only 0.5% less than the theoretically-exact value. There are some small variations found in the distribution of plastic strain (figure 4.5a). The predictions using the micro show less variation. The mainframe solution for sticking friction, figure 4.5b, gives a pattern of deformation similar to that found with the micro.

As a further check on the results from the micro, the variation of plastic strain along a circumferential section may be compared to that obtained using a much finer mesh of 842 elements on the mainframe. In general good agreement is found (figure 4.6).

These results clearly demonstrate the feasibility of adapting non-linear programs to run on small machines. These provide valuable demonstration and

Fig. 4.5 Distorted grids and distribution of generalised strain for (a) zero friction and (b) high friction at 20% reduction in height. Results using a CDC7600 mainframe computer.

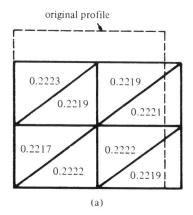

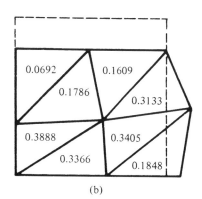

(a)   (b)

tuition examples, allowing a student to observe the construction of the program in full detail and even to check selected calculations manually. The programs can be fully interactive and give a good background to real, practical, analysis.

4.3   A 'LARGE' FORTRAN-BASED SYSTEM FOR NON-LINEAR FINITE-ELEMENT PLASTICITY

4.3.1   Hardware selection

The following features are suggested as desirable requirements for the microcomputer system:

- (i)   16-bit machine
- (ii)   Programming language should be FORTRAN
- (iii)   Operating system should be standard
- (iv)   Large RAM should be available
- (v)   Disc drive, preferably one floppy and one hard disc
- (vi)   Single user system
- (vii)   Good resolution graphics
- (viii) System should include printer and plotter

The 16-bit machines offers some advantages over 8-bit machines in terms of accuracy and speed, and the more powerful 32-bit models can be recommended if funds permit.

The use of FORTRAN as the programming language not only aids future transfer of the program from one micro to another but makes the task of transference from a mainframe computer easier.

The machine RAM must be at least 256k with simple facilities for additional memory. The availability of a hard disc drive allows the large amounts of data produced by non-linear FE analyses to be easily and rapidly stored ready for further processing if necessary. Graphical display of the results is particularly important. A reasonable screen resolution is important for assessing the output.

Fig. 4.6 Generalised plastic strain predictions using various computers and meshes for the circumferential section AB: —— CDC7600, 842 TRI3 elements; ---- CDC7600, 8 elements; –·– Olivetti M24, 8 elements.

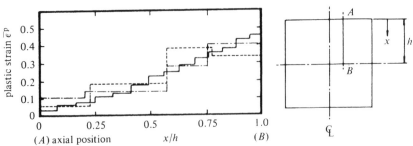

Provisions must be made for producing hard copies of the results, so any system should be complemented with a suitable printer and plotter.

Numerous 16-bit machines are available and though the majority are developed for business use, not scientific applications, several can, in principle, be used. The final choice will involve reliability, after-sales support, country of manufacture, delivery time, location of suppliers and the total cost. A number of systems would meet the requirements but the following system was set up for the work described here:

Microcomputer	Future Computer FX30
Operating system	Concurrent CPM-86
Main processor	Intel 8088 (8 MHz)
Working RAM	768k
Programming language	SSS FORTRAN
Disc storage	784k floppy, 10M hard disc
Graphics	PLUTO graphics system
Peripherals	Dot matrix printer, A4 plotter

The concurrent CPM-86 operating system was included to enable more than one task to be undertaken simultaneously. This was particularly useful for program development when frequent changes were made. The programming language, 'Small Systems Services' FORTRAN (SSS FORTRAN) is equivalent to ANSI-66 FORTRAN. The normal directly addressable space in RAM for a 16-bit machine of the type used here is 64k. The additional RAM could be accessed by using specially prepared routines written in assembler language or by using a different FORTRAN compiler as described in section 4.3.5. One floppy disc was augmented with a 10 Mbyte Winchester disc. An A4 flat bed plotter and a dot matrix printer were also included. The components of the system are illustrated schematically in figure 4.7, and further details can be found in reference [4.12].

Fig. 4.7 Components of the microcomputer system hardware and schematic layout.

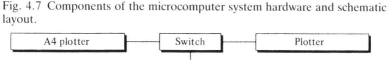
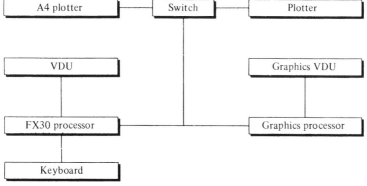

4.3.2   Program transfer

Two small-strain programs were transferred from the mainframe computers. These were the programs developed by Hartley [4.11] and Al-Sened [4.13] for the study of 2-D metalforming problems. The former used only constant-strain triangular elements with three nodes (TRI3), while the latter was extended to use eight-node isoparametric quadrilateral elements (QUAD8). The major contents of the programs remained unaltered but the means of data input/output and graphics all require modification. The SSS FORTRAN compiler requires a minimum of 128k of RAM for the initial program compilation. The operating system requires 40k and the executable file of either program needs 96k. So for running any sensible problem within the machine RAM, thus avoiding many data transfers, at least 200k is necessary to allow for reasonable amounts of data storage. Constants and blank common blocks are limited to 64k blocks. This imposes some restriction on the number of elements that could be considered in a model if the program is to run entirely within the machine RAM. Although this restriction can be overcome by various methods the examples in the immediately following sections are all conducted within the above limitations. Later sections (4.3.4 and 4.3.5) describe the alternative methods and further examples are given.

4.3.3   Applications of the non-linear system

4.3.3.1   Axi-symmetric upsetting

Simple upsetting of an axi-symmetric billet between flat parallel platens is used here for initial assessment of the microcomputer programs. The process was analysed using both TRI3 elements and QUAD8 elements. The results are compared with those obtained using the original mainframe programs.

Previous experimental results and solutions using the mainframe FE programs [4.14] are available. For the present analysis high friction (no lubricant, $m = 0.7$) is assumed at the die and platen interface. (See Chapter 6 for a description of the technique used to include the effects of interface friction.) The FE mesh representing the aluminium workpiece is shown in figure 4.8.

The internal deformation predicted by the microcomputer program using TRI3 elements is shown in figure 4.9. It can be seen that the profiles clearly indicate significant barrelling. The edge profiles exhibit some oscillations at high reductions in height, attributable to the coarse mesh. A finer mesh would give more precise results. The predictions are almost identical to those produced by the mainframe program. Figure 4.10 shows the distribution of generalised stress values, at 50% reduction in height of $m = 0.7$. The average percentage variation in the stress values compared to those produced by the mainframe program is 0.06%.

The computing times for these examples reveal an interesting trend. Figure 4.11 shows the computing time plotted as a function of the total number of degrees of freedom for the two solutions corresponding to the use of (a) the mainframe computer and (b) the microcomputer. The percentage increase in computing time with the number of degrees of freedom is roughly the same for both computers, but the micro takes about 40 times as long for this type of solution.

The predictions for simple upsetting with the QUAD8 elements are closely similar to those obtained with TRI3 elements. In general fewer of these elements

Fig. 4.8 Initial FE meshes for upsetting.

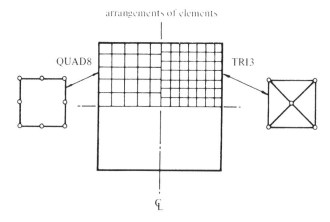

Fig. 4.9 Predicted grid distortion in simple upsetting using a microcomputer with TRI3 elements at 50% reduction with high friction.

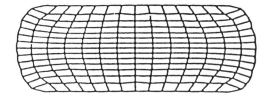

Fig. 4.10 Generalised stress (N/mm^2) distribution predicted with TRI3 elements at 50% reduction using the microcomputer.

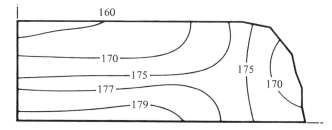

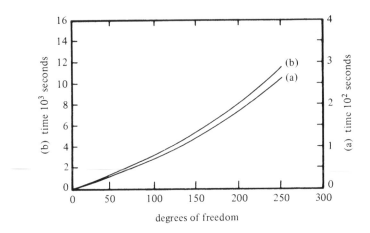

Fig. 4.11 Computing time per increment for (a) the mainframe computer and (b) the microcomputer.

are needed to achieve the same degree of accuracy. Although the use of QUAD8 elements does not reduce the computational time, the reduction in the number of elements generally reduces the file space needed to contain the data and results.

4.3.3.2   Cold heading

To assess the use of both types of element in a more complicated process, cold heading of an axi-symmetric billet is considered (figure 4.12). The specimen has a diameter of 50 mm, a free-length to diameter ratio of 1 and a gripped length of 50 mm. The incremental displacement of the upper punch for each step is 1% of the original free height of the billet. Figure 4.12 shows the initial FE meshes.

Fig. 4.12 Initial FE meshes for cold heading.

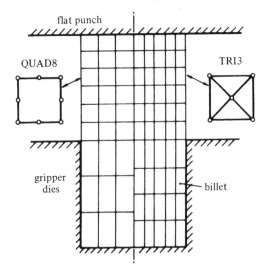

Symmetry of the die about the centre line of the billet means that the solution need only be carried out over half the billet. This symmetry determines the boundary conditions on the centre line, where shear traction and normal displacement are both zero. Two models are used for the theoretical comparisons. In the mesh used for the first model, 220 TRI3 elements with 254 degrees of freedom and 220 integration points are used. In the second model, 30 QUAD8 elements with 224 degrees of freedom and 120 integration points are used. Again, $m = 0.7$ is used for high friction. The effect of friction in the gripper dies is not considered.

The distorted grids are shown in figure 4.13. The distorted grid lines are somewhat smoother with the QUAD8 elements than with the TRI3 elements and in particular the local deformation at the inner corner below the head is more accurately modelled by the less rigid QUAD8 elements.

4.3.4   Overcoming the FORTRAN compiler limitations

To allow more elements to be used in the analysis it is necessary to overcome the restrictions on the amount of normally addressable RAM that can be accessed. The major problem in the non-linear analysis is storing the banded stiffness matrix during the solution routine. The direct Gaussian elimination technique has been retained throughout this work (see Appendix 5). A front solver technique would reduce the storage required but would also result in increased computing times.

When not located in the RAM a large stiffness matrix can be stored in one of two ways:

(i)   The matrix can be sorted in the out-of-bounds memory within the machine memory but not within the current RAM being used for the main program. This is designated EPFEPAOM

Fig. 4.13 Predicted grid distortion in cold heading at 50% reduction and high friction, (a) using QUAD8 elements and (b) using TRI3 elements. Microcomputer results.

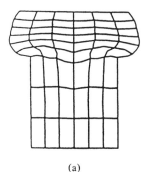
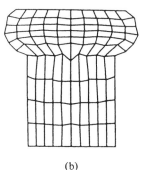

(a)   (b)

(ii) The matrix can be stored in a random access file on the hard disc. This is designated EPFEPMSD

These two methods together with the original version (EPFEPM) are compared in terms of computing time for various numbers of elements in the FE mesh. Once again the familiar process of simple upsetting with high friction will be used as the benchmark test. Only QUAD8 elements will be considered. In the original version EPFEPM, a maximum of 30 elements could be used. This could be increased to about 250 for the EPFEPAOM technique and quite easily 1000 elements when the EPFEPMSD technique is used.

Figure 4.14 shows clearly the expected result that the programs run more quickly if data transfer to any medium can be avoided. When the RAM storage limit is reached data transfer to the hard disc is the most efficient.

4.3.5 Introducing an improved FORTRAN compiler and the 8087 maths co-processor

4.3.5.1 System improvements

The system developed allows the analysis of many practical plane-strain and axi-symmetric metalforming problems; for many 2-D processes 100 QUAD8 elements are adequate except where severe geometric changes are involved. To reduce running times to an acceptable level for these, two further enhancements to the system can be introduced:

(i)   the addition of an 8087 mathematics co-processor

(ii) the addition of a new FORTRAN compiler, Pro-FORTRAN

The 8087 maths co-processor is specifically designed to deal with mathematics operations and does so much quicker than the 8088 processor. It does not replace

Fig. 4.14 Computing times per increment for three different data storage and transfer techniques. Microcomputer results.

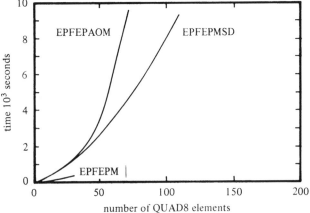

the latter chip but operates alongside it. To make full use of the maths processor the appropriate library routines in the FORTRAN compiler must be specified. Although not done here some programs may also need re-structuring to speed up the mathematics processing [4.15].

The Pro-FORTRAN compiler allows the RAM memory to be accessed in 64k pages. This access to large blocks of data will allow the data transfer to be done much quicker than using the EPFEPAOM method described earlier which could transfer data only in very small units. The paging type of access is achieved by incorporating additional routines written in assembler language within the FORTRAN compiler.

Incorporating the 8087 mathematics co-processor reduced the running times of the examples shown earlier to about 70% of the original processing time. The Pro-FORTRAN compiler reduced this by a further 75%, so the running times were typically a fifth of the original. This reduction was largely due to the reduced time in data transfer in RAM locations.

4.3.5.2   Analysis of upsetting and heading with refined meshes

To demonstrate typical results that can be achieved with realistic FE models on a micro, two examples, upsetting and heading, are again considered.

The upsetting process is analysed using 440 TRI3 elements, and distorted grids and displacement vectors are shown in figure 4.15. These are very similar to the earlier results but show improved free surface profiles and more internal detail. The running time for this mesh was 1850 seconds per increment.

Fig. 4.15 Microcomputer predicted distorted grids and displacement vectors in simple upsetting.

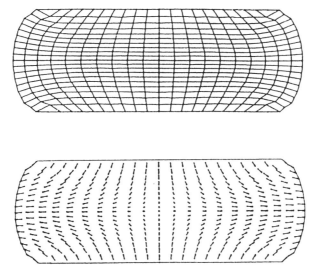

Heading is analysed using 105 QUAD8 elements. Figure 4.16 shows improved results compared to those found earlier. The running time in this case was 1440 seconds per increment.

The introduction of the 8087 co-processor and the Pro-FORTRAN compiler means that realistic metalforming analyses can be conducted within 24 hours on a microcomputer. This may still seem lengthy to the user who, having seen trivial results produced immediately on a VDU, expects the computer to produce all results instantly. Nevertheless, the present system offers a practical means of obtaining very detailed solutions to metalforming problems relatively quickly.

4.4   SUMMARY

In this chapter we have seen examples of metalforming process simulation using a small-strain FE formulation. This type of formulation is also referred to as 'small-deformation' or as 'infinitesimal-strain'. The main point here is that this approach is strictly applicable only to situations in which the plastic strain is of the same order of magnitude as the elastic strain. When this is not the case, the small-strain formulation cannot calculate accurate values of the components of strain and stress, even though, as shown by the upsetting and heading examples in this chapter, the generalised values of strain and stress and the overall pattern of flow may be reasonable.

For a complete and much more accurate model of metalforming plasticity, a more rigorous formulation, the finite-strain approach, must be used. This approach is described in the following chapters.

Fig. 4.16 Microcomputer predicted distorted grids and displacement vectors in cold heading.

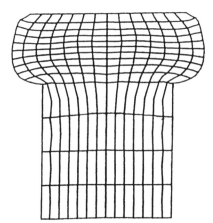

References

[4.1] Avison, D.E. *Microcomputers and their Commercial Applications*, Blackwell Scientific Press (1983).

[4.2] Samish, F. *Choosing a Microcomputer*, Granada (1983).

[4.3] Lane, J.E. *Choosing Programs for Microcomputers*, National Computer Centre (1980).

[4.4] Bell, W.T. The role of microcomputers in civil and structural engineering. *Engineering Software II, Proc. 2nd Int. Conf. on Engineering Software*, ed. R.A. Adey, CML Pubs., pp. 852–61 (1981).

[4.5] O'Connell, C., Sheviak, J.K., Browne, A.R. and Johnson, S.V. *Directory of Microcomputer Software for Mechanical Engineering Design*, Marcel Dekker (1985).

[4.6] Killingbeck, J.P. Microcomputer calculations in physics. *Rep. Prog. Phys.* **48**, 54–99 (1985).

[4.7] Brown, D.K. *An Introduction to the Finite Element Method using Basic Programs*, Surrey University Press (1984).

[4.8] Moir, P.J. and Stoker, J.R. Theta: A desktop computer solution for non-linear time dependent heat flow problems. *Engineering Software II, Proc. 2nd Int. Conf. on Engineering Software*, ed. R.A. Adey, CML Pubs., pp. 796–805 (1981).

[4.9] Stelzer, J.F. Consideration and strategies in developing finite element software for desktop computers. *Eng. Comp.* **1**, 106–24 (1984).

[4.10] Backx, E. and Rammant, J.P. Structural dynamic interactive analysis in Basic on micros. *Engineering Software II, Proc. 2nd Int. Conf. on Engineering Software*, ed. R.A. Adey, CML Pubs., pp. 842–51 (1981).

[4.11] Hartley P. Metal flow and homogeneity in extrusion-forging, Ph.D. thesis, University of Birmingham, UK (1979) (unpublished).

[4.12] Hussin, A.A.M. Transference of mainframe finite-element elastic–plastic analysis to microcomputers, and its application to forging and extrusion, Ph.D. thesis, University of Birmingham, UK (1986) (unpublished).

[4.13] Al-Sened, A.A.K. Development of simulation techniques for cold forging sequences, Ph.D. thesis, University of Birmingham, UK (1984) (unpublished).

[4.14] Hartley, P., Sturgess, C.E.N. and Rowe, G.W. Influence of friction on the prediction of forces, pressure distributions and properties in upset forging. *Int. J. Mech. Sci.* **22**, 743–53 (1980).

[4.15] Startz, R. *8087 Applications and Programming for the IBM PC and other PCs*, R.J. Brady Company, Prentice-Hall (1983).

5 Finite-strain formulation for metalforming analysis

5.1   INTRODUCTION

The first applications of the FEM were concerned with structural problems, and so it is not surprising that when the FEM began to be applied to the modelling of plastic deformation, this was viewed simply as an extension of non-linear elastic behaviour, with the elastic stress/strain matrix in the FE formulation replaced by an appropriate plastic one, as described in Chapter 3. When the amounts of plastic deformation and material displacement are small, this is a valid approach and is often used in the study of plastic failure of structures such as pressure vessels. The small-strain technique also has the advantage of being easily understood, thus providing a good introduction to the principles and underlying FE plasticity, and of being easily incorporated into available elastic FE programs.

As the examples in the previous chapter show, the small-strain approach can give good approximations to the overall pattern of deformation in certain simple forming processes. However, this technique cannot accurately predict the values of important workpiece parameters, such as the components of strain and stress, particularly if the metalforming process involves appreciable material rotation. This is due partly to the nature of elastic–plastic deformation, for which correct definitions of stress and strain increment must be chosen, and partly to the large total and incremental strains involved in metalforming, which require special numerical techniques for their evaluation.

This chapter re-examines the FEM in the light of these considerations in order to derive a formulation that is able to model large-strain elastic–plastic deformation. The next chapter considers how this formulation can be implemented in a computer program to study practical metalforming operations.

The theory so far has been presented mainly in matrix notation, since this is the most familiar approach. However, the ideas that are considered in this chapter are more clearly, and more naturally, expressed using tensors. For those unfamiliar with suffix notation, or with the concept of tensors, Appendix 8 gives a brief introduction to the subject.

The FE formulation is derived here for a general 3-D deformation: the corresponding 2-D expressions (for plane-strain, plane-stress or axi-symmetric geometries) may be obtained, if necessary, as a special case of the 3-D ones.

It should be noted that throughout this chapter and the next, all expressions refer to a single element of the FE mesh. However, for reasons of clarity, the subscripts denoting this have been omitted.

5.2 GOVERNING RATE EQUATIONS

5.2.1 Rate of potential energy for updated-Lagrangian increment

Since metalforming processes may involve large amounts of deformation, the FE analysis must be divided into a number of steps, or increments, each considering just a small part of the total deformation.

Consider an element of the body at the start of an increment of deformation. Let the reference Cartesian co-ordinates of a particle at point P within the element be x_i. During the increment, the element deforms so that at some later instant the reference co-ordinates of the particle, now at point P', are x_i'. These co-ordinates will be functions of the initial co-ordinates and of time.

The total rate of increase in potential energy of an element of the deforming material is the rate of increase in the internal deformation energy of the material, less the rate of decrease of the potential energy due to the movement of the points of application of external applied forces. The average rate of increase of the internal energy per unit volume is the dyadic product (that is, the tensor equivalent of the scalar product of vectors) of the deformation-rate tensor and the average stress tensor.

In the updated-Lagrangian approach considered here, the state variables are specified with respect to the reference configuration at the start of the increment, so the rate of change of internal energy is required per unit *initial* volume. The deformation rate is therefore the derivative of the instantaneous velocity of a particle with respect to its initial reference co-ordinates.

Thus, if the displacement of the particle is $u_i = x_i' - x_i$, the deformation rate is:

$$\dot{u}_{i,j} = \frac{\partial \dot{x}_i'}{\partial x_j} \tag{5.1}$$

For the same reason, the correct stress to use in this context is the *nominal* stress s_{ij} (see Appendix 9 for a description of the different stress tensors).

The FEM assumes that the forces acting on the body are concentrated at the nodes of an element. If f_{lm} is the mth reference component of force acting at node I, and d_{Im} is the corresponding component of the displacement of this node from some convenient starting point, the average rate of increase of potential energy $\dot{I}$ of the element during time dt is:

$$\dot{I} = \int (s_{ij} + \tfrac{1}{2}\dot{s}_{ij}\,dt)\dot{u}_{j,i}\,dV - (f_{Im} + \tfrac{1}{2}\dot{f}_{Im}dt)\dot{d}_{Im} \tag{5.2}$$

where the integration is carried out over the volume V of the element. Henceforth, the usual summation convention will be extended to include summation over all the nodes of an element when an upper-case subscript is repeated in a term of an expression.

5.2.2   Minimisation of rate of potential energy

The velocity field is such that the rate of increase of potential energy of the element is a minimum. By the calculus of variations, the condition for the functional $\dot{I}$ to have a stationary value with respect to the functions $\dot{u}_{j,i}$ and $\dot{d}_{Im}$ is that:

$$\delta(\dot{I}) = \int (s_{ij} + \tfrac{1}{2}\dot{s}_{ij}\,dt)\delta(\dot{u}_{j,i})\,dV - (f_{Im} + \tfrac{1}{2}\dot{f}_{Im}\,dt)\delta(\dot{d}_{Im}) = 0 \tag{5.3}$$

in which $\delta()$ denotes an arbitrary variation in the function enclosed in the parentheses.

Since the element is in equilibrium at the start of the increment, the first-order terms in equation 5.3 cancel to give:

$$\delta(\dot{d}_{Im})\dot{f}_{Im} = \int \delta(\dot{u}_{j,i})\dot{s}_{ij}\,dV \tag{5.4}$$

5.2.3   Incorporation of strain rate

To complete the governing rate equations, it is necessary to express the stress rate in terms of strain rate by means of the appropriate elastic–plastic constitutive law. The strain rate is defined to be the symmetric part of the deformation-rate tensor:

$$\dot{\epsilon}_{ij} = \tfrac{1}{2}(\dot{u}_{i,j} + \dot{u}_{j,i}) \tag{5.5}$$

The constitutive relationship normally involves the rate of true or *Cauchy* stress σ_{ij}, but since the integration in equation 5.4 is carried out over the *initial* volume of the element, the correct measure to use in this instance is the rate of Kirchhoff stress τ_{ij}, which is simply equal to the Cauchy stress multiplied by the ratio of the current to the initial volume. Furthermore, the rate of stress used in the constitutive relationship must vanish during rigid-body rotation, when $\dot{\epsilon}_{ij}$ vanishes. For this reason, the time derivative of Kirchhoff stress is measured in the Cartesian co-ordinate system, which is co-incident with the reference system at the start of the increment, and which rotates with the element. This derivative is often called the rotationally-invariant or *Jaumann* derivative $\dot{\tau}^*_{ij}$. In Appendix 10, it is shown that this is related to the rate of nominal stress by:

$$\dot{s}_{ij} = \dot{\tau}^*_{ij} - \sigma_{kj}\dot{\epsilon}_{ik} - \sigma_{ik}\dot{\epsilon}_{kj} + \sigma_{ik}\dot{u}_{j,k} \tag{5.6}$$

The elastic–plastic constitutive law therefore takes the form:

$$\dot{\tau}^*_{ij} = D_{ijkl}\dot{\epsilon}_{kl} \tag{5.7}$$

Substitution of equations 5.6 and 5.7 into equation 5.4, using the symmetric properties of the Kirchhoff stress-rate and the strain-rate tensors, gives the basic rate expression:

$$\delta(\dot{d}_{lm})\dot{f}_{lm} = \int \left[\delta(\dot{\epsilon}_{ij})(D_{ijkl} - 2\sigma_{ik}\delta_{jl})\dot{\epsilon}_{kl} + \delta(\dot{u}_{j,i})\,\sigma_{ik}\dot{u}_{j,k} \right] dV \tag{5.8}$$

where δ_{jl} is the Kronecker delta introduced in Chapter 3.

5.2.4   Importance of correct choice of stress rate

In FE formulations designed for small-scale plasticity, the constitutive relationship is assumed to be between strain rate and the rate of *nominal* stress. This leads to a much simpler FE governing expression but is invalid for formulations intended for metalforming applications, in which gross deformation can occur. It has been shown [5.1] that the use of the nominal stress rate in the flow expression leads to significant errors in the resulting stiffness equations unless the current modulus of plasticity is much greater than the current yield stress, a situation rarely met in practical metalforming processes. The small-strain technique, therefore, cannot accurately model plastic deformation however small the increments are made.

In addition, a small-strain formulation may be inaccurate, even for small-strain elastic deformation, if the magnitudes of the displacements are large, as the following example shows.

Two FE analyses of the elastic deformation of a single cubic element are performed (figure 5.1), in which simple tension is combined with rotation. In one of the FE analyses, the small-strain constitutive relationship, involving nominal stress rate, is used while, in the other, the formulation is based upon the finite-strain governing equation 5.8.

The element is assumed to start with a tensile stress in the x_1 direction; the stress and forces acting on the element at any later stage in the rotated co-ordinate system can easily be deduced from Hooke's law. The reference components may therefore be calculated using Mohr's circle and compared with the FE results.

Even when the incremental tensile strain is as small as 0.0016%, and the angle of rotation is as small as one degree per increment, figure 5.2 shows that the small-strain approach predicts incorrect values of stress and force in the element, whereas the· predictions of the finite-strain solution agree with the analytical values, even up to 90 degrees rotation.

5.3   GOVERNING INCREMENTAL EQUATIONS

5.3.1   Modification of rate expression

In the present formulation, the fundamental variables are the changes in the various parameters during each increment of the deformation, so the various time derivatives in equation 5.8 need to be replaced by the corresponding increment, during time step $\triangle t$, using:

$$\dot{d}_{lm}\triangle t \simeq \triangle d_{lm} \tag{5.9}$$

$$\dot{f}_{lm}\triangle t \simeq \triangle f_{lm} \tag{5.10}$$

Fig. 5.1 Combined extension and rotation of a single element.

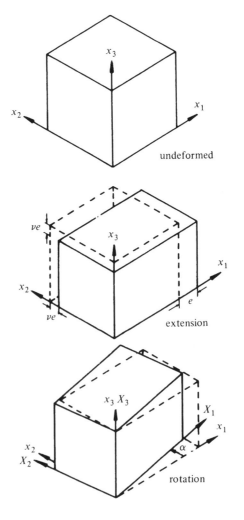

$$\dot{u}_{i,j}\Delta t \simeq \Delta u_{i,j} = \frac{\partial(\Delta u_i)}{\partial x_j} = \frac{\partial(\Delta x_i')}{\partial x_j} \qquad (5.11)$$

$$\dot{\epsilon}_{ij}\Delta t \simeq \Delta \epsilon_{ij} \qquad (5.12)$$

where $\Delta u_{i,j}$ is called the deformation gradient. The approximations in equations 5.9–5.12 will be valid provided the deformation increment is not too large (for example, smaller than 2 or 3%, which is larger than the value usually chosen for FE analysis).

5.3.2 Effect of rotation

By multiplying equation 5.5 by the time increment and substituting from equation 5.11, it is possible to express the increment of strain as the symmetric part of the deformation gradient. However, if the change in strain is defined in this way

Fig. 5.2 Comparison of analytical values of stress and force with FE predictions using small-strain and finite-strain formulations: —— theoretical; ○ FE – small-strain; □ FE – finite-strain.

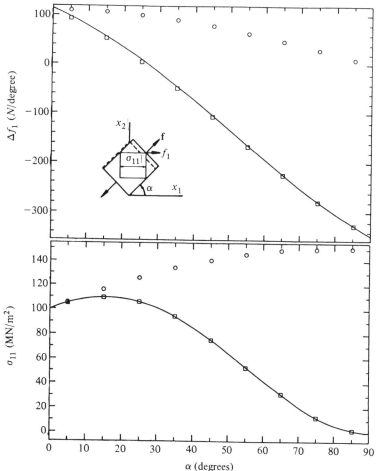

(the infinitesimal definition) it is easily demonstrated that a non-zero value is calculated when the material is undergoing rigid-body rotation, a situation for which the strain increment should, by definition, be zero.

More generally, whenever there is a rotational component to the deformation, the infinitesimal definition of incremental strain predicts an anomalous change in volume. Since in metal deformation the volume can change only elastically, the erroneous volume strain will result in the calculation of very large elastic values of hydrostatic stress.

5.3.3   LCR expression for strain increment

To avoid these problems, an alternative method of calculating the strain increment is adopted, which incorporates a correction for any rotational component of the deformation. This leads to the definition of the *linearised co-rotational (LCR)* strain increment [5.2]:

$$\triangle \epsilon_{ij} = \frac{1}{2}\left[r_{ki}\triangle u_{k,j} + r_{kj}\triangle u_{k,i} - \frac{1}{2}(\langle r_{ik}\rangle\langle\triangle u_{j,k}\rangle + \langle r_{jk}\rangle\langle\triangle u_{i,k}\rangle) \right] \quad (5.13)$$

where the angle brackets denote the skew-symmetric part of the enclosed tensor, for example:

$$\langle\triangle u_{i,j}\rangle = \frac{1}{2}(\triangle u_{i,j} - \triangle u_{j,i}) \quad (5.14)$$

r_{ik} is the rotational part of the unique decomposition of the real and non-singular deformation mapping into orthogonal and symmetric components. Thus:

$$r_{ik}\, q_{kj} = \frac{\partial(\triangle x_i')}{\partial x_j} + \delta_{ij} \quad (5.15)$$

where:

$$r_{ki}\, r_{kj} = \delta_{ij} \text{ and } q_{ij} = q_{ji} \quad (5.16)$$

In practice, the values of the rotational matrices are estimated during the previous step of the analysis, so the right-hand side of equation 5.13 is a linear function of displacement gradients. It can be seen from this equation that the LCR increment of strain is the same as the infinitesimal value when the material is not rotating. Furthermore, the tensor $r_{ij} - \delta_{ij}$ is approximately skew-symmetric providing the incremental angles of rotation of the material do not exceed about ten degrees. Thus for pure rotations up to this magnitude, equation 5.13 leads to approximately zero values of strain increment [5.2].

The importance of using the correct definition of strain increment is clearly demonstrated by figure 5.3. This shows two FE predictions (one with LCR strain, the other using the infinitesimal definition) of the force and stress during the combined extension and rotation of a single element considered earlier, and compares these with analytical values.

With the LCR definition of strain increment, and the deformation gradient defined in equation 5.11, equation 5.8 may be rewritten in its incremental form:

$$\delta(\triangle d_{lm})\triangle f_{lm} = \int \Big[\delta(\triangle \epsilon_{ij})(D_{ijkl}-2\delta_{jl}\sigma_{ik})\triangle \epsilon_{kl}$$

$$+ \delta(\triangle u_{j,i})\sigma_{ik}\triangle u_{j,k}\Big]dV \tag{5.17}$$

5.4   ELASTIC–PLASTIC FORMULATION

5.4.1   Yield criterion

So far, no special assumptions have been made about the nature of the material being deformed, and hence about the form of the constitutive matrix D_{ijkl}. It is now necessary to obtain the particular constitutive matrix that may be applied

Fig. 5.3 Comparison of analytical values of stress and force with FE predictions using infinitesimal and LCR definitions of strain increment: —— theoretical; ○ FE – infinitesimal-strain; □ FE – LCR strain.

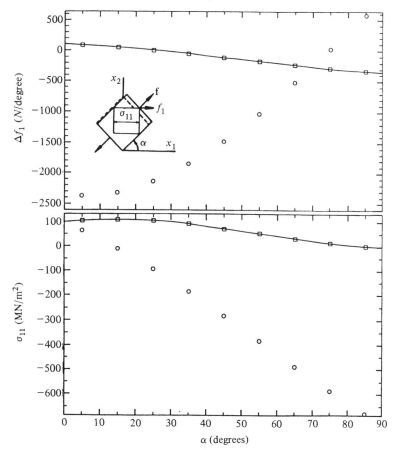

to the deformation of metals. Clearly, the general theory may just as easily be applied to the deformation of other materials, such as rocks or soils, provided that the appropriate constitutive matrix is defined.

Most common metals appear to obey the von Mises yield criterion described in Chapter 3. Previously, we have assumed that yield stress depends only upon the value of plastic strain. Now we consider the more general statement that a region will deform plastically when its generalised stress $\bar{\sigma}$ reaches a critical value determined by the accumulated plastic strain $\bar{\epsilon}^{\mathrm{p}}$, the strain rate $\dot{\bar{\epsilon}}^{\mathrm{p}}$ and the temperature T, that is when:

$$\bar{\sigma}^2 = \frac{3}{2}\sigma'_{ij}\,\sigma'_{ij} = Y^2(\bar{\epsilon}^{\mathrm{p}}, \dot{\bar{\epsilon}}^{\mathrm{p}}, T) \tag{5.18}$$

5.4.2   Elastic–plastic flow rule

As mentioned in Chapter 3, a basic assumption of a non-viscous elastic–plastic formulation is that an increment of strain may be divided into its elastic (recoverable) and plastic (irrecoverable) parts. This assumption appears to be valid providing the increments are not too large [5.3]. Normality of the plastic strain increment to the yield locus in stress/strain space [5.4], and the use of the generalised form of Hooke's law for the elastic component leads as before to the Prandtl–Reuss flow expression [5.5]:

$$\triangle \epsilon_{ij} = \frac{1}{E}((1 + \nu)\triangle\sigma_{ij} - \nu\delta_{ij}\triangle\sigma_{kk}) + \triangle\lambda\sigma'_{ij} \tag{5.19}$$

5.4.3   Elastic–plastic constitutive relationship

The Prandtl–Reuss equations may be rearranged [5.6] to obtain the incremental form of equation 5.7. This is described in Appendix 3. From equation A3.18 the elastic–plastic constitutive matrix D_{ijkl} is defined to be:

$$D_{ijkl} = 2G \left(\delta_{ik}\,\delta_{jl} + \delta_{ij}\,\delta_{kl} \left(\frac{\nu}{1-2\nu} \right) - \frac{3\sigma'_{ij}\,\sigma'_{kl}}{2\bar{\sigma}^2 \left(1 + \dfrac{Y'}{3G} \right)} \right) \tag{5.20}$$

In practice, the last term of equation 5.20 is omitted if an element is deforming elastically.

5.4.4   Effect of plastic incompressibility

In rigid-plastic and similar FE formulations, an explicit constraint must be imposed upon the solution in order to enforce volume constancy during plastic deformation. There are several methods of doing this, one example being the penalty-function technique [5.7]. All of these can be shown to be equivalent to multiplying the bulk-strain terms in the governing equations by a large number.

With an elastic–plastic formulation however, plastic incompressibility is an implicit assumption of the Prandtl–Reuss flow rule (equation 3.18).

As the governing equations stand, this tends to enforce plastic volume constancy at every point of these elements. Since each element has only a limited number of degrees of freedom, the effect is to over-constrain the deformation [5.8].

In the actual workpiece, the condition of volume constancy will apply throughout the plastic region, thus introducing one constraint for every yielded particle, each of which has three independent components of displacement. This ideal ratio of 3:1 for the number of degrees of freedom to the number of volume constraints is rarely met in an FE model, unless special techniques are used. For example, in an FE mesh there are three degrees of freedom for each node. The ratio of degrees of freedom to volume constraints will therefore depend upon the number of constraints per element and upon the ratio of the number of nodes to the number of elements in the mesh. Both of these quantities depend upon the type and arrangement of the elements. Thus in a regular mesh of 3-D eight-node linear elements, the numbers of nodes and elements are approximately equal, provided that the number of elements is large, and there are seven volume constraints per element. The constraining ratio is therefore 3:7. A regular 3-D 20-node quadratic element has sixteen volume constraints and, when the number of elements is large, there are four times as many nodes as elements in a regular mesh. The constraining ratio in this case is therefore 3:4. In neither of these examples does the constraining ratio approach the ideal value.

Since the nodal displacements in an FE mesh have to satisfy so many additional constraining conditions imposed by the requirements of volume constancy, the FE solution tends to predict unrealistic modes of flow, particularly when plasticity is fully developed. This requires some relaxation of the constraints.

5.4.5   Element-dilatation technique

The element-dilatation technique, first proposed by Nagtegaal, Parks and Rice [5.9], resolves the problem of over-constraint by separating the expression that determines the volume change out of the governing equations. This term may then be modified to produce the required number of volume constraints per element.

From equation 5.20 it can be shown that:

$$\delta(\triangle \epsilon_{ij}) D_{ijkl} \triangle \epsilon_{kl} = \delta(\triangle \epsilon'_{ij}) D_{ijkl} \triangle \epsilon'_{kl} + \kappa \cdot \delta(\triangle \epsilon_{pp}) \triangle \epsilon_{qq} \qquad (5.21)$$

in which the deviatoric strain increment is defined by:

$$\triangle \epsilon'_{ij} = \triangle \epsilon_{ij} - \frac{1}{3} \delta_{ij} \triangle \epsilon_{pp} \qquad (5.22)$$

and κ is the elastic Bulk Modulus $E/3(1-2v)$ (Appendix 7). It can be seen that the last term of equation 5.21 represents the contribution of the bulk or volume strain to the element stiffness.

The bulk-strain increment is defined to be an independent function $\triangle\phi$, called the dilatation increment, of position:

$$\triangle\epsilon_{jj} = \triangle\phi(x_i) \tag{5.23}$$

The form of the dilatation function depends upon the number of volume constraints that are required per element. For example, a regular mesh of a large number of eight-node linear elements requires one extra equation to be satisfied per element to give the desired ratio of degrees of freedom to volume constraints. Thus the dilatation function must have only one coefficient and so be constant throughout the element. The sole volume constraint is then the condition:

$$\int\triangle\epsilon_{jj}\,dV = \int\triangle\phi\cdot dV \tag{5.24}$$
$$= V\cdot\triangle\phi$$

or:

$$\triangle\phi = \frac{1}{V}\int\triangle\epsilon_{jj}\,dV \tag{5.25}$$

Alternatively, an extensive regular mesh of 20-node quadratic elements requires four volume constraints per element, so the dilatation function in this case must contain four independent coefficients defining a tri-linear function of local co-ordinates X_i:

$$\triangle\phi = \triangle\phi_0 + \triangle\phi_iX_i \tag{5.26}$$

(The local co-ordinates of a hexahedral element are defined to be plus or minus one at the corners of the element. The origin of this co-ordinate system is at the centroid of the element.)

Multiplying equation 5.23 by each of the local co-ordinates (or by one) and integrating with respect to the volume V^* of the element in the local co-ordinate system gives four constraining equations to be satisfied for each element:

$$\triangle\phi_0 = \frac{1}{8}\int\triangle\epsilon_{jj}dV^* \tag{5.27}$$
$$\triangle\phi_i = \frac{3}{8}\int X_i\triangle\epsilon_{jj}dV^*$$

Using equation 5.23, equation 5.17 may be re-written as:

$$\delta(\triangle d_{lm})\triangle f_{lm} = \int\left(\delta(\triangle\epsilon_{ij})(D_{ijkl} - 2\delta_{jl}\sigma_{ik})\triangle\epsilon_{kl} + \delta(\triangle u_{j,i})\sigma_{ik}\triangle u_{j,k}\right.$$
$$\left. - \kappa\cdot\delta(\triangle\epsilon_{ii})\triangle\epsilon_{jj} + \kappa\cdot\delta(\triangle\phi)\triangle\phi\right)dV \tag{5.28}$$

Since most of the examples of 3-D FE analyses presented in this work use eight-node linear elements, it will henceforward be assumed that one volume constraint is to be introduced into the expressions governing the deformation of each element. Most of the following theory would still apply if more than one volume constraint were required, though the resulting equations would be slightly more complex.

Substituting equation 5.25 into equation 5.28 produces:

$$\delta(\triangle d_{lm})\triangle f_{lm} = \int\Bigg(\delta(\triangle \epsilon_{ij})(D_{ijkl} - 2\delta_{jl}\, \sigma_{ik})\triangle \epsilon_{kl} + \delta(\triangle u_{j,i})\sigma_{ik}\triangle u_{j,k}$$

$$-\kappa\cdot\delta(\triangle \epsilon_{ii})\triangle \epsilon_{jj}\Bigg)\,dV + \frac{\kappa}{V}\int \delta(\triangle \epsilon_{ii})\,dV\cdot\int \triangle \epsilon_{jj}\,dV \tag{5.29}$$

5.5   ELEMENT EXPRESSIONS

5.5.1   Interpolation of nodal displacement

In order to evaluate the volume integrals in equation 5.29 it is necessary to express the strain increment and the deformation gradient at each point in terms of the nodal displacements. The incremental displacement of a particle of the element with initial co-ordinates x_j may be written in the form:

$$\triangle u_i = N_I(x_j)\triangle d_{Ii} \tag{5.30}$$

where N_I is the interpolation function of the Ith node of the element which was introduced in Chapter 2 (equation 2.10). If:

$$N_{I,i} = \frac{\partial N_I(x_j)}{\partial x_i} \tag{5.31}$$

then from equation 5.11:

$$\triangle u_{i,j} = N_{I,j}\triangle d_{Ii} \tag{5.32}$$

and from equation 5.13:

$$\triangle \epsilon_{ij} = B_{ijIm}\triangle d_{Im} = \frac{1}{2}(\bar{B}_{ijIm} + \bar{B}_{jiIm})\triangle d_{Im} \tag{5.33}$$

where B_{ijIm} is the LCR strain increment/nodal displacement increment matrix and:

$$\bar{B}_{ijIm} = r_{mi}N_{I,j} - \frac{1}{4}\langle r_{ik}\rangle(\delta_{jm}N_{I,k} - \delta_{km}N_{I,j}) \tag{5.34}$$

The exact nature of the interpolation functions depends upon the number and arrangement of the nodes of the element. Usually, they may be expressed most simply as functions of a local co-ordinate system defined for each element. In this case, the Cartesian spatial derivatives of the shape functions in equations 5.32 and 5.34 must be evaluated from the local spatial derivatives using the chain rule.

5.5.2   Incremental element-stiffness equations

Substitution of equations 5.32 and 5.33 into equation 5.29 produces:

$$\delta(\triangle d_{Im})\triangle f_{Im} = \delta(\triangle d_{Im})\iint\Big(B_{ijIm}(D_{ijkl} - 2\delta_{jl}\sigma_{ik})B_{klJn}$$

$$+ N_{I,i}\,\delta_{mn}\sigma_{ik}N_{J,k} - \kappa\cdot B_{iiIm}B_{jjJn}\ \Big)\triangle d_{Jn}\mathrm{d}V \qquad (5.35)$$

$$+ \delta(\triangle d_{Im})\ \Big(\frac{\kappa}{V}\int B_{iiIm}\mathrm{d}V\cdot\int B_{jjJn}\mathrm{d}V\ \Big)\triangle d_{Jn}$$

from which the arbitrary variations in nodal incremental displacement may be cancelled to give the element stiffness equations:

$$\triangle f_{Im} = (K^{\epsilon}_{ImJn} + K^{\sigma}_{ImJn} + K^{\phi}_{ImJn})\triangle d_{Jn} \qquad (5.36)$$

where:

$$K^{\epsilon}_{ImJn} = \int B_{ijIm}D_{ijkl}B_{klJn}\mathrm{d}V \qquad (5.37)$$

is the element *deformation stiffness matrix*, i.e. the usual infinitesimal-strain matrix,

$$K^{\sigma}_{ImJn} = \int\big(N_{I,i}\delta_{mn}\sigma_{ik}N_{J,k} - 2B_{ijIm}\delta_{jl}\sigma_{ik}B_{klJn}\big)\mathrm{d}V \qquad (5.38)$$

is the element *stress-increment correction matrix* and:

$$K^{\phi}_{ImJn} = \kappa\ \Big(\frac{1}{V}\int B_{iiIm}\mathrm{d}V\cdot\int B_{jjJn}\mathrm{d}V - \int B_{iiIm}B_{jjJn}\mathrm{d}V\ \Big) \qquad (5.39)$$

is the element *dilatation correction matrix*.

The volume integrals in equations 5.37–5.39 are most conveniently evaluated by Gaussian quadrature. To this end, the integrands may be multiplied by the Jacobian of the mapping of local to global co-ordinates to enable the integration to be carried out with respect to the volume of the element in local co-ordinates.

As described in Chapter 2, the global stiffness matrix is obtained by assembling the stiffness matrices of all the elements in the mesh.

References

[5.1]   Lee, E.H. *The basis of an elastic–plastic code.* SUDAM rep. no. 76–1, Stanford University (1976).

[5.2]   Pillinger, I., Hartley, P., Sturgess, C.E.N. and Rowe, G.W. A new linearized expression for strain increment for the finite-element analysis of deformations involving finite rotation. *Int. J. Mech. Sci.* **28**, 253–62 (1986).

[5.3]   Lee, E.H. and McMeeking, R.M. Concerning the elastic and plastic components of deformation. *Int. J. Solids Struct.* **16**, 715–21 (1980).

[5.4]   Drucker, D.C. A more fundamental approach to plastic stress–strain relations. *Proc. 1st National Congress on Applied Mechanics*, ed. E. Sternberg, ASME, pp. 487–91 (1951).

[5.5] Hill, R. *The Mathematical Theory of Plasticity.* Clarendon Press (1950).

[5.6] Yamada, Y., Yoshimura, N. and Sakurai, T. Plastic stress–strain matrix and its application for the solution of elastic–plastic problems by the finite element method. *Int. J. Mech. Sci.* **10**, 343–54 (1968).

[5.7] Price, J.W.H. and Alexander, J.M. Specimen geometries predicted by computer model of high deformation forging. *Int. J. Mech. Sci.* **21**, 417–30 (1979).

[5.8] Nagtegaal, J.C. and de Jong, J.E. Some computational aspects of elastic–plastic large strain analysis. *Computational Methods in Nonlinear Mechanics*, ed. J.T. Oden, North-Holland, pp. 303–39 (1980).

[5.9] Nagtegaal, J.C., Parks, D.M. and Rice, J.R. On numerically accurate finite element solutions in the fully plastic range. *Comp. Meth. Appl. Mech. Eng.* **4**, 153–77 (1974).

6 Implementation of the finite-strain formulation

6.1 INTRODUCTION

The previous chapter was concerned with producing a correct statement of the elastic–plastic element stiffness equations and their assembly into the global stiffness equations. As they stand, these expressions simply describe how the forces applied to the nodes of a discretised model of the workpiece change as those nodes are displaced by small amounts from their starting positions. It is now necessary to consider how these equations can be used within an FE analysis for the study of practical metalforming operations.

6.2 PERFORMING AN FE ANALYSIS – AN OVERVIEW

It is instructive to consider the entire process of performing an FE analysis of a metalforming operation (figure 6.1). Of the four main parts of this process, the FE calculation will always be undertaken by a computer program: the description of the metalforming operation and the pre-processing and post-processing stages may be integrated into the FE package, may be separate computer programs, or even be performed by hand.

The starting point in any such analysis is the metalforming operation itself. The first stage in the analysis is therefore concerned with obtaining a complete description of the operation in geometric or numerical form. This description will include information about the initial geometry of the workpiece, the shape of the dies and how the relative position and orientation of the dies and workpiece change during the deformation, the previous history of the workpiece and the dies, and the particular metal being formed.

Although not often considered when discussing FE procedures, this particular stage is just as much a part of the analysis process as the FE calculation itself, for the data used by the numerical calculation cannot be determined until the different aspects of the metalforming operation have been rigorously described. This preliminary stage is usually performed manually, but in principle the process could be automated to a certain extent by means of an expert system [6.1].

Fig. 6.1 Schematic representation of the FE analysis process.

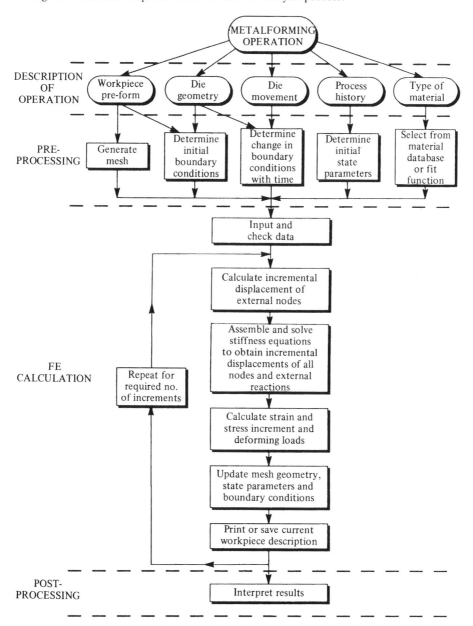

6.3   PRE-PROCESSING

The pre-processing procedures take the description of the metalforming operation and express it as numerical data that may be input into the FE program.

6.3.1   Mesh generation

An FE mesh is generated by dividing the volume of the workpiece into a number of elements, joined together at nodes. The mesh is then defined to the FE program by specifying the nodal co-ordinates and by listing the nodes belonging to each element, usually in some standard sequence that determines the topology of the element.

As a general rule, and perhaps with the exception of the constant-strain triangle in 2-D and the tetrahedron in 3-D, the commonly-available elements are all suitable for performing metalforming analyses. Although the different elements may lead to different results [6.2], providing the correct formulation is used the differences will not be large. This is in contrast to elastic FE analyses in which it may be of vital importance to choose the correct type of element for a particular problem. That this should be so is perhaps not surprising: the accuracy of the metalforming analysis is mainly a result of obtaining a correct variational principle governing metal flow, valid for any discretisation, and not in choosing a particular set of element degrees of freedom and interpolation functions to model a given strain and stress distribution.

Up to a certain point, the accuracy of the solution improves as more elements are included in the mesh, but in practice the constraints of available computer memory or time will put a limit to the number that can be used. It is desirable, therefore, to use a large number of smaller elements in the regions of the workpiece which are expected to undergo severe deformation. The generation of meshes with non-uniform element size is made easier if a selection of different element types is available within the FE program, though this is not essential.

The mesh can be generated manually, and the numerical information obtained typed into the computer, but for any except the smallest meshes this is very time-consuming and prone to error. The alternatives are either to write a computer program to generate the mesh or to use a commercial mesh-generation package, if a suitable one is available. The latter option would certainly be preferable if it is intended to examine a large number of different workpiece geometries. The disadvantages of this approach are that it may take a long time to learn to use commercial mesh generators properly, that it is sometimes difficult or impossible to obtain a mesh with precisely the required variation of element type and resolution, and that these programs are expensive.

Commercial programs may contain many thousands of lines of computer code, and yet by taking advantage of the characteristics of a particular geometric configuration, it is possible to write a program containing only a few hundred lines that can produce similar or better results. Such programs will generally be much easier to use and can of course be tailored to produce exactly the mesh required in any particular instance. If all the workpieces to be modelled are members of a small number of geometric classes, it will be possible to compile a suite of such specially-written programs.

6.3.2   Boundary conditions

There are two methods of defining mechanical boundary conditions – by means of nodal constraining conditions and by modelling die surfaces. Nodal constraining conditions specify how nodes are to move (or not to move) throughout the deformation. These would be used, for example, to make a face of the FE mesh act as a plane of symmetry of the workpiece (figure 6.2).

A constraining condition will determine which of the three Cartesian components of displacement are to be unconstrained (i.e. be obtained as solutions of the stiffness equations), and which are to have some specified (frequently zero) incremental value. Since nodes will not always be constrained upon planes that are aligned with the global axis system, it is necessary to be able to specify a nodal constraining condition with respect to a locally rotated set of axes (Appendix 6).

In practice, it is often found that many nodes are subject to the same constraining condition. It is therefore convenient to define each constraining condition once only, and then to specify the constraint applied to a given node using the constraining-condition number.

Nodal constraining conditions apply throughout the deformation, but the boun-

Fig. 6.2 Modelling planes of symmetry using constraining conditions.

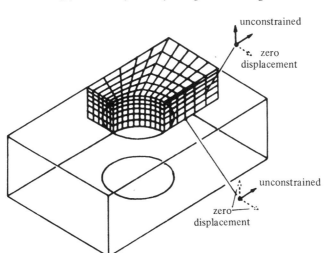

dary conditions resulting from contact between the workpiece and the dies will change as the metalforming operation proceeds. To determine the boundary conditions at any part of the outside of the FE mesh at any stage it is necessary to determine which nodes are in contact with the dies. The shape and position of the dies must therefore be made known to the FE program.

The easiest way of doing this is to model each die by means of a set of primitive geometric surfaces [6.3]. In this way the determination of nodal contact is much simplified. All but the most complex die surfaces can be described using very few primitive shapes. Figure 6.3 shows examples of some of these – other simple geometric surfaces may also be found useful.

The description of the die surface will usually contain information about the

Fig. 6.3 Primitive geometric surfaces used to model dies.

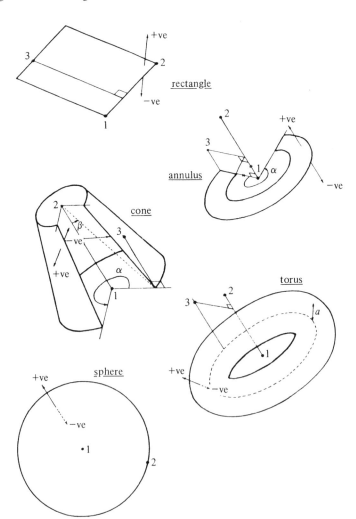

frictional restraint resulting from the particular lubricant used (if any). The heat transfer coefficient of this layer may also be specified.

As an alternative to using geometric primitives, the dies could be modelled by means of additional finite elements. In addition to permitting contact between the workpiece and the dies to be determined, this would also allow the deformation of the dies to be investigated, but with added difficulties due to the complex nature of the contact between the two surfaces. This approach will not be considered further here.

6.3.3   Material properties

These include the elastic coefficients (Young's Modulus, Poisson's Ratio), the thermal parameters (conductivity, specific heat) and the yield stress. For the purposes of the FE calculation, the elastic properties may be taken to be constant. However, it may be necessary to specify the dependence of the thermal properties upon temperature, and the yield stress will certainly vary with strain, and probably with strain rate and temperature as well.

The instantaneous values of these variables may either be calculated from some specified function, or be calculated by interpolation between empirically-derived data. The former process is much simpler but it means the FE analysis is restricted to those materials for which suitable functions have been written into the computer program.

The method of data interpolation is more flexible but may require a large amount of information to be specified, particularly if the yield function is to be considered as a function of three variables. The process of interpolation in this last case is not trivial, though it is a standard mathematical technique and library routines are available. Fortunately only the values of yield stress and its derivative with respect to strain are required, not the derivatives with respect to strain rate and temperature.

6.3.4   Deformation

One way of specifying the deformation for the FE calculation program has already been described, namely by prescribing values for the components of incremental displacement of particular external nodes (Section 6.3.2).

In general, it is not known beforehand exactly how a particular node, even an external one, will move during the metalforming operation. What is known is how the dies move. So in addition to specifying the shape and initial position of the die surfaces, it is necessary to provide information to the FE program about how these positions change during the metalforming operation deformation.

The simplest way is to specify the incremental displacement of each of the die surfaces. The specification of incremental rotations may also be necessary (figure 6.4) in order to model, for example, the rolling process.

This information allows the FE program to determine two things: where each primitive boundary surface is at the start of each increment, and the nature of the constraints to be applied to any nodes in contact with these surfaces (figure 6.5).

By specifying the displacements of boundary surfaces and boundary nodes with respect to an arbitrary increment of the deformation, it is possible, if a simple approach is required, to ignore the influence of time entirely. On the other hand, time can just as easily be included in the formulation by specifying a time step for each increment. This time step may be constant, or it may be calculated in accordance with the sinusoidal variation of deformation with time typical of a mechanical press. It is also possible to model more complicated press characteristics.

Fig. 6.4 Incremental movement of boundary surface.

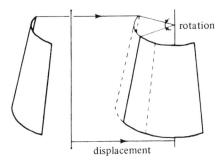

Fig. 6.5 Constraint of node in contact with moving boundary surface. P is position of node at the start of the increment; P' is the position of node at the end of the increment for sticking-friction conditions.

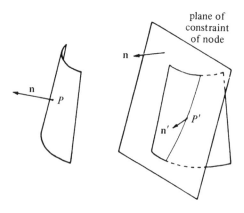

6.3.5   Initial state parameters

In order to evaluate the FE stiffness equations, and hence to calculate the pattern of nodal displacements, it is necessary to know the geometry of each element and the values of stress, strain, strain rate, temperature and incremental rotation throughout the FE mesh. The description of the initial mesh geometry has been considered earlier; the initial values of the other quantities must also be specified to the FE program. The temperature distribution resulting from the pre-heating of the workpiece, or the stress and strain distribution arising from a prior stage in the forming sequence may both be modelled quite easily. Initial values of the state variables may also be calculated as a result of a re-meshing procedure (Section 6.6.1).

6.4   FE CALCULATION

The main parts of the FE program are shown in figure 6.1. These parts will now be considered individually in more detail.

6.4.1   Input of data

The pre-processing stage of the FE analysis will have produced a file of numerical information precisely defining the deformation process. This information must now be read into the FE program. It is sensible at this stage to check the information for error.

However, the extent of the checking is open to interpretation. Thus, the internal consistency of the data, such as whether an element has the correct number of nodes, or whether all the nodes have been defined, is easy to verify, and it is fairly straightforward to pin-point values which could lead to machine overflow or other errors of calculation.

It is not so easy to determine whether the information supplied is 'sensible'. Thus an FE program designed to look at forging in the automobile industry could quite reasonably consider that any co-ordinate measured in metres rather than millimetres must be a mistake. This will cause problems if the program is later used by someone to model the forming of a marine propellor shaft. Clearly, the program could be easily modified, but it illustrates the difficulty of trying to impose sensible limits on the values of the data without unduly restricting the application of the program.

Providing such unusual values do not jeopardise the FE calculation, it is probably best just to issue warnings that the data is suspect and carry on regardless. It goes without saying that such warnings, as is the case with all error messages produced by the FE program, should endeavour to describe fully not just what is wrong, but what exactly has caused it and, if possible, what can be done to put it right.

However much error checking takes place, it is customary to echo the data as soon as they are read in to provide a record of how the program has interpreted the data file.

6.4.2   Displacement of surface nodes

The FE analysis is carried out incrementally, and at the start of each increment it is necessary to determine the values of any prescribed components of incremental nodal displacement. A component of displacement of a node may be prescribed either because a constraining condition has been specified for that node, or because the node is in contact with a die surface, or for both these reasons.

At this stage in the FE calculation therefore, the program must check the position of each external node with respect to each of the specified primitive boundary surfaces. Any node which appears to have passed through a boundary surface must be re-positioned on the surface (figure 6.6).

If the die surface is stationary, the node should be constrained to move within a plane that is tangential to the surface at point P (the point of contact of the node with the surface at the start of the increment – figure 6.5); if the die surface is moving, the node is constrained to move within a plane passing through the point P' that P will occupy at the end of the increment. The normal $\bar{n}$ of this plane of constraint is the average of n and n', where n is the normal to the surface at P at the start of the increment, and n' is the normal to the surface at P' at the end of the increment.

This rather complicated method of constraint is necessary in order to be able to deal with rotation as well as displacement of boundary surfaces. To take the two extremes, it can readily be seen that the plane of constraint is a tangent to the displaced boundary surface if it does not rotate, and that if a cylindrical surface is subject to a simple rotation about its axis, the plane of constraint is a chord to the surface passing through P and P' (figure 6.7).

The situation just described, where the node is constrained to move within a plane, will apply whenever a free-surface node comes into contact with just one

Fig. 6.6 Checking for contact between FE mesh and a boundary surface: (a) at the end of an increment, (b) after checking for contact with dies.

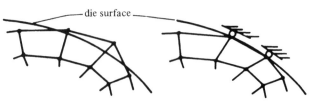

(a)   (b)

primitive boundary surface. If the node was originally subject to a constraining condition (figure 6.8) or if the node is found to be in contact with more than one boundary surface (figure 6.9), then the constraint applied to the node will need to satisfy, if it is possible, all the conditions imposed upon it. This may result in the node being made to move along a line. of intersection or, in the extreme case, being fixed in position upon the die surfaces.

In general though, a node in contact with a die is allowed to move freely parallel to the die surface throughout the increment. No restriction is otherwise placed upon this movement; any frictional restraint applying to the die surface is imposed by a separate technique (Section 6.4.3.1).

One final point should be noted in connection with the boundary-surface technique. This is that once a node has made contact with a die surface, the method outlined above will tend to keep it in contact throughout the rest of the deformation. The method should therefore also check whether the stress acting

Fig. 6.7 Constraint of a node in contact with (a) displaced die, (b) rotating die.

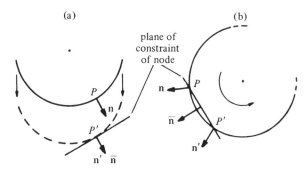

Fig. 6.8 Contact of die with node subject to previous constraining condition.

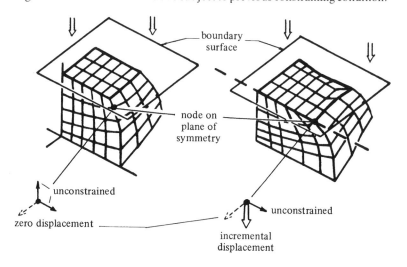

normally to the surface of the mesh at a given node is tensile, indicating that there is a tendency for the node to pull away from the die. If this is the case, the node should not be constrained upon the die surface during the next increment, even if it appears to be in contact with it.

6.4.3   Assembly and solution of the stiffness equations

The basic principles of the assembly and solution of the stiffness equations have been discussed in Chapter 2. However, several other aspects of the solution technique need to be considered in a practical metalforming program.

6.4.3.1   Incorporation of frictional restraint

The frictional restraint acting at the interface between the workpiece and the dies is an important factor in determining the pattern of flow, strain and stress in the component. In metalforming operations, the condition of macroscopically-rigid sliding surfaces is no longer met, and the familiar proportionality between the frictional force opposing the relative movement and the normal reaction no longer applies.

Instead, for a given die/workpiece interface and lubricating condition, it is found that, over most of the region of contact, the shear stress at the interface is an essentially-constant fraction m of the shear yield stress of the metal undergoing deformation. In practice this is an approximation, since the lubricant is

Fig. 6.9 Constraint of nodes in contact with arbitrary planes p^1, p^2 and p^3.

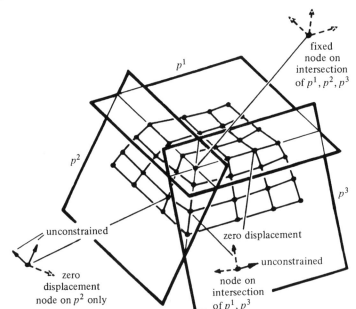

affected by the interface pressure, but the assumption is good enough for most purposes.

Thus, at one extreme, if the friction factor m equals zero, there is no tagential force acting at the interface and the sliding of the workpiece against the die is unimpeded. On the other hand, if m equals one, the plastic deformation of the workpiece at the interface must be due solely to a process of shear parallel to the die surface. Under these conditions, there can be no relative movement between the die and the material actually in contact with it.

As might be expected, these two idealised conditions of zero and sticking friction are not met in practice and the friction factor m is found to approach, but not equal, the two limiting values. The actual value is commonly determined from laboratory trials using the ring test [6.4].

If the direction of sliding of the surfaces of the workpiece with respect to the die were known from the start, the imposition of a frictional traction force would be a simple matter. In a few simple geometries the direction is indeed known, and in others experiments may be carried out to determine this information, but this is not possible if the FE program is to be generally applicable and fully predictive, for example in the ring test itself.

A method of applying a frictional restraint without prior knowledge of the pattern of flow is provided by the friction layer technique [6.5]. This requires a fictitious layer of elements to be created at the interfaces between the workpiece and the dies (figure 6.10a). The boundary technique ensures that the interface nodes can only move tangentially to the die surface, and the extra friction-layer nodes are prevented from moving in this direction. (If the dies are moving, of course, both sets of nodes may have the same superimposed motion.)

The stiffness matrix of each friction-layer element is then multiplied by a factor, the Stiffness-Matrix Multiplier (SMM), that is proportional to $m/(1-m)$ (figure 6.10b). Since the friction-layer nodes cannot move parallel to the die surface, the effect is to apply a shear force to the interface nodes acting in the opposite direction to their movement. As m tends to zero, the SMM, and hence this shear force, also tends to zero. As m tends to one, the SMM tends to infinity and the friction-layer elements become very stiff, essentially preventing any tangential movement of the interface nodes and thus sticking them to the die.

The friction layer is fictitious. It is not actually modelling a lubricant but is merely a mechanism for modelling the effects of such a lubricant. Its existence is also transitory. Each friction-layer element can be created when the stiffness matrix of the associated interface element is evaluated. The stiffness matrices of the interface element and the friction-layer element may be assembled together and a Gaussian reduction used to eliminate the friction-layer nodes from these equations (for the displacements of these nodes are known) before the equations of the interface element are assembled into the global matrix. Thus the friction-layer nodes need never form part of the main FE calculation.

6.4.3.2   Solution techniques

For reasons of economy, it is important to be able to use large step sizes in the FE analysis of metalforming operations. Many of the examples in this book have used nominal deformation increments of 1 or 2%. Since the FE stiffness equations are non-linear, the determination of the nodal displacement increments requires the use of some sort of iterative solution procedure.

For example, in the *initial-stiffness* technique [6.6], which is illustrated in figure 6.11, the set of incremental nodal displacements (represented in the diagram by $\triangle d^{(1)}$) are obtained as usual by solving the global stiffness equations for the set of applied incremental forces ($\triangle f^{(1)}$) incorporating into the solution any pre-scribed values of nodal displacement. The increments of strain, and hence stress ($\triangle \sigma^{(1)}$) throughout the mesh may then be calculated, and the incremental form of equation 5.4:

$$\delta(\dot{d}_{lm})\dot{f}_{lm} = \int \delta(\dot{u}_{j,i})\dot{s}_{ij}\, dV \qquad (6.1)$$

used to determine the values of incremental nodal force ($\triangle f^{(1)*}$) that are in equilibrium with the calculated nodal displacements.

Fig. 6.10 Friction-layer technique.

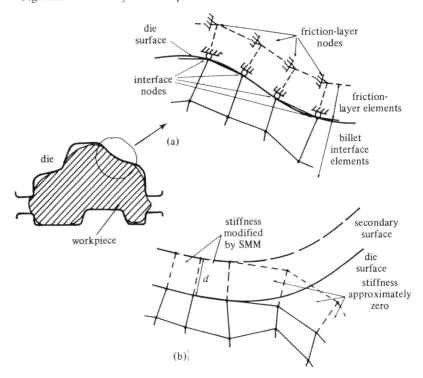

Because of the non-linearity of the stiffness equations, these nodal equilibrium forces will not, in general, be equal to the actual applied values, but the difference between them, the out-of-balance or residual forces ($\triangle f^{(2)}$), may be used to calculate a set of corrections to nodal displacement ($\triangle d^{(2)}$) by solving the original stiffness equations again, this time setting any prescribed components of nodal displacement to zero. The increment of nodal force ($\triangle f^{(2)*}$) in equilibrium with the displacement correction can then be found, and compared with the first set of residual forces ($\triangle f^{(2)}$). The difference may then be used to calculate a second set of corrections to the nodal displacements ($\triangle d^{(3)}$), and so on. This process may be repeated until the values of the residual forces or displacement corrections are smaller than some required limit.

The advantage of this technique is that the global stiffness matrix needs to be calculated and inverted only once, at the start of each increment, so each sub-

Fig. 6.11 Schematic representation of initial-stiffness iteration.

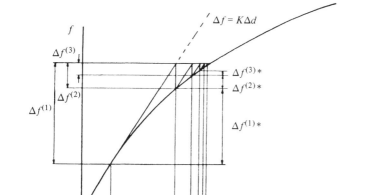

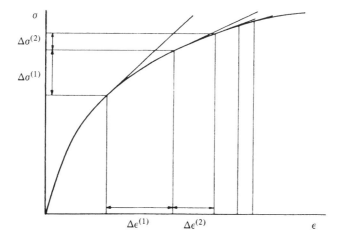

sequent iteration may be performed in a fraction of the time required for a full solution of the stiffness equations. The disadvantage is that many iterations may be required for the convergence of the solution. Occasionally, the solution may not converge at all.

In an alternative form of iteration, the global stiffness matrix is re-calculated at the end of each iteration; although this means more computation at each step, convergence is usually quicker.

A good example of this approach is provided by the *secant-modulus* method of solution, illustrated in figure 6.12. This is a second-order Runge–Kutta technique in which the stiffness matrix is evaluated using the material stress and strain at the start of the increment (represented by σ^0 and ϵ^0 in the diagram).

Fig. 6.12 Schematic representation of secant-modulus technique.

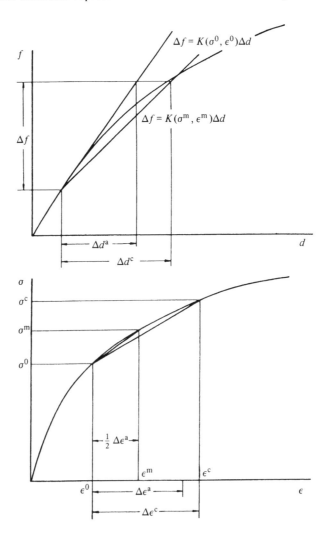

The equations are then solved for the set of applied forces $\triangle f$ to obtain a first approximation to the set of incremental nodal displacements ($\triangle d^a$). These may be used to estimate the stress and strain half-way through the step (σ^m, ϵ^m). A new stiffness matrix is then constructed, based upon these mid-increment values, and solved to give a better estimate of the incremental nodal displacements ($\triangle d^c$) and the strain (ϵ^c) and stress (σ^c) at the end of the increment.

6.4.3.3   Yield-transition increments

A more extreme form of non-linearity occurs when an element of the workpiece yields and starts to deform plastically, for at such times the elastic constitutive matrix in equation 5.37 needs to be replaced by the elastic–plastic one, with a correspondingly-large change in the magnitude of the deformation modulus.

In order to overcome this problem, it is possible to reduce the size of yield-transition increments automatically [6.7], so that elements always yield at, or immediately before, the end of a step.

6.4.4   Updating of workpiece parameters

As shown in Chapter 5, the FE stiffness equations depend upon the current geometry of the mesh and the current distributions of stress, strain, strain rate, temperature and element rotational values. These quantities must therefore be updated at the end of each increment. In addition, since the boundary conditions for the solution of the equations change throughout the deformation, these must also be re-determined before the next increment of the analysis, and the values of force acting upon the surface of the workpiece need to be calculated in order to be able to assess die stresses and forming loads.

Updating the mesh geometry is simply a matter of adding the incremental nodal displacement values to the nodal co-ordinates. Updating the other quantities will be considered in more detail.

6.4.4.1   Strain components and rotational values

The incremental-strain tensor may be evaluated at any point within an element using the expression:

$$\triangle\epsilon_{ij} = \frac{1}{2}\left(r_{ki}\triangle u_{k,j} + r_{kj}\triangle u_{k,i} - \frac{1}{2}(<r_{ik}><\triangle u_{j,k}> + <r_{jk}><\triangle u_{i,k}>)\right) \tag{6.2}$$

which was introduced in Chapter 5. The deformation gradient $\triangle u_{i,j}$ may itself be obtained from the known values of $\triangle d_{li}$, the incremental nodal displacements for that element, using:

$$\triangle u_{i,j} = N_{I,j}\triangle d_{li} \tag{6.3}$$

As explained in Chapter 5, the rotational matrix r_{ij} is defined by the expression:

$$r_{ik}q_{kj} = \triangle u_{i,j} + \delta_{ij} \tag{6.4}$$

Because r_{ij} is an orthogonal matrix (equation 5.16), multiplying each side of this expression by its transpose gives:

$$r_{im}q_{ml}r_{ik}q_{kj} = \delta_{mk}q_{ml}q_{kj} = (\triangle u_{i,l} + \delta_{il})(\triangle u_{i,j} + \delta_{ij}) \tag{6.5}$$

Denote the right-hand side of equation 6.5 by the matrix h_{lj}. Then since q_{ij} is symmetric:

$$q_{lk}q_{kj} = h_{lj} \tag{6.6}$$

h_{lj} may be evaluated quite simply from the known increments of nodal displacement $\triangle d_{li}$ using equation 6.3, and equation 6.6 solved iteratively using Newton's method, to obtain q_{kj}:

$$q_{kj}^{(i)} = \frac{1}{2}\left(q_{kj}^{(i-1)} + p_{kj}^{(i-1)}\right) \tag{6.7a}$$

where:

$$p_{lk}^{(i-1)}q_{kj}^{(i-1)} = h_{lj} \tag{6.7b}$$

and:

$$q_{kj}^{(0)} = w_{kj} \tag{6.7c}$$

in which $q_{kj}^{(i)}$ denotes the ith approximation to q_{kj}. In practice, this process of iteration converges rapidly since the components of $\triangle u_{i,j}$ are much smaller than one.

The rotational matrix r_{ij} may then be evaluated for use in equation 6.2, and subsequently in the stiffness formulation for the next increment, using equation 6.4.

6.4.4.2   Deviatoric stress

The elastic–plastic constitutive matrix (equation 5.20) should only really be used to evaluate increments of stress from increments of strain if the increment size is very small. This is because the Prandtl–Reuss flow expression (equation 5.19) on which this matrix is based, is strictly only defined in terms of infinitesimal deformation steps (equation 3.20a). If finite-sized steps are used, Nagtegaal and de Jong [6.8] have shown that the FE solution obtained may be inaccurate or unstable.

Instead, the *mean-normal* method may be used to calculate the increments of deviatoric stress. This technique was first suggested by Rice and Tracey [6.9] and extended to include strain-hardening properties by Tracey [6.10].

The technique is illustrated in figure 6.13 with reference to stresses plotted in the synoptic plane. Point A represents the stress state at the start of the increment on the initial yield locus, and B represents the final stress state after some degree of strain hardening. The form of the Prandtl–Reuss flow expression given in equation 5.19:

$$\triangle \epsilon_{ij} = \frac{1}{E}((1+\nu)\triangle\sigma_{ij} - \nu\delta_{ij}\triangle\sigma_{kk}) + \triangle\lambda\sigma'_{ij} \tag{6.8}$$

assumes that the increment of plastic strain is parallel to the deviatoric stress at the start of the increment. The mean-normal method makes the more realistic assumption that the plastic strain increment is parallel to the deviatoric stress half-way through the step. It can be seen from figure 6.13 that the latter stress (OC) is parallel to $\hat{\sigma}'_{ij}$ (OD) where:

$$\hat{\sigma}_{ij} = \sigma^0_{ij} + \frac{1}{2} \triangle \sigma^e_{ij} \tag{6.9}$$

and from the generalised form of Hooke's law:

$$\triangle \sigma^e_{ij} = 2G(\triangle \epsilon_{ij} + \delta_{ij} \, (\frac{\nu}{1-2\nu}) \triangle \epsilon_{pp}) \tag{6.10}$$

The modified form of the Prandtl–Reuss equations given in equation 6.8 is therefore:

$$\triangle \epsilon_{ij} = \frac{1}{E} \, ((1+\nu) \triangle \sigma_{ij} - \nu \delta_{ij} \triangle \sigma_{kk}) + \frac{\triangle m}{2G} \hat{\sigma}'_{ij} \tag{6.11}$$

which may be rearranged to give:

$$\triangle \sigma_{ij} = \triangle \sigma^e_{ij} - \triangle m \hat{\sigma}'_{ij} \tag{6.12}$$

Equation 6.12 allows the stress increment to be calculated once the proportionality factor $\triangle m$ is known. Application of von Mises's yield criterion to the state of stress at the end of the increment shows that this factor may be evaluated by solving the equation:

$$p(\triangle m)^2 + q \cdot \triangle m + r = Y^2 \, (\bar{\epsilon}^{0p} + \frac{\sqrt{p}}{3G} \triangle m, \dot{\bar{\epsilon}}^P, T^f) \tag{6.13}$$

Fig. 6.13 Mean-normal method of calculating finite changes in stress.

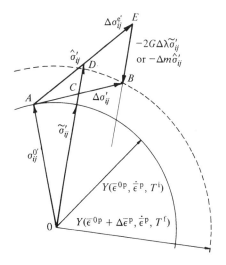

where:

$$p = \frac{3}{2}\hat{\sigma}'_{ij}\hat{\sigma}'_{ij} \qquad (6.14)$$

$$q = -3\hat{\sigma}'_{ij}(\sigma^{0\prime}_{ij} + \triangle\sigma^{\varepsilon\prime}_{ij}) \qquad (6.15)$$

$$r = \frac{3}{2}(\sigma^{0\prime}_{ij} + \triangle\sigma^{\varepsilon\prime}_{ij})(\sigma^{0\prime}_{ij} + \triangle\sigma^{\varepsilon\prime}_{ij}) \qquad (6.16)$$

The solution is carried out iteratively (figure 6.14), and is found to be rapidly convergent. The procedure is particularly efficient since the coefficients of the quadratic function need only be evaluated once at the start of the iteration. It can be seen that increment of generalised plastic strain is evaluated automatically as part of the mean-normal method.

The increment of stress obtained by this method depends only upon the change in strain in the element. However, rotation of a stressed element will also cause the values of the stress components to change, even if there is no change in strain. Therefore, when updating the Cartesian components of total (Cauchy) stress, it is necessary to include any additional stress increment resulting from the element rotation.

The importance of using an accurate method of calculating stress is illustrated in figure 6.15. This shows two FE predictions of the compression of a cube of

Fig. 6.14 Iterative solution of mean-normal equations.

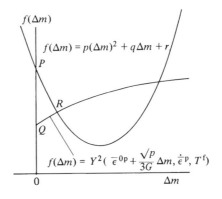

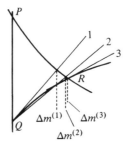

aluminium with zero inter-facial friction. The deformation ought, therefore, to be homogeneous. Figure 6.15a shows the unstable deformation resulting when the stress is calculated by a simple tangent-D-matrix approach; figure 6.15b shows the results obtained when the mean-normal method is used.

6.4.4.3   Hydrostatic stress

Changes in hydrostatic stress may be calculated by multiplying the changes in bulk strain by the bulk modulus. However, it is useful to have an independent means of evaluating the hydrostatic component of stress. Providing there are external surfaces of the FE mesh that are not in contact with dies, hydrostatic stress may be calculated from equilibrium principles. This is the indirect method proposed by Alexander and Price [6.11].

Since the present technique assumes zero body forces, principles of force equilibrium lead to the expression:

$$\frac{\partial \sigma_{kk}}{\partial x_i} = -3\frac{\partial \sigma'_{ij}}{\partial x_j} \qquad (6.17)$$

The deviatoric stress may be calculated at arbitrary points within the mesh using the mean-normal method, so the distribution, and hence the spatial gradients, of deviatoric stress may, in principle, be evaluated throughout the body. Integration of equation 6.17 along a given line therefore allows the change in hydrostatic stress to be calculated between its two end-points.

Since the hydrostatic stress at any free surface is known to be equal to the negative of the deviatoric stress normal to that surface, a single starting point,

Fig. 6.15 Zero-friction compression of cube: comparison of (a) tangent-D-matrix and (b) mean-normal methods of calculating stress.

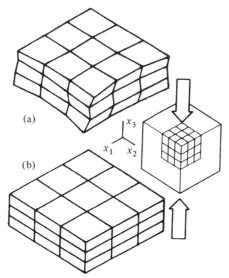

or possibly several, can usually be found for the integration, so that the hydro-static stress can be determined anywhere within the workpiece.

As is often the case with the calculation of the gradients of a field variable, the calculation of the spatial derivatives of deviatoric stress is very susceptible to small errors in these stress components. If this is found to cause unacceptable errors in the hydrostatic stress distribution, then the estimates of the stress derivatives can be improved by performing a smoothing operation of some sort on the original stress values [6.12].

6.4.4.4   External forces

Minimising the instantaneous rate of change of potential energy for an element (cf. equation 5.3 which is based upon the *average* rate in an infinitesimal step) leads to the following expression:

$$\delta(\dot{d}_{Im})f_{Im} = \int \delta(\dot{u}_{j,i})\sigma_{ij}dV \tag{6.18}$$

It should be noted that it is perfectly correct to use the Cauchy stress in this context because we are not considering any change in the stress state. Since Cauchy stress is symmetric, this may be written as:

$$\delta(\dot{d}_{Im})f_{Im} = \int \delta(\dot{\epsilon}_{ij})\sigma_{ij}dV \tag{6.19}$$

Using the interpolation functions N_I, this gives the relationship:

$$\delta(\dot{d}_{Im})f_{Im} = \delta(\dot{d}_{Im}) \int \frac{1}{2}(\delta_{im}N_{I,j} + \delta_{jm}N_{I,i})\sigma_{ij}dV \tag{6.20}$$

or:

$$f_{Im} = \int \frac{1}{2}(\delta_{im}N_{I,j} + \delta_{jm}N_{I,i})\sigma_{ij}dV \tag{6.21}$$

Equation 6.21 allows the forces acting upon any node of an element to be calculated from the previously-calculated values of element stress. If a node belongs to more than one element, the actual force applied to that node will be the sum of the values obtained for all these elements. Usually only the forces acting at the surface nodes are of interest: the forces acting at internal nodes should, because of equilibrium, all be zero. The integrals in equation 6.21, like the element-stiffness integrals, may be evaluated by Gaussian quadrature.

6.4.4.5   Strain rate

Since the yield stress may, in general, be a function of plastic strain rate, this needs to be calculated throughout the analysis. The plastic strain rate is simply the current incremental plastic strain divided by the time interval for the current increment.

The disadvantage with this simple approach is that the value of strain rate is only available after the calculation of deviatoric stress (Section 6.4.4.2), and this calculation itself requires that the current strain rate be known.

An alternative then is to assume that the rate of plastic strain is the same as the rate of total (that is, elastic plus plastic) strain. In practice, there will be little difference between them, and it is a great advantage to be able to calculate the strain rate as soon as the increments of nodal displacement are known. In addition to simplifying the mean-normal calculation, it also avoids the discontinuous change in strain rate which would otherwise occur when an element begins to yield.

The strain rate can most conveniently be calculated as part of the procedure which checks for transition between the elastic and plastic states and scales the increment size accordingly (Section 6.4.3.3) since this takes place immediately after the solution of the stiffness equations and must calculate the strain increment for its own purposes. As pointed out in Section 6.3.4, the actual time step associated with each increment of deformation may change throughout the analysis.

6.4.4.6   Temperature

The calculation of the temperature changes in the workpiece involves, among other things, the determination of the heat flowing throughout the mesh. Since this will itself depend upon the temperature distribution, the thermal calculation carried out during each increment of the deformation may need to be further divided into a number of steps. Let this number be N. Later, we shall examine a way of determining how many thermal steps are required.

The change in temperature δT of a given element during a thermal step will depend upon the change in energy of that element during the step:

$$\delta T = \frac{1}{cV}\left(\delta Q^{\mathrm{d}} + \delta Q^{\mathrm{f}} + \delta Q^{\mathrm{c}} \right) \qquad (6.22)$$

where c is the thermal capacity per unit volume of the material, V is the volume of the element, δQ^{d} is the increase in energy of the element due to the work of deformation, δQ^{f} is the increase in the energy due to frictional work at external boundary faces, and δQ^{c} is the change in energy due to the conduction of heat between the element under consideration and adjacent elements or dies. The last quantity may, of course, be positive or negative. Consider each of these contributions in turn.

Work of deformation

The work of deformation per unit volume is the product of the average element generalised stress $\bar{\sigma}$ for the *whole* deformation increment, and the corresponding change in element plastic strain $\triangle\bar{\epsilon}^{\mathrm{p}}$.

Thus:

$$\delta Q^{\mathrm{d}} = \frac{\alpha}{N} V \cdot \bar{\sigma} \cdot \triangle \bar{\epsilon}^{\mathrm{p}} \qquad (6.23)$$

in which α is the proportion of the work of deformation converted into heat (an empirical quantity, usually in the range 0.85–0.95) [6.13].

Friction

The frictional-work contribution depends upon the tangential force acting at any faces of the element that are in contact with die surfaces, and the relative movement of these surfaces. Suppose that F faces of the element are in contact with dies and let the index α refer to one of these faces. Since the shear stress at the boundaries is equal to the shear yield stress multiplied by the friction factor (Section 6.4.3.1) and, by the von Mises yield condition, the shear yield stress equals the yield stress in tension divided by the square root of three, then by assuming that half of the heat generated flows into the element, the frictional-work contribution may be written as:

$$\delta Q^{\mathrm{f}} = \frac{1}{N} \sum_{\alpha=1}^{E} \frac{m^{\alpha} A^{\alpha} Y \triangle d^{\alpha}}{2\sqrt{3}} \qquad (6.24)$$

where m^{α} is the friction factor associated with face α of the element, A^{α} is its area, $\triangle d^{\alpha}$ is the displacement of this face relative to the die surface during the whole of the deformation increment, and Y is the average yield stress of the element during the increment, a function of plastic strain, strain rate and temperature.

Conduction of heat

The contribution to the increase in energy of the element due to the conduction of heat through its faces is proportional to the duration of the thermal step $\triangle t/N$ where $\triangle t$ is the time interval of the whole deformation increment.

The temperature T at any point will be a function of its co-ordinates. Let x_j^0 be the global reference co-ordinates of the centroid of the element. If we assume the temperature at this centroid is the average element temperature, then applying the standard heat-flow expression to the element gives:

$$\delta Q^{\mathrm{c}} = k \cdot V \frac{\partial^2 T}{(\partial x_i)^2} \frac{\triangle t}{N} \quad \text{at } x_j^0 \qquad (6.25)$$

where k is the thermal conductivity of the material.

In order to evaluate the right-hand side of this expression, we shall assume that 3-D hexahedral elements are being used. The treatment for other types of element will be broadly similar.

Define a local curvilinear axis system X^i, with origin at the centroid of the element, and with axes that pass through the centroids of the elements adjacent to the six faces of the element under consideration. Let the non-zero local co-ordinate of each of these adjacent centroids be plus or minus one (figure 6.16). If a particular face of the element under consideration is on the surface of the mesh, there is no adjacent element, so in this case let the axis pass through the centre of the face and have co-ordinate ±1 at this point.

The six neighbouring points that, together with the centroid of the element under consideration, define the curvilinear axes, are labelled with a number between one and three and a plus or a minus sign, according to the ray of the local axis on which they are situated. Thus the reference co-ordinates of the neighbouring point (centroid or centre of face) with local co-ordinates $(0,-1,0)$ will be x_i^{2-}, and so on.

Since each of the three local axes is defined by three points (the centroid of the element under consideration and two neighbouring points) it will, in general, be parabolic in shape, and the value of any state variable may be approximated by a quadratic function of each of the three local co-ordinates.

Thus the co-ordinates x_i of any point within the element are given approximately by the expression:

$$x_i = x_i^0 + a_{ij}X^j + \frac{1}{2}b_{ij}(X^j)^2 \qquad (6.26)$$

where a_{ij}, b_{ij} are constants determined by the geometry of the local configuration.

Similarly, the temperature T may be approximated by:

$$T = T^0 + r_j X^j + \frac{1}{2}s_j(X^j)^2 \qquad (6.27)$$

where T^0 is the temperature at the centroid.

Fig. 6.16 Local curvilinear axis system for calculation of heat flow into an element.

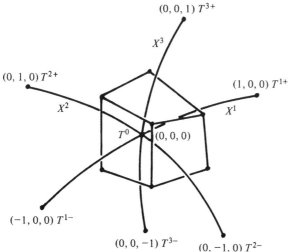

$(0, 0, 1)\ T^{3+}$

X^3

$(0, 1, 0)\ T^{2+}$

$(1, 0, 0)\ T^{1+}$

X^2

X^1

T^0 $(0, 0, 0)$

$(-1, 0, 0)\ T^{1-}$

$(0, 0, -1)\ T^{3-}$ $(0, -1, 0)\ T^{2-}$

The coefficients in equations 6.26 and 6.27 can be determined by evaluating these functions at each of the six neighbouring points:

$$a_{ij} = \frac{1}{2} \, (x_i^{j+} - x_i^{j-}) \tag{6.28a}$$

$$b_{ij} = x_i^{j+} + x_i^{j-} - 2x_i^0 \tag{6.28b}$$

$$r_j = \frac{1}{2} \, (T^{j+} - T^{j-}) \tag{6.29a}$$

$$s_j = T^{j+} - T^{j-} - 2T^0 \tag{6.29b}$$

Now by the chain rule:

$$\frac{\partial^2 T}{(\partial x_i)^2} = \frac{\partial^2 T}{(\partial X^j)^2} \left(\frac{\partial X^j}{\partial x_i} \right)^2 + \frac{\partial T}{\partial X^j} \frac{\partial^2 X^j}{\partial x_i \partial X^k} \frac{\partial X^k}{\partial x_i} \tag{6.30}$$

since $\partial^2 T / \partial X^j \partial X^k = 0$ if $j \neq k$.

Differentiating equation 6.26 with respect to the local co-ordinates produces:

$$\frac{\partial x_i}{\partial X^j} = a_{ij} + b_{ij} X^j \tag{6.31}$$

and by differentiating equation 6.27 we have:

$$\frac{\partial T}{\partial X^j} = r_j + s_j X^j \tag{6.32a}$$

and

$$\frac{\partial^2 T}{(\partial X^j)^2} = s_j \tag{6.32b}$$

Combining equations 6.29, 6.30 and 6.32, and evaluating at the centroid of the element under consideration:

$$\frac{\partial^2 T}{(\partial x_i)^2} = (T^{j+} + T^{j-} - 2T^0) \left(\frac{\partial X^j}{\partial x_i} \right)^2 + \frac{1}{2} \, (T^{j+} - T^{j-}) \frac{\partial^2 X^j}{\partial x_i \partial X^k} \frac{\partial X^k}{\partial x_i} \quad \text{at } x_i^0$$

$$= C^{j+} \, (T^{j+} - T^0) + C^{j-} \, (T^{j-} - T^0) \tag{6.33}$$

where:

$$C^{j+} = \delta_{jk} \left(\frac{\partial X^k}{\partial x_i} \right)^2 + \frac{1}{2} \frac{\partial^2 X^j}{\partial x_i \partial X^k} \frac{\partial X^k}{\partial x_i} \quad \text{at } x_i^0 \tag{6.34a}$$

$$C^{j-} = \delta_{jk} \left(\frac{\partial X^k}{\partial x_i} \right)^2 - \frac{1}{2} \frac{\partial^2 X^j}{\partial x_i \partial X^k} \frac{\partial X^k}{\partial x_i} \quad \text{at } x_i^0 \tag{6.34b}$$

The matrices of first-order partial derivatives in equations 6.34 are obtained by inverting the matrix of partial derivatives in equation 6.31; the second-order partial derivatives are obtained by differentiating these first-order derivatives with respect to the local co-ordinates. When evaluated at the centroid of the element, which has local co-ordinates $(0,0,0)$, these partial derivatives are found to be functions of just the quanitities a_{ij} and b_{ij} defined by equations 6.28. The coefficients C^{j+} and C^{j-} therefore depend only on the local geometry and so only need to be evaluated once for each increment of deformation, however many steps there may be in the thermal calculation. Returning to the heat-flow equation, the increase in energy of the element due to conduction of heat in one thermal step is:

$$\delta Q^c = \frac{kV\triangle t}{N} \left(S^{j+} C^{j+} (T^{j+} - T^0) + S^{j-} C^{j-} (T^{j-} - T^0) \right) \qquad (6.35)$$

The temperatures in this expression refer to the start of the particular thermal step, and are updated at the end of each such step.

The coefficients S^{j+} and S^{j-} in equation 6.35 allow the expression to be used even if the element has faces that are on the outside of the mesh.

If neighbouring point $j+$ or $j-$ is an adjacent element centroid:

$$S^{j+} = 1 \quad \text{or} \quad S^{j-} = 1 \qquad (6.36a)$$

if neighbouring point $j+$ or $j-$ is the centre of a face in contact with a die:

$$S^{j+} = \frac{l^{j+} k^{j+}}{l^{j+} k^{j+} + k} \quad \text{or} \quad S^{j-} = \frac{l^{j-} k^{j-}}{l^{j-} k^{j-} + k} \qquad (6.36b)$$

where k^{j+}, k^{j-} are the heat transfer coefficients of the die/billet interfaces, and l^{j+}, l^{j-} are the distances from the centre of face $j+$ or $j-$ to the centroid of the element.

Finally, if neighbouring point $j+$ or $j-$ is a free-surface face:

$$S^{j+} = 0 \quad \text{or} \quad S^{j-} = 0 \qquad (6.36c)$$

The expressions in equation 6.36b allow for the fact that the centre of the element face will not be at the same temperature as the die (T^{j+} or T^{j-}). Equation 6.36c ensures that there is no heat flowing through free-surface faces. The method just described can easily be modified, if required, to include the heat flowing due to convection at free surfaces. Similarly, it is possible to adapt the method to calculate the flow of heat through the dies themselves [6.14].

During each of the N thermal steps of the deformation increment, the temperature change in each element is calculated using equation 6.22 and then the distribution of temperature throughout the mesh is updated for the next thermal step.

The heat flowing at any instant between two isolated points of a medium is proportional to the difference in temperature of those points, and acts in such a way as to reduce this temperature difference. In the absence of any other effects, the temperature difference will tend to zero with time. Since the temperature difference becomes smaller all the time, a step-wise calculation will over-estimate the heat flowing between the two points. If these steps are too large, then not only will the corresponding estimates of temperature be inaccurate, but there is a possibility that the over-estimates of the heat flowing will actually lead to predictions of the temperature difference that diverge and change in sign between successive steps.

For similar reasons, this type of instability, though of a much more complicated nature, can also occur in the FE thermal calculation if the size of the thermal step is too large. The critical step size is very difficult to assess accurately, but by making some broad simplifying assumptions, it is quite easy to estimate a lower bound to this quantity. Fortunately it does not matter in an FE program that this value is an under-estimate since the computational effort expended upon the thermal calculation is small compared with that required for the construction and solution of the FE stiffness equations. It is of no great consequence, therefore, if more thermal steps are used than are strictly necessary.

The problem then is, given a time interval of $\triangle t$ for an increment, to estimate the number of thermal steps, N, that this must be divided into in order to ensure firstly that the calculated temperature changes converge to finite values, and secondly that the errors in these values are smaller than some specified limit.

Equation 6.22 may be re-written in the form:

$$\delta T = \left(P + \frac{k}{c} \left[\ S^{j+}C^{j+}(T^{j+} - T^0) + S^{j-}C^{j-}(T^{j-} - T^0) \ \right] \ \right) \frac{\triangle t}{N} \qquad (6.37)$$

in which:

$$P = \frac{N}{cV\triangle t} \left(\delta Q^d + \delta Q^f \right) \qquad (6.38)$$

For simplicity, assume that the temperatures of the neighbouring elements and dies are constant throughout the time interval $\triangle t$. If:

$$Z = \frac{k}{c} \ (S^{j+}C^{j+}T^{j+} + S^{j-}C^{j-}T^{j-}) + P \qquad (6.39)$$

and:

$$X = \frac{k}{c}(S^{j+}C^{j+} + S^{j-}C^{j-}) \qquad (6.40)$$

then the estimated change in the temperature of the element during the *i*th thermal step is:

$$T^{0(i)} - T^{0(i-1)} = \left(Z - XT^{0(i-1)} \right) \frac{\triangle t}{N} \qquad (6.41)$$

or:

$$T^{0(i)} = Z\frac{\triangle t}{N} + (1 - \frac{X\triangle t}{N})T^{0(i-1)} \qquad (6.42)$$

hence, the estimate at the end of N thermal steps is:

$$T^{0(N)} = -\frac{1}{X}(Z - XT^{0(0)})(1 - \frac{X\triangle t}{N})^N + \frac{Z}{X} \tag{6.43}$$

and the estimated change in the temperature $\triangle T$ of this element is given by:

$$\triangle T^{0\text{est}} = \frac{1}{X}(Z - XT^{0(0)})\left[1 - (1 - \frac{X\triangle t}{N})^N\right] \tag{6.44}$$

The correct value of the change in temperature in this simplified model is found by integrating equation 6.37 with the substitutions in equations 6.39 and 6.40:

$$\triangle T^0 = \frac{1}{X}(Z - XT^{0(0)})(1 - e^{-X\triangle t}) \tag{6.45}$$

From equation 6.44, it can be seen that the estimated temperature difference will converge if:

$$1 - \frac{X\triangle t}{N} > -1 \tag{6.46}$$

(Note, the left-hand side will always be less than $+1$.)

Since various simplifying assumptions have been made, the number of steps required for a convergent solution is taken to be an order of magnitude larger than that defined by equation 6.46, i.e. it is required that:

$$N > 10X\triangle t \tag{6.47}$$

The absolute error in the estimated temperature difference may be found by comparing equations 6.44 and 6.45. If τ is the specified maximum permitted error in the temperature difference, then it is additionally required that:

$$\frac{|Z - XT^{0(0)}|}{X}\left[e^{-X\triangle t} - (1 - \frac{X\triangle t}{N})^N\right] < \tau \tag{6.48}$$

A value of N is therefore chosen which satisfies equations 6.47 and 6.48 in every element of the FE mesh.

6.4.5 Output of results

The FE analysis is capable of producing information about many different aspects of the deformation. During each increment, the FE program calculates values of fundamental parameters such as nodal co-ordinates, and the distributions of stress and strain components, temperature and strain rate, as well as derived quantities such as nodal velocity, generalised stress and strain, hydrostatic stress, principal stress components, work rate and external loads. Whether the results are displayed numerically or in graphical form, the amount of information produced can be vast.

If only a limited portion of this information is required, or if the results are only required at a few stages during the deformation, it is certainly possible to include numerical and graphical routines in the FE program. Generally, however, it is better to save a complete description of the current state of the FE model, in compact machine-readable form, during every increment of the analysis, and to use a post-processing program to display the results as required.

6.5   POST-PROCESSING

Post-processing simply means extracting from the large amounts of information produced by an FE program just those quantities that are of interest and either displaying these in some form or using them in an additional non-FE calculation. In practice, the information will be stored in a computer file, and so the post-processing will be performed by a computer program, usually interactively.

There are several advantages in using a post-processing program. Firstly, as mentioned previously, the output of the FE analysis becomes more manageable because only the particular information required need appear on paper. There is also the flexibility of being able to display the information in a variety of numerical and graphical forms – tables, charts, graphs, contour plots etc.

Secondly, since the FE program will have saved results throughout the analysis, it is possible to examine in detail the deformation at any increment. Indeed, by interpolating between the results obtained for successive increments, it is possible to study any stage of the deformation.

Thirdly, it is not necessary to know beforehand which quantities will be of interest in a particular process. Fourthly, since the post-processing information can be saved indefinitely, different aspects of the metalforming process can be studied at a later date, perhaps in the light of subsequent experimental findings.

Lastly, the post-processing information can be used as input to other computer programs, such as CAD and control packages, and numerical-application programs. One such application is to use the FE stress and strain components throughout the deformation to predict when and where ductile fracture is likely to occur during the forging [6.15].

There are many commercial packages available for the analysis of information produced by FE prgrams. Most of these will have been designed for use in conjunction with elastic FE analyses, but some may be sufficiently flexible to be able to display the information relevant in the study of metalforming. Alternatively, it is not difficult to write very simple programs that can extract and output FE information in the required form.

6.6   SPECIAL TECHNIQUES

6.6.1   Processes involving severe deformation

Metalforming operations can involve total natural strains of unity or more. Depending upon the exact nature of the deformation, this may mean that, by

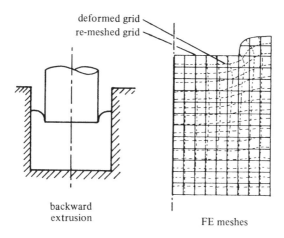

Fig. 6.17 Re-meshing of distorted FE grid during the analysis of backward extrusion.

the end of the FE analysis, the original mesh becomes so distorted that the interpolation polynomials are incapable of modelling the geometry of the elements and the associated state variables.

If this is the case, a regular FE mesh must be re-defined within the current boundaries of the workpiece at intervals throughout the analysis (figure 6.17); on each such occasion the values of stress, strain, strain rate, temperature and incremental angles of rotation need to be transferred from the old to the new model using the element interpolation functions (figure 6.18).

Fig. 6.18 Interpolation of nodal values during re-meshing.

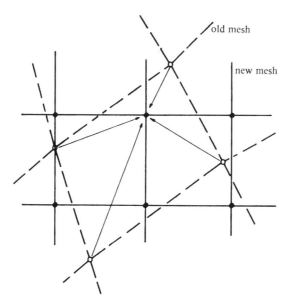

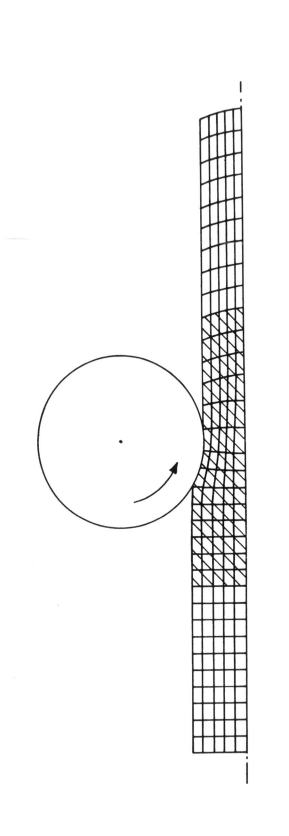

Fig. 6.19 FE analysis of steady-state rolling: region for which matrices solved (shaded).

For simple geometries, such as 2-D axi-symmetric or plane-strain deformations, it may be possible to perform the re-meshing automatically; more complex examples may require the FE mesh to be re-created manually, with perhaps the interpolation of variables being performed by a specially written computer program.

When the FE mesh is re-defined in this way, the pattern of deformation is made much clearer if the geometry of the original mesh is updated along with that of the current FE mesh at the end of each increment – the *super-grid technique* [6.16]. This gives the impression that the original mesh has been used throughout the analysis and so more readily accords with experience gained from experimental observation. Some examples of this technique will be given later.

6.6.2   Steady-state processes

In order to use an incremental FE program to study steady-state forming processes, such as rolling, extrusion or drawing, the analysis must examine the process from the instant when the workpiece first enters the dies or roll gap until such time as the spatial distributions of the process parameters are unchanging. This may require a large number of increments, and if so, the length of the workpiece which needs to pass between the rolls or through the dies would also be large. The FE analysis of a steady-state process can therefore require considerable computational effort.

However, the effort can be reduced. At any instant, there may be large parts of the workpiece which are not affected by the deformation, either having passed completely through the region of deformation or having not yet reached it. Thus only the part of the workpiece actually undergoing deformation needs to be modelled in the FE analysis. Naturally, this part of the workpiece will change during the deformation and so will need to be constantly updated.

One way of doing this is to produce an FE mesh that encompasses the entire length of the workpiece and to include an algorithm in the FE program that assembles and solves the stiffness matrices of only those elements in, or just adjacent to, the deforming region (figure 6.19). The nodes elsewhere in the mesh can be assigned default values of displacement if required. The method of selecting elements may be based on an examination of the deformation occuring in the mesh, or may use some previously-determined criterion for gauging the extent of the deformation region.

A slightly more complicated procedure, but one which reduces the amount of computer storage required, is to use an FE mesh that models only the part of the workpiece immediately surrounding the deformation region. As the deformation proceeds, a type of re-meshing can be carried out whereby elements are removed from the downstream end of the workpiece and new ones are created at the upstream end (figure 6.20).

6.6.3 Three-dimensional analyses

Unless otherwise stated, the principles outlined in this chapter, and the previous one, are applicable equally to those FE analyses which are intended to study general 3-D flow problems, and those which take advantage of the simplifications possible when a metalforming process involves only 2-D flow.

The simplifications mean that the element stiffness equations in 2-D programs have fewer degrees of freedom and are much simpler to construct. In addition, a full 3-D analysis requires a mesh with many more elements than are necessary for a comparable 2-D analysis because elements are required throughout the volume. These considerations mean that the space needed to store the global stiffness matrix and associated arrays is very much larger in 3-D programs than in 2-D ones. Not surprisingly, it also takes considerably longer to assemble and solve the 3-D matrix equations.

Depending upon the particular computer used to run the FE program, the large storage requirements of 3-D programs may mean that special techniques have to be used. For example, on some machines it may be impossible to store the entire stiffness matrix in main memory, even taking into account the sparse and symmetric nature of this matrix.

The solution here is to combine the processes of assembly and conversion to upper-triangular form so that equations can be transferred to secondary disc storage as soon as they no longer play any part in the triangularisation procedure. The equations may then be read back from disc during the back-substitution phase of the solution. Using this *frontal-solution* technique [6.3], only a small part of the stiffness matrix needs to be in main memory at any instant (figure 6.21).

Fig. 6.20 FE grid (a) before, and (b) after re-meshing and re-numbering during an analysis of steady-state rolling.

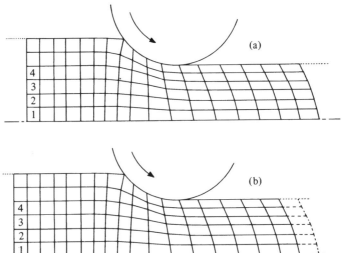

The disadvantage of the frontal-solution technique, apart from the added complexity, is the large amount of time spent reading from, and writing to disc storage devices.

Many modern mainframe computers have what is called a virtual memory. Often the size of arrays permitted under such systems can be essentially unlimited. The main memory of these computers is not, of course, unlimited, and may indeed be quite small. The apparent size of the virtual memory is a result of the operating system using disc devices as a secondary store, just as the frontal technique does. The difference is that a computer with virtual memory moves blocks of information back and forth between main memory and the disc devices quite automatically, a process called paging, so that the computer user is never aware of the fact.

The problems of storage will probably not arise in these cases, although the computer operating system may require that special action is taken when using very large arrays.

Obviously, a great deal of effort is expended by the computer manufacturers to ensure that transfer of information is performed as efficiently as possible, but it is nevertheless important to remember that, even on virtual machines, 3-D

Fig. 6.21 Diagrammatic representation of frontal-solution technique (shaded regions represent information stored in computer memory): (a) equations relating to node I fully assembled, (b) equations relating to node I removed to disc storage.

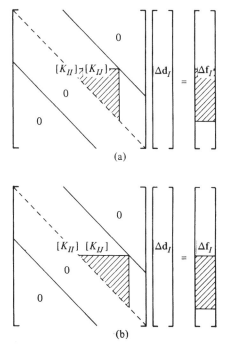

FE programs will probably spend a great deal of their time waiting for disc read or write operations.

It should also be remembered that although the operating system performs paging as efficiently as it can, it cannot determine with any certainty when it is selecting a page of information for transfer to disc, whether that page will be needed in main memory again quite soon. It is possible, therefore, for pages to be transferred to and from the disc devices quite unnecessarily.

Thus while the explicit disc read and write operations used by the frontal technique will not be performed as efficiently as those carried out directly by the operating system, the FE program exploits the nature of the calculation being performed and only writes the information to disc when necessary. It is possible therefore that it is more efficient to use the frontal technique, even on computers with virtual memory. This is a question that can only be decided, on a particular machine, by comparing actual program timings.

Whichever method is employed, and whatever the machine used, 3-D FE analyses take a long time to compute. This may or may not be a problem. Dedicated machines can be left running for the number of hours, days, perhaps weeks, required for the analysis. The only problem then is one of machine reliability.

On time-sharing systems, there may be a limit to the length of time taken by any one job submitted to the computer. The length of time will frequently be shorter than that required for a complete 3-D FE analysis of a metalforming process.

In such situations, it will be necessary to divide the analysis up into a sequence of jobs, each performing a specified number of increments of the calculation. At the end of each stage of the analysis, the program must save in a file all the information that will be required in order to enable the next job to continue the analysis from where it was halted.

This technique could also be useful even when using a dedicated computer to avoid the loss of interim solutions. In this case, the FE program would save the restart-information at regular intervals during the analysis. If a malfunction occurs, the analysis could be re-started from the last file of information saved.

References

[6.1] Lu, S.C-Y. A consultative expert system for finite element modelling of strip drawing. *Proc. 13th Nth. American Manufacturing Res. Conf.*, SME, pp. 433–41 (1985).

[6.2] Mallett, R.L. *Finite-element selection for finite deformation elastic–plastic analysis*. SUDAM rep. no. 80–4, Stanford University (1980).

[6.3] Pillinger, I. The prediction of metal flow and properties in three-dimensional forgings using the finite-element method. Ph.D. thesis, University of Birmingham, UK (1984) (unpublished).

[6.4]   Male, A.T. The friction of metals undergoing plastic deformation at elevated temperature. Ph.D. thesis, University of Birmingham, UK (1962) (unpublished).

[6.5]   Hartley, P., Sturgess, C.E.N. and Rowe, G.W. Friction in finite-element analyses of metalforming processes. *Int. J. Mech. Sci.* **21**, 301–11 (1979).

[6.6]   Zienkiewicz, O.C., Valliappan, S. and King, I.P. Elasto-plastic solutions of engineering problems 'Initial Stress', finite element approach. *Int. J. Num. Meth. Eng.* **1**, 75–100 (1969).

[6.7]   Yamada, Y., Yoshimura, N. and Sakurai, T. Plastic stress–strain matrix and its application for the solution of elastic-plastic problems by the finite element method. *Int. J. Mech. Sci.* **10**, 343–54 (1968).

[6.8]   Nagtegaal, J.C. and de Jong, J.E. Some computational aspects of elastic–plastic large strain analysis. *Computational Methods in Nonlinear Mechanics*, ed. J.T. Oden, North-Holland, pp. 303–39 (1980).

[6.9]   Rice, J.R. and Tracey, D.M. Computational fracture mechanics. *Numerical and Computer Methods in Structural Mechanics*, ed. A.R. Robinson and W.C. Schnobrich, Academic Press, pp. 585–623 (1973).

[6.10] Tracey, D.M. Finite element solutions for crack-tip behaviour in small-scale yielding. *Trans. ASME, J. Eng. Mater. Techn.*. **98**, 146–51 (1976).

[6.11] Alexander, J.M. and Price, J.W.H. Finite element analysis of hot metal forming. *Proc. 18th Int. Machine Tool Des. Res. Conf.*, ed. J.M. Alexander, MacMillan, pp. 267–74 (1977).

[6.12] Liu, C. Modelling of strip and slab rolling using an elastic–plastic finite-element method. Ph.D. thesis, University of Birmingham, UK (1985) (unpublished).

[6.13] Mahrenholtz, O., Westerling, C. and Dung, N.L. Thermomechanical analysis of metal forming processes through the combined FEM/FDM. *Proc. 1st Int. Workshop on the Simulation of Metal Forming Processes by the Finite Element Method (SIMOP-I)*, ed. K. Lange, Springer, pp. 19–49 (1986).

[6.14] Rebelo, N. and Kobayashi, S. A coupled analysis of viscoplastic deformation and heat transfer. *Int. J. Mech. Sci.* **22**, 699–705 & 707–18 (1980).

[6.15] Clift, S.E., Hartley, P., Sturgess, C.E.N. and Rowe, G.W. Fracture initiation in plane-strain forging. *Proc. 25th Int. Machine Tool Des. Res. Conf.* pp. 413–19 (1985).

[6.16] Al-Sened, A.A.K., Hartley, P., Sturgess, C.E.N. and Rowe, G.W. Forming sequences in axi-symmetric cold-forging. *Proc. 12th Nth. American Manufacturing Res. Conf.*, SME, pp. 151–8 (1984).

7 Practical applications

7.1 INTRODUCTION

This chapter will examine the application of FE metalforming techniques to a wide range of industrially-relevant processes.

The main reasons for conducting computer simulations of metalforming processes are to:

(i) reduce development lead times by minimising the number of experimental trials required (get closer to 'right first time')

(ii) reduce development costs, particularly those incurred by the manufacture of expensive dies for experimental trials

Both of these considerations result in increased industrial competitiveness and flexibility through the ability to introduce new products quickly and cheaply.

Estimates vary of the proportion of forged parts in the UK that have axi-symmetric geometries, but this figure probably lies somewhere in the region of 60 to 70%. The development of tooling for axi-symmetric parts does not present anywhere near the difficulties that are associated with the design of dies for non-symmetric components, but even so, computer simulation can significantly speed up the design process for axi-symmetric parts. The savings in time and money will be even greater when non-symmetric parts are to be formed.

The emphasis here will therefore be on those aspects of the results that are most important in an industrial context, such as:

(i) the prediction of operational parameters (force/stroke/time history, die stressing)

(ii) the prediction of product properties (distribution of hardness, residual stresses) and defects (die-filling problems, folds, ductile fracture)

(iii) the implications for the design of dies, and of sequences of dies in multi-stage processes

The practical considerations (difficulties) involved in the simulations (modelling of boundary conditions, length of computation) will also be discussed.

The examples in this chapter are set out according to the type of metalforming process involved. The main types of process that will be examined are forging, extrusion and rolling. Where appropriate, the FE results will be compared with the predictions of simple metalforming theory and experimental tests.

7.2   FORGING

7.2.1   Simple upsetting

Simple upsetting (also known as dumping or cheesing) of a cropped billet is frequently performed prior to more complex forging operations in order to increase the width of the workpiece, and to induce a certain level of plastic work. Simple compression is also widely used in the laboratory, along with the common tensile test, in order to obtain stress/strain data.

   Whereas the simple upsetting of upright cylinders (and billets under plane-strain conditions) is relatively well understood, the compression of workpieces under more complex forming geometries has received less attention. It is in these situations, in which the simple empirical guidelines and rules of thumb derived for axi-symmetric and plane-strain examples no longer apply, that finite-element techniques can be of most value, providing as they do information unobtainable by any other means. As a first example of this, we can consider block compression.

   Figure 7.1 shows the initial shape of the FE mesh used to model the simple upsetting of a rectangular block of commercially-pure aluminium with high inter-facial friction [7.1].

Fig. 7.1 Upsetting of rectangular block – finite-element discretisation.

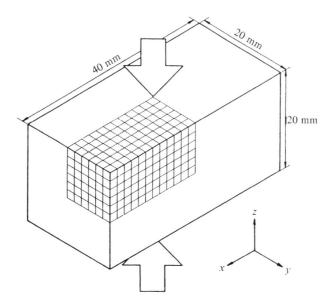

432 eight-node iso-parametric 3-D elements were used in the FE mesh. Due to the symmetry, this represented one eighth of the workpiece. The elements were arranged in a simple rectangular configuration.

The FE analysis used the finite-strain 3-D elastic–plastic formulation described in Chapters 5 and 6 of this monograph. The material properties of commercially-pure aluminium were used in the calculation, with the yield stress increasing with the amount of plastic strain. The analysis assumed isothermal conditions and a slow forming speed, so strain-rate effects could be ignored.

The frictional conditions were modelled by means of the beta-stiffness technique using a value of $m = 0.7$. For most of the analysis, the height of the mesh was reduced by 2% of its original value during each increment. During the early stages, while there were regions that were still elastic, the program automatically selected smaller incremental deformation steps as described in Chapter 6. The analysis was stopped when the height of the mesh had been

Fig. 7.2 Predictions of shape of finite-element mesh at various stage during upsetting: (a) whole mesh; (b) horizontal section through centre of workpiece; (c) longitudinal vertical section through centre.

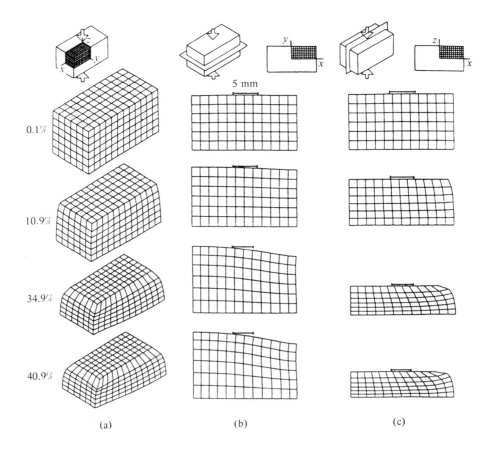

(a)   (b)   (c)

reduced to 40% of its original value, which required 39 increments. Each increment took 190 seconds of CPU time on a CDC 7600 computer to calculate.

Figure 7.2 shows the distorted FE mesh and two centre-line sections at various stages during the deformation. The roll-round of the billet onto the dies is clearly seen. This figure also illustrates the formation of concavities on the vertical faces of the workpiece near the external corner, and the associated 'earing' of this corner. This feature is also observed experimentally. The prediction of this sort of detail would be extremely difficult by any other theoretical technique.

In figure 7.3, the FE predicted profiles of horizontal and transverse sections through the workpiece are compared with experimentally measured shapes. Agreement between the two is quite close. The discrepancies are partly due to the coarseness of the FE mesh, and partly due to using a frictional restraint in the FE simulation that was apparently larger than that in the laboratory trials. This can be seen from figure 7.3b, where the arrow indicates the current position of the original corner of the experimental workpiece at this stage of the defor-

Fig. 7.3 Comparison of finite-element and experimental profiles at 40% reduction in height: (a) one quarter of horizontal section through centre of workpiece; (b) one quarter of transverse vertical section through centre.

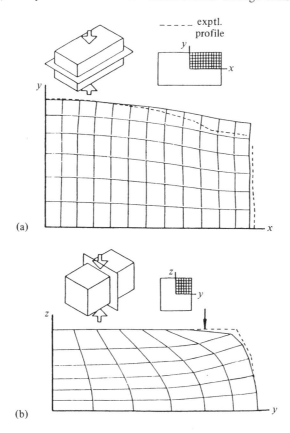

Fig. 7.4 Comparison of finite-element and experimental distributions of Vickers Pyramid Hardness Number (VPN) across one quarter of the transverse vertical section through the centre of the workpiece at 40% reduction in height.

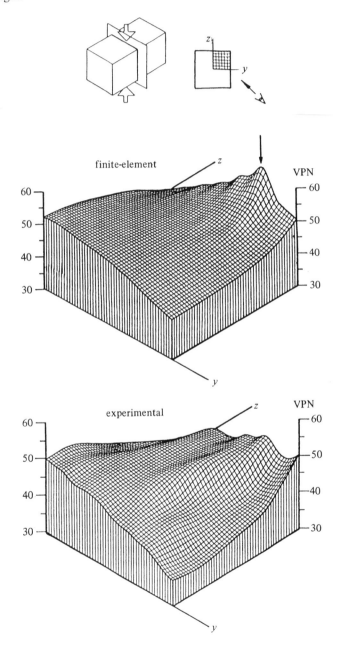

mation. Clearly, the amount of interfacial sliding was greater in the experiment than in the FE analysis. The effect of a greater frictional restraint was for there to be greater roll-round of material in the FE case.

Despite differences in the frictional conditions, the FE predictions of the internal patterns of deformation agree very well with experimental observations. This is illustrated by figure 7.4, which compares FE and experimental distributions of Vickers Pyramid Hardness Number (VPN) on the transverse vertical centre-line section through the workpiece at about 40% reduction in height. (For convenience of presentation, the deformed shapes of the sections have been mapped onto a square base.) The FE hardness values were obtained from the predicted values of yield stress Y using the relationship VPN $\simeq 0.3Y$. The overall pattern and level of the distribution is very similar in the two cases, except for the prediction by the FE program of a peak value of hardness on the top face of the billet near the edge. This is again a result of the differences in the frictional conditions and the larger amount of roll-round calculated by the FE analysis.

The predicted distribution of pressure across one quarter of the top of the billet at 40% reduction in height is given in figure 7.5. This shows that the pressures decreases towards the centre of the face. Experimental work carried out by Nagamatsu and Takuma [7.2] has also provided evidence for the formation of a 'friction valley' in the upsetting of billets with this type of geometry. The finite-element analysis calculated an overall value of 200 kN for the deforming load at this stage of the process. This compares with a value of 210 kN measured experimentally.

Fig. 7.5 Distribution of pressure (P) across one quarter of top surface of finite-element mesh at 40% reduction in height.

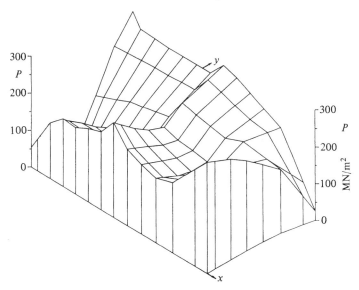

7.2.2   Upset forging

Upset forging is commonly used to make axi-symmetric components from a bar. A typical component is shown in figure 7.6.

Fig. 7.6 Geometry of component produced by upset forging.

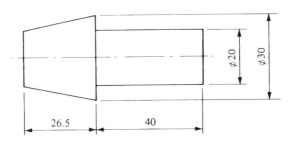

The FE analysis of the forging of this component performed by Eames *et al.* [7.3] again used the finite-strain 3-D elastic–plastic formulation described in Chapters 5 and 6. The forging was performed slowly under isothermal conditions. The yield stress of the material (aluminium alloy HE30TF) varied only with strain. The beta-stiffness technique was used to model the frictional conditions between the dies and the billet with $m = 0.1$ to simulate lanolin lubricant. The predicted deformation is shown in figure 7.7b. This analysis was also performed with $m = 0.7$, representing no lubrication. The forces for both conditions are shown in figure 7.7a.

Since this example is axi-symmetric, only a single-layer of 145 eight-noded iso-parametric hexahedral elements needed to be used in the analyses. These were arranged to form a thin wedge. The nodes on the outer faces of this wedge were constrained to move on radial planes to maintain the conditions of axial symmetry. The lower part of the mesh was contained within a closely fitting cylindrical die, while the upper part was located within a conical die.

In this process there is very little sliding contact between the workpiece and the dies, the effect of the upsetting being for the material to roll onto the inside of the conical die. As a result, there is little difference between the forces (both experimental and FE) obtained for the two levels of friction. It can be seen that the FE results are in reasonable agreement with the experimental values.

7.2.3   Heading

In the heading process, a portion of a bar is held between clamps or grippers and the exposed portion is compressed by a flat or shaped punch to spread the material into some sort of cap attached to the undeformed stem. It is used to make a very wide variety of components, including fastenings, rivets and bolts. Heading can of course be one part of a multi-stage forming process.

Figure 7.8 shows the finite-element mesh used by Hussin *et al.* [7.4] to model an example of a heading operation, in which a cylindrical bar of radius 25.4 mm and length 50.8 mm is gripped for half its length and the remaining portion upset by a flat punch. The FE analysis used a small-strain 2-D (axi-symmetric) elastic–plastic formulation. The calculation assumed isothermal conditions and no strain-rate effects, the yield stress of the commercially-pure aluminium varying only with plastic strain.

The beta-stiffness technique was used to model the frictional conditions between the punch and the workpiece, and two analyses were performed, one using a value of friction factor $m = 0.1$, the other using $m = 0.7$. These two levels of friction were intended to correspond to heading with a good lubricant (such as lanolin) and heading with no lubricant at all. In both analyses, 105 eight-noded iso-parametric quadrilateral elements were used to model half the

Fig. 7.7 (a) Comparison of finite-element and experimental forging forces (kN) during upset forging: ● experimental high friction; —— finite-element high friction; □ experimental low friction; ---- finite-element low friction.

(b) Comparison of distorted finite-element meshes and experimental profiles at various stages in upset forging process with lanolin lubricant.

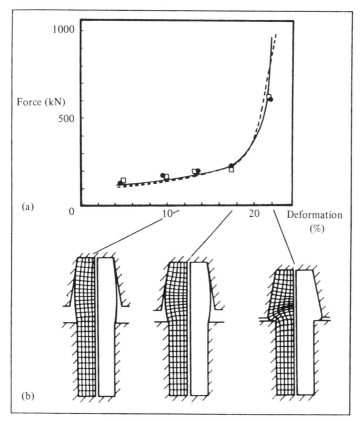

section of the workpiece, arranged as in figure 7.8. Larger elements were used in the gripper region of the mesh because it was expected that little or no deformation would occur here.

During the FE analysis, the exposed portion of the mesh was reduced by 1% of its original height during each increment of the calculation up to a maximum of 60%. The FE calculation was performed on a Future Computer FX30 described in Chapter 4.

Figure 7.9 shows the distorted FE meshes, drawn for the full cross-section, for the two levels of friction, at intervals of 10% reduction in height. The differences in the pattern of deformation between the two frictional conditions can be clearly seen. It can also be seen that in the later stages of the deformation, the FE program predicts deformation within the gripper region of the mesh. This is not observed experimentally, indicating that a finer mesh should have been used to model the exposed part of the workpiece and at least the upper part of the gripper region. Remeshing in this region at various stages of the simulation would also have helped to reduce this effect.

Figure 7.10 compares FE predictions of Vickers Pyramid Hardness Number with experimentally-measured values for the two levels of friction at a deformation of 47% reduction in height. The FE values were obtained from the distribution of yield stress as described in section 7.2.1. Although the FE and experimental values differ by about 5–10% over most of the section, the overall pattern is similar. There is a larger discrepancy between the two sets of values in the lower part of the head portion. This is again a result of a coarse mesh and the FE

Fig. 7.8 Heading – initial finite-element mesh.

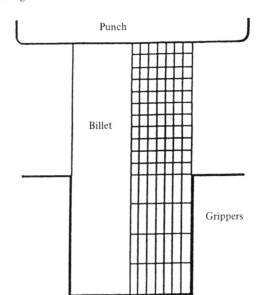

program predicting that the region of deformation extends down into the gripped part of the mesh.

Although the patterns of internal deformation obtained with the two levels of friction are strikingly different, the forming loads in the two cases are very similar. Figure 7.11 compares the forces measured experimentally with the FE results. Agreement is very good. The kink in the FE-predicted load/deformation curve for $m = 0.7$ at 42% reduction is due to the folding round of an element onto the punch surface.

7.2.4   Plane-strain side-pressing

Plane-strain side-pressing is the name given to the transverse forging of billets

Fig. 7.9 Finite-element simulation of heading – FE meshes between initial state and 60% reduction in height of head at intervals of 10%: (a) $m = 0.1$; (b) $m = 0.7$.

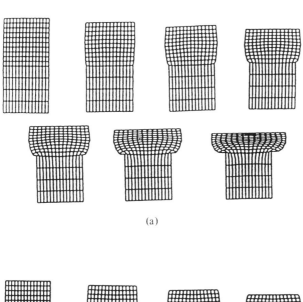

(a)

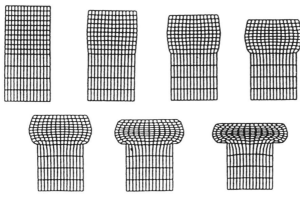

(b)

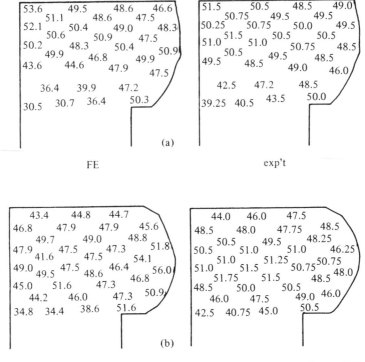

Fig. 7.10 Comparison of finite-element and experimental values of Vickers Pyramid Hardness Number (VPN) across section of billet during heading at 47% reduction in height: (*a*) m = 0.1; (*b*) m = 0.7.

Fig. 7.11 Comparison of finite-element and experimental results for the variation of forging force (kN) during heading operation as a function of reduction in height of head: ◆ experimental (*m* = 0.1); ● experimental (*m* = 0.7); —— prediction.

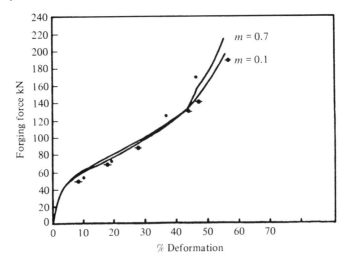

in which no deformation is assumed to occur in the longitudinal direction. A typical example would be in the early stage of the forging of a circular bar into a turbine blade. Here, friction between the shaped dies and the workpiece ensures that there is little increase in the length of the billet. The process is discussed here principally in order to illustrate the ability of finite-element techniques to predict when and where the workpiece is likely to fracture during plastic deformation. This ability is of great importance, particularly when materials of low ductility, such as those chosen for aerospace applications, are used.

Three geometries of side-pressing between flat dies were examined by Clift *et al.* [7.5]. In the first of these, the workpiece had a circular cross-section with diameter 20 mm. In the other two cases, parallel flats were machined on the billets before forging. For one of these specimens, the ratio H/W of the machined height of the billet to the width of the machine flat was 2.03, and for the other specimen the ratio H/W was 1.33 (figure 7.12).

The large-displacement 3-D elastic–plastic formulation described in Chapters 5 and 6 was used for the FE analysis. The program was constrained to model plane-strain behaviour by preventing the movement of nodes in the longitudinal direction. The forging was assumed to be carried out slowly under isothermal conditions, so there were no strain-rate or temperature effects in the model. The yield stress was calculated as a function of plastic strain in order to simulate the work-hardening properties of 7075 aluminium. This fairly brittle material was chosen for the trials so that fracture would be observed at an early stage of the deformation.

The beta-stiffness technique was used to model the frictional conditions between the workpieces and the dies, a value of friction factor $m = 0.25$ being selected to simulate unlubricated dies.

Approximately 80 eight-node iso-parametric hexahedral elements were used to model one quarter of the cross-section of the three types of workpiece. The initial FE meshes are shown in figure 7.12. Because these were plane-strain examples, only a single layer of elements was required in each case.

Figure 7.13 shows the sites of fracture initiation observed experimentally for the three geometries of specimen considered here. The specimen that was originally circular started to fracture in the centre of the workpiece at approximately 16% reduction in height. In contrast, fracture in the specimens with the machined flats started at the outer surface of the billets, at points that were near the original positions of the corners of the specimens. The levels of deformation for fracture initiation in these two cases were 18% for the specimen with $H/W = 2.03$ and 16% for the specimen with $H/W = 1.33$. Figure 7.14 shows the FE meshes for the three geometries at the levels of deformation at which fracture started in each case.

The prediction of fracture requires the monitoring of some fracture-initiation

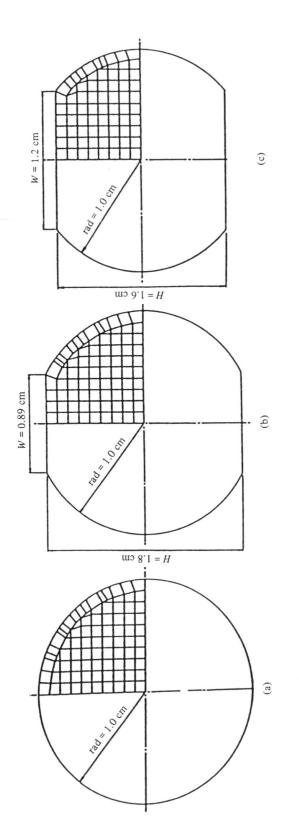

Fig. 7.12 Three geometries of plane-strain side-pressing showing initial finite-element meshes: (a) full circular section; (b) circular section with machined flats, ratio of height of billet H to width of flats $W = 2.03$; (c) circular section with machined flats, $H/W = 1.33$. The material used is 7075 Al.

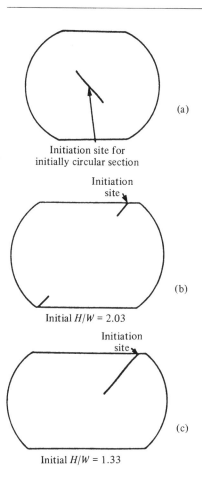

Initiation site for
initially circular section

Initial H/W = 2.03

Initial H/W = 1.33

Fig. 7.13 Site of fracture initiation in experimental plane-strain side-pressing specimens: (a) initially circular section; (b) initial H/W = 2.03; (c) initial H/W = 1.33.

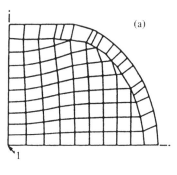

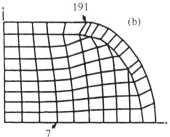

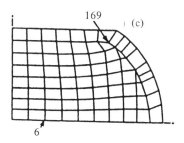

Fig. 7.14 Distorted finite-element meshes for the three plane-strain side pressing examples at the levels of deformation at which fracture initiated experimentally: (a) initially circular section; (b) initial H/W = 2.03; (c) initial H/W = 1.33. Numbers indicate nodes referred to in figure 7.15.

function at all points of a body, and determining where and at what level of deformation the value of this function exceeds some critical empirical quantity. Many different functions have been proposed [7.6], most of which depend upon the state of stress, and some of which involve integration along the strain paths of points within the body. Since FE techniques produce highly detailed information about the distribution of stress, strain and other quantities throughout the deformation, they are ideally suited to the task of predicting the onset of fracture.

One fracture-initiation measure that has proved to be particularly successful is called the generalised plastic work criterion. This integrates the plastic work per unit volume from the start of the deformation up to the current stage:

$$\begin{matrix} \text{Total generalised} \\ \text{plastic work} \\ \text{per unit volume} \end{matrix} = \int_0^{\bar{\epsilon}_f} \bar{\sigma} d\bar{\epsilon} \qquad (7.1)$$

The values of total generalised plastic work per unit volume are plotted in figure 7.15 as a function of deformation for the three cases examined. The curves

Fig. 7.15 Variation of the maximum value of total generalised plastic work per unit volume up to the levels of deformation at which fracture initiated experimentally for the three geometries of plane-strain side-pressing: (1) maximum value for $H/W = 2.03$, node 191; (2) maximum value for $H/W = 1.33$, node 169; (3) maximum value for circular section, node 1. Location of nodes given in figure 7.14.

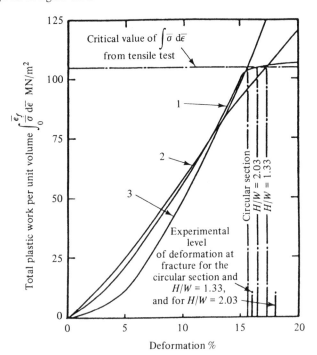

drawn are for the nodes in the FE meshes at which this quantity has its greatest values during the deformation. The position of these nodes is indicated in figure 7.14. It can be seen that the FE program predicts that the maximum values of this fracture function occur at the positions where fracture is observed to start in the laboratory trials. Moreover, the curves in figure 7.15 attain the critical value for the onset of fracture at levels of deformation which are quite close to the values measured in the experimental trials.

Further experimental studies [7.7] have shown transition from centre cracking to corner cracking at a value of H/W approximately equal to 2.5. This transition value is also predicted by FE analyses using the generalised plastic work criterion.

7.2.5   Rim-disc forging

Rim-disc forging is a form of backward extrusion in which a punch is pushed into the centre of a workpiece held within a container to form a rim around a central depression. It differs from backward extrusion (considered later) in that the deformation is non-steady state throughout.

A small-strain axi-symmetric (2-D) elastic-plastic FE program was used to study this process [7.8]. The strain-hardening properties of commercially-pure aluminium were modelled and the deformation was assumed to be performed slowly at room temperature. The beta-stiffness friction technique was used to simulate the frictional conditions between the billet and the die and punch. A value of friction factor $m = 0.04$ was selected to model good lubrication.

Fig. 7.16 Initial finite-element mesh for rim-disc forging.

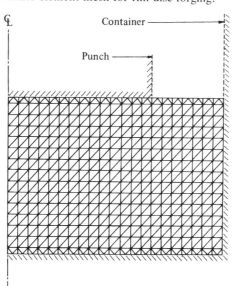

The FE mesh represented one half of the cross-section of the workpiece which had a diameter of 39 mm and a height of 13.3 mm (figure 7.16). A regular arrangement of 546 constant strain triangular elements were used for the analysis which was performed with increments of 0.25% reduction in height of the central portion of the billet.

Figure 7.17 shows the distorted FE meshes at various stages of the deformation. The diagonal band of high shear can be seen in figure 7.17d leading from the corner of the punch downwards at an angle of approximately 45°. This forms the boundary of a conical region of lightly deformed material beneath the punch. The pattern of deformation agrees broadly with experimental observations of gridded specimens, though the shape of the extruded free surface shows the effect of the restraint to deformation imposed by the sharp corner of the punch in the FE simulation. The pattern of flow in the FE analysis can be improved by re-meshing at intervals during the deformation. An example of this is given in reference [7.9] and illustrated in figure 7.18.

The distributions of generalised strain predicted by the FE program are shown in figure 7.19. This indicates a region of low deformation at the bottom of the billet a short distance away from the corner in addition to that directly under the punch. Examination of the grain structure on etched surfaces of sectioned workpieces confirms these predictions. The 'dead zone' near the bottom corner

Fig. 7.17 Finite-element predictions of internal distortion during rim-disc forging: (a) 10% reduction in height of central portion; (b) 16% reduction; (c) 21% reduction; (d) 26% reduction.

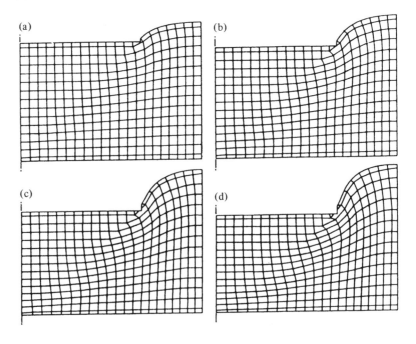

Fig. 7.18 Internal deformation in backward extrusion predicted by the small deformation program, with $m = 0.6$, at each 5% deformation up to 60%. The results were obtained using re-meshing and the supergrid technique.

of the workpiece has sometimes been found to be the site of defects in industrial components of this type.

The comparison of the FE and experimental distributions of VPN hardness illustrated in figure 7.20 again shows that the FE program predicted the correct pattern of deformation, though at some points it over-estimated the values of strain. This is almost certainly because of the inability of the FE model to simulate correctly the flow around the corner of the punch which has led to excessive levels of deformation in the body of the workpiece.

Fig. 7.19 Distribution of generalised strain in rim-disc forging predicted by finite-element analysis at 26% reduction: (A) 0.61–1.03; (B) 0.46–0.60; (C) 0.31–0.45; (D) 0.16–0.30; (E) 0.11–0.15.

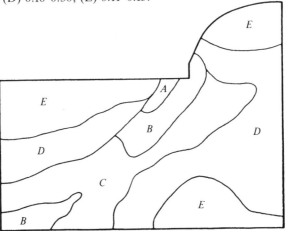

Fig 7.20 Comparison of finite-element predicted values and experimentally-measured values (in parentheses) of VPN hardness in rim-disc forging at 26% reduction.

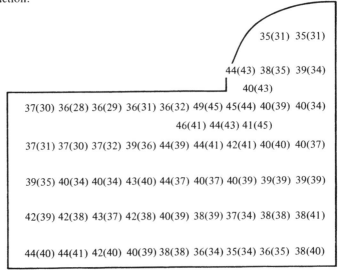

7.2.6   Extrusion-forging

Extrusion-forging, as its name implies, embodies two forms of deformation in one process. This combination of different types of deformation is typical of many operations and is illustrated in an idealised form in figure 7.21. In this example, a billet is compressed between a flat platen and a platen with a central orifice. It is useful because by varying certain geometric parameters, the relative importance of the two deformation mechanisms, extrusion and forging, can be altered.

The results shown here were obtained by Kato *et al.* [7.10, 7.11] and illustrate the influence of the initial geometry of the billet and the size of the central orifice. The finite-strain 3-D program was used for these analyses, constrained to reproduce plane-strain deformation. Sticking friction was assumed on both

Fig. 7.21 Idealised illustration of flow in extrusion forging: (a) before deformation; (b) during deformation.

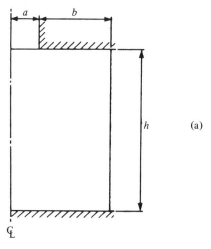

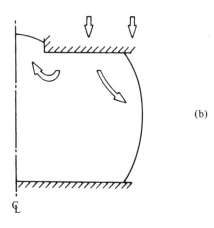

platens and only a small amount of deformation was undertaken in order to examine how the flow patterns developed before significant geometric changes occurred. Commercially-pure aluminium was used as the model material. Rate and temperature effects were not included.

The results of the analyses showed that three distinct modes of flow could be identified. Mode I, figure 7.22a, is typified by a rigid-core region around the plane of symmetry. This is predominantly a forging operation. Mode II, figure 7.22b, clearly shows mixed deformation as the forging of material between the platens is accompanied by extrusion into the central orifice. A neutral point, indicating a flow divide, can be seen along the upper surface in contact with the platen. Mode III, figure 7.22c, is very similar to simple upsetting since material

Fig. 7.22 Typical flow patterns for (a) mode I, (b) mode II and (c) mode III types of deformation.

(a)

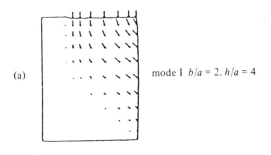

mode I $b/a = 2, h/a = 4$

(b)

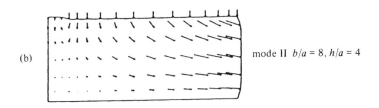

mode II $b/a = 8, h/a = 4$

(c)

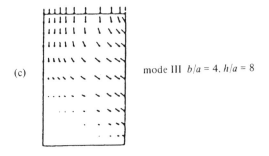

mode III $b/a = 4, h/a = 8$

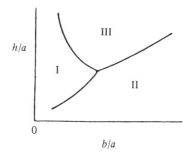

Fig. 7.23 Flow pattern 'classification' diagram.

beneath the orifice moves down towards the lower platen along with the rest of the material.

It is possible to produce a 'classification' diagram as shown in figure 7.23, which relates modes of flow to initial geometry. This is somewhat of a simplification, as the boundaries of the regions in the diagram are not actually as distinct as they are shown to be.

7.2.7   'H'-section forging

Full 3-D analyses require many elements and take a correspondingly large amount of computer time to perform, but it is sometimes possible to use a two-dimensional simplification to examine a particular feature or stage of a complex 3-D forging.

Fig. 7.24 Finite-element mesh and initial positions of die boundary surfaces for 'H'-section forging.

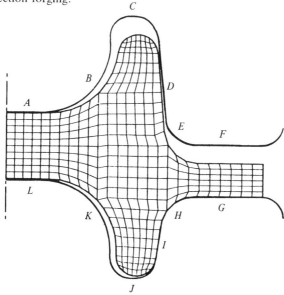

Figure 7.24 shows an example of this. This depicts an FE mesh modelling one half of an 'H'-section of a long bar. This represents one stage in a sequence of operations, so there is already some flash on the pre-form (to the right in the figure). During the forging of this component it is found that there is very little change in the length of the bar, so the behaviour of the central portion can be modelled satisfactorily assuming plane-strain conditions.

The finite-strain 3-D elastic–plastic formulation described earlier in this monograph was used to perform the analysis. The 336 eight-node iso-parametric hexahedral elements were arranged in a single layer as shown in figure 7.24. The nodes were prevented from moving perpendicularly to the plane of this figure in order to model plane-strain deformation. Because of the offset parting line, the 'H'-section possessed only one-fold symmetry, and so the FE mesh represented one half of the section.

The initial positions of the dies are also shown in figure 7.24. These were modelled in the FE program by means of 18 primitive boundary surfaces. A friction factor of $m = 0.2$ was assumed between the dies and the billet, the frictional restraint being imposed in the FE calculation by the beta-stiffness technique.

During each increment of the analysis, the upper boundary surfaces were moved downwards in order to reduce the thickness of the central part of the section by a nominal value of 1.05% of its original height. The time step for the increment was chosen to give a nominal strain rate of 1.0 s^{-1} in this region.

The model material was Al 1100, the yield stress of which depended upon plastic strain, strain rate and temperature. The following thermal properties were used in the analysis:

$$
\begin{aligned}
\text{thermal conductivity} &= 242 \text{ J/msK} \\
\text{thermal capacity/volume} &= 2.43 \text{ MJ/m}^3\text{K} \\
\text{die-interface conductivity} &= 35 \text{ kJ/m}^2\text{K}
\end{aligned}
$$

Since this was a warm forging, the billet was assumed to be at a temperature of 700 K and the dies at a temperature of 600 K at the start of the deformation.

The complete FE analysis of this deformation up to a reduction of 26% in the height of the central portion of the section required approximately five hours of CPU time on a Honeywell DPS8 mainframe computer.

Figure 7.25 shows how the plastic region develops during the early stages of the deformation. At first, the billet is only in contact with the dies in the regions A, G–I and L marked in figure 7.24, so the initial deformation is principally that of simple compression of the central part of the section. The shear bands can be seen clearly in figure 7.25a. Gradually, the region of plastic deformation grows until at 2% reduction it extends right across the section and into the flash region (F, G). Even at this stage, however, the upper and lower lobes of the section are still deforming elastically.

Fig. 7.25 Development of plastic zone (shaded) during early stage of 'H'-section forging: (a) 0.05% reduction; (b) 0.065%; (c) 0.2%; (d) 0.5%; (e) 1%; (f) 2%.

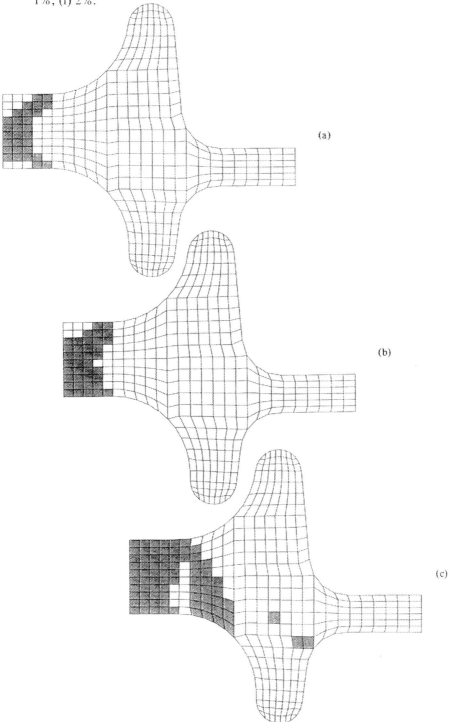

(a)

(b)

(c)

Fig. 7.25 (*continued*)

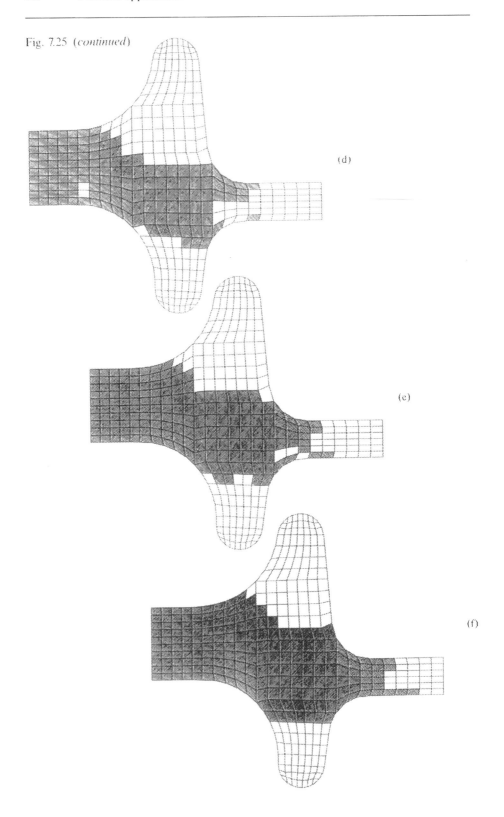

(d)

(e)

(f)

Fig. 7.26 Flow velocity vectors at various stages in forging of 'H'-section: (a) 2% reduction; (b) 3%; (c) 10%; (d) 12%; (e) 15%; (f) 26%.

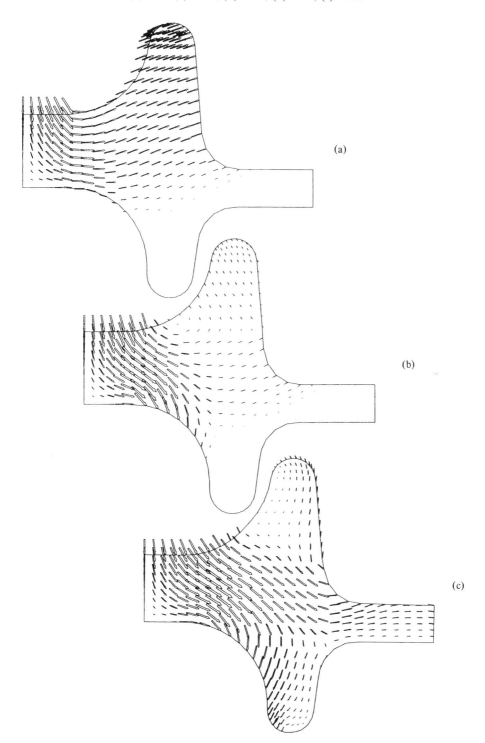

(a)

(b)

(c)

Fig. 7.26 (*continued*)

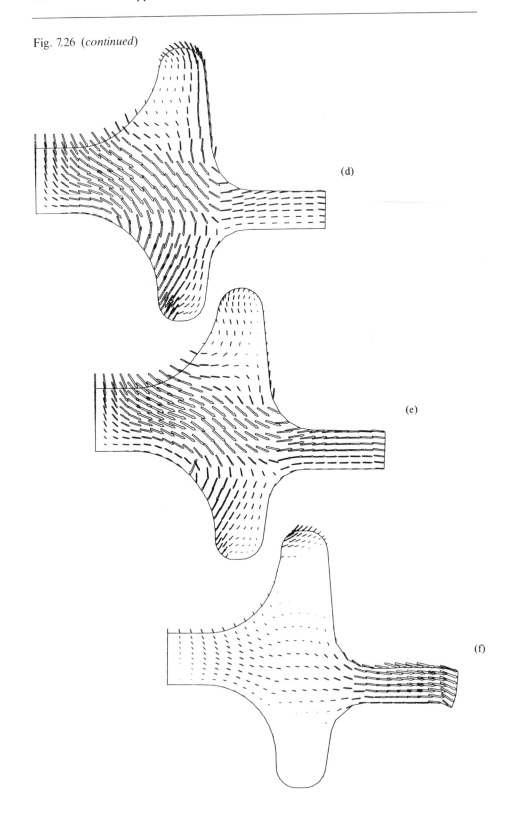

(d)

(e)

(f)

The velocity vectors in figure 7.26 illustrate the patterns of flow in the 'H'-section in more detail. At 2% reduction for instance (figure 7.26a), it can be seen that the upper lobe is being moved across by the outward flow of material from the central region, whereas the lower is prevented from moving by its contact with die surfaces H and I. There is therefore very little flow of material in this part of the workpiece.

At about 3% reduction (figure 7.26b), the upper lobe of the 'H'-section comes into contact with the die at surface D and its outward movement is halted. Material from beneath the central indenter now flows mainly into the bottom lobe, and begins to fill up the space near surface K. There is also a small amount of flow into the upper lobe. This situation persists until about 10% reduction (figure 7.26c), by which time the downward movement of die surface D has begun to reverse the flow in the upper lobe. Material is still filling the space in the bottom, but is now also flowing into the flash region.

This behaviour is somewhat intensified at 12% reduction (figure 7.26d), but by 15% reduction in height (figure 7.26e), the situation has changed again. At this stage, the upper die is also in contact with the workpiece at surface B, and this has reversed the flow in the upper lobe once again, but this time only partially. The right-hand side of the lobe is still being pushed down by the die wall, but material is flowing upwards on the left-hand side. The complex pattern of flow is shown in greater detail in figure 7.27, in which a flow divide on the right-hand side can be seen clearly, together with a rotational pattern on the left-hand side.

Finally, figure 7.26f shows that at the end of the deformation, the lower lobe has completely filled, whereas there is still some flow of material into the gap at the top. Most of the flow however is into the flash.

The distribution of plastic strain at the end of the deformation is illustrated in figure 7.28. This shows how most of the deformation has occurred in an approximately horizontal band across the section; there is relatively little strain in the two lobes. Much of their movement has occurred as almost rigid-body movement. The region of highest strain is near the lower right-hand corner of the die (surface H).

The temperature contours at 22% reduction shown in figure 7.29 indicate that the principal determinant of temperature in this forging example is the effect of die chilling on the workpiece. The hottest part of the workpiece is half-way between the lower and upper lobes, but the temperature gradient between the centre and the outside of the billet is not uniform. Those parts of the billet that have been in contact with the dies the longest time (e.g. A, D, G–I and L) show the greatest amount of cooling. An interesting consequence of this is that the temperature difference across the flash thickness has led to a difference in the flow stress, and to an asymmetrical flow of material into the flash gutter.

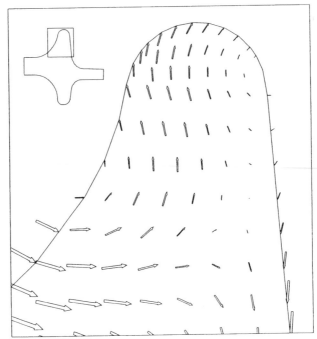

Fig. 7.27 Close-up of flow velocity vectors in upper lobe of 'H'-section forging at 15% reduction.

Fig. 7.28 Contours of generalised plastic strain in 'H'-section forging at 26% reduction.

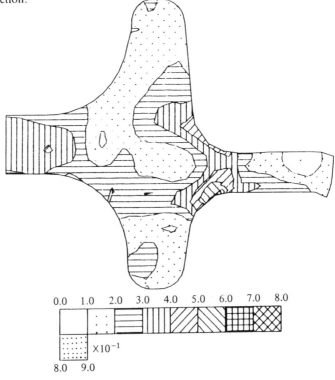

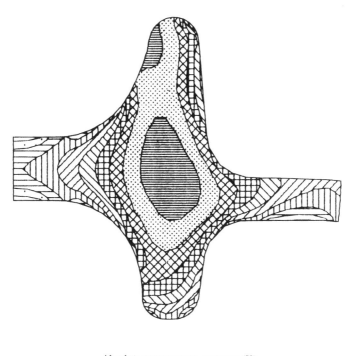

Absolute temperature contours (K)

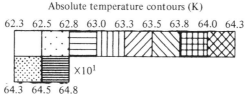

Fig. 7.29 Contours of temperature (K) in 'H'-section forging at 22% reduction.

7.2.8 Forging of a connecting rod

Industrially, connecting rods are usually forged from steel at a temperature of approximately 1200 °C. The forging is typically carried out in six stages, starting from the original bar stock and ending with a coining operation before the final clipping of the flash. The finite-element analyses described here model a laboratory simplification of one of these stages (from final preform to mould forging) using commercially-pure aluminium deformed at room temperature [7.12].

The initial preform geometry and the desired shape of the billet at the end of this stage of the deformation are shown in figure 7.30. The symmetry of the component meant that only one quarter of the preform needed to be modelled. The initial FE mesh is shown in figure 7.31. This contained 600 eight-noded 3-D linear isoparametric elements.

The FE analysis was carried out using the 3-D large-displacement elastic–plastic approach described in Chapters 5 and 6 of this monograph. The commercially-pure aluminium used for the model was assumed to deform isothermally and no strain-rate effects were included. Sixteen geometric primitive surfaces were used to model the complex shape of the dies. For simplicity, two complete analyses were performed, one assuming zero-friction conditions between the billet and the dies and the other assuming sticking friction. During each increment of the

Fig. 7.30 (a) preform and (b) required final shape of connecting rod forging.

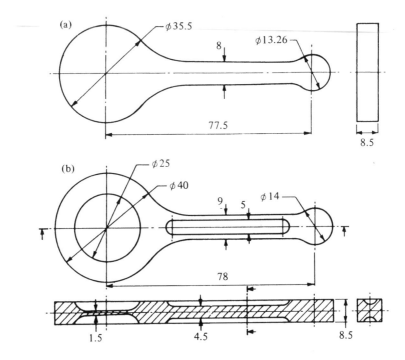

Fig. 7.31 Finite-element idealisation of connecting rod preform.

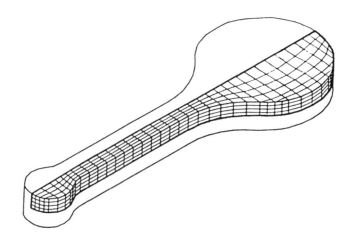

analysis, the thickness of the billet at the centre of the big end was reduced by no more than 0.5% of its original value. Each increment took approximately 120 s of CPU time on a CDC 7600 computer, a complete analysis taking about $5\frac{1}{2}$ hours.

Figure 7.32 compares the experimentally measured change in length of the billet during the deformation with the values predicted by the two FE analyses. The predictions of the sticking-friction and zero-friction analyses are markedly different. Despite the pronounced scatter in the experimental results, it can be seen that the sticking-friction results agree very well with the actual measurements.

The patterns of deformation predicted by the two FE analyses are compared in greater detail in figures 7.33–7.35. The vertical sections through the deformed FE meshes (figures 7.33 and 7.34) show that the pattern of deformation in the zero-friction case is much more homogenous than that obtained with sticking friction. The velocity vectors illustrated in figure 7.35 clearly indicate other essential differences in the two modes of flow. Figure 7.35a shows that when there is no frictional restraint, flow on the horizontal centre-line plane has a predominant longitudinal component, so that material extruded from under the big-end indenter tends to elongate the billet. In contrast, with sticking friction, material flows radially outwards from under the big-end indenter, tending to increase the amount of flash around this region. There is practically no longitudinal flow along the connecting arm of the billet. Instead, flow in this region is almost entirely in a transverse direction and results from the formation of the indentation in this arm.

However, one feature of the experimental forging that was predicted by the FE analysis with sticking friction was the reduction in the width of the flash at

Fig. 7.32 Experimental measurements ($\triangle$) and finite-element predictions (——) of changes in length (δL) of connecting rod forging plotted against percentage reduction in height, R.

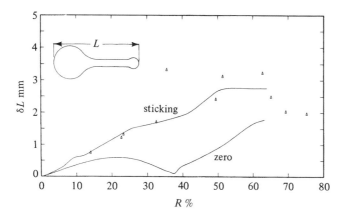

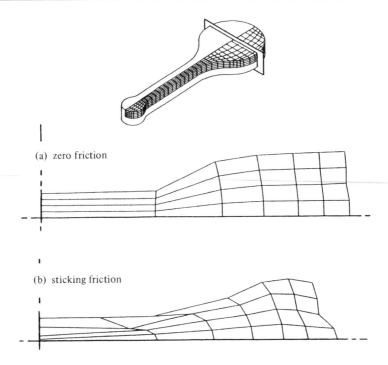

(a) zero friction

(b) sticking friction

Fig. 7.33 Transverse vertical sections through big end of deformed FE connecting rod meshes at about 60% reduction assuming (a) zero friction and (b) sticking friction.

Fig. 7.34 Longitudinal vertical sections along centre-line of connecting rod showing FE meshes at about 60% deformation assuming (a) zero friction and (b) sticking friction.

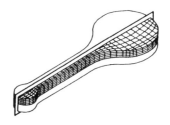

(a) zero friction

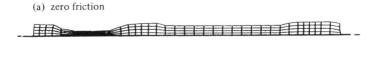

(b) sticking friction

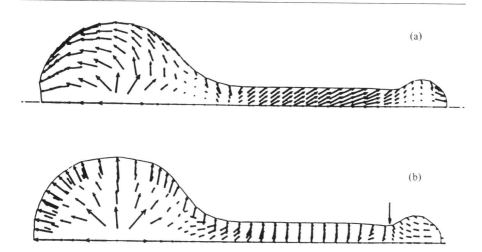

(a)

(b)

Fig. 7.35 Material velocity vectors on horizontal centre-line section of connecting rod at about 60% deformation predicted by FE analyses assuming (a) zero friction and (b) sticking friction.

the point where the small end joins the connecting arm. This can be clearly seen in figure 7.35 at the location marked.

Figure 7.36 compares experimentally measured values of hardness on the vertical plane of symmetry of the connecting rod with the values predicted by the sticking-friction FE analysis. Despite the coarseness of the FE mesh, the two distributions are in good agreement.

Fig. 7.36 Comparison of experimental values (top) and sticking-friction finite-element predictions (bottom) of Vickers Pyramid Hardness distributions on longitudinal vertical plane through the centre-line of the connecting rod. Hardness ranges: (kgf/mm^2): a, 23–35; b, 35–45; c, 45–55; d, 55+.

Experimental result

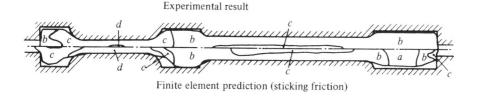

Finite element prediction (sticking friction)

7.3   EXTRUSION

7.3.1   Forward extrusion

Forward extrusion is used to produce a large variety of continuous sections. It may also form part of a more complex forging operation, such as the formation of a central boss in a gear blank. It is in this context that the process is examined here.

Figure 7.37 shows the example studied by Hussin *et al*. A short billet is positioned within a container prior to extrusion through a conical die. The billet is 25.4 mm in diameter and 20.32 mm long. The die orifice has a diameter of 15.86 mm, corresponding to an area reduction of 61%. The included die angle is 124 degrees.

A small-strain 2-D (axi-symmetric) elastic–plastic finite-element analysis was used to study this process. The material (commercially-pure aluminium) was assumed to be extruded slowly and under isothermal conditions, with the yield stress depending only upon strain.

The beta-stiffness friction model was used, with a value of friction factor $m = 0.7$ employed to correspond with unlubricated contact between the billet and the container and punch.

The initial FE mesh consisted of 109 eight-noded iso-parametric quadrilateral elements arranged in a simple rectangular grid (figure 7.37). Since the billet was symmetrical about the centre line, only half of the cross-section was modelled.

During each increment of the analysis, the punch was displaced by 1% of the original length of the billet. Because of the severe deformation occurring in the extrusion process, particularly at the corner of the orifice, the mesh was reformed after every 5% deformation. Information about the internal flow distribution was used to update the nodal co-ordinates of a reference mesh at the end of each increment to indicate the pattern of lines that would be obtained on a gridded experimental specimen (super-grid technique).

The distorted FE reference grids are shown at stages throughout the deformation in figure 7.38. The areas of gross deformation at the corners of the orifice can be seen clearly, as can zones of much lower deformation near the internal corners of the container. This is bounded by a diagonal band of shear, making an angle of approximately 45 degrees with the container wall.

Fig. 7.37 Initial finite-element mesh for forward extrusion.

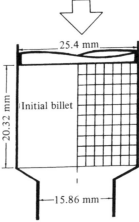

The values of VPN hardness predicted by the FE program are compared with the experimentally-measured values in figure 7.39. The FE hardness numbers were calculated from the strain values as described in Section 7.2.1. The two sets of values agree quite well and support the evidence of the deformation patterns in that they indicate regions of high deformation at the corner of the orifice, and lower deformation beneath the punch.

Figure 7.40 compares the FE and experimental results for the variation of the extrusion force during the deformation. The FE program tended to over-estimate the loads, but the slope of the predicted curve is in excellent agreement with the experimental behaviour.

Fig. 7.38 Finite-element predictions of the distortion of an initially-regular grid at various stages of forward extrusion (super-grid technique).

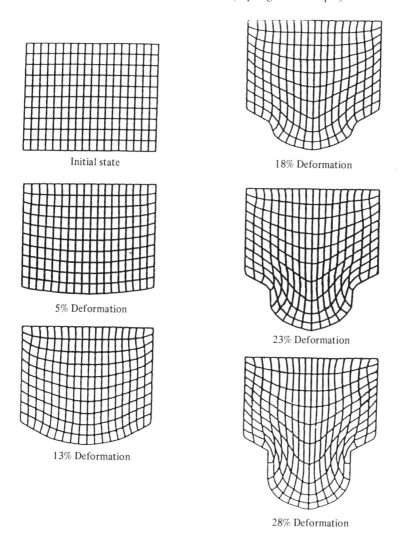

Initial state

18% Deformation

5% Deformation

23% Deformation

13% Deformation

28% Deformation

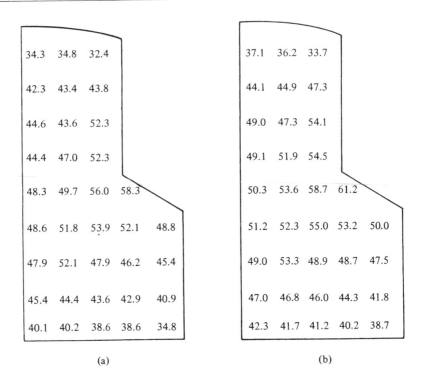

Fig. 7.39 Comparison of VPN hardness values at 28% reduction of original billet length: (a) finite-element values; (b) experimentally-measured values.

Fig. 7.40 Comparison of the finite-element predictions and measured values of the variation of forging force (kN) during forward extrusion: □ experimental; —— prediction.

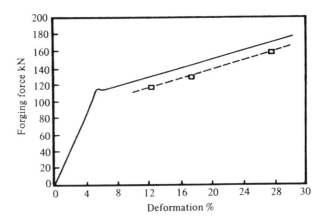

7.4 ROLLING

7.4.1 Strip rolling

In strip rolling, the width of the workpiece is very much greater than its thickness. As a result, the frictional restraint to transverse movement of material is large and there is little sideways spread. Strip rolling can therefore be reasonably modelled as a plane-strain deformation.

The FE analyses carried out by Liu *et al.* [7.13, 7.14] used the finite-strain 3-D elastic–plastic formulation described earlier in this monograph. The rolling was assumed to be carried out under conditions that allowed the effects of strain rate and temperature on the properties of the material to be ignored. The strain-hardening characteristics of copper were used in the analysis. The friction between the roll and the strip was modelled by means of the beta-stiffness technique using a value of friction factor $m = 0.2$.

In this example, the strip was 1.56 mm thick and the roll has a diameter of 78 mm. The thickness of the strip was reduced by a nominal value of 22.76% in one pass. Due to symmetry, only one half of the strip needed to be modelled. A strip length of 31.4 mm was used for the analysis to ensure that steady-state conditions would be reached before all the strip has passed between the rolls. This required a total of 88 eight-noded iso-parametric hexahedral elements arranged in a single layer with four elements in the thickness direction. The nodes of the mesh were prevented from moving in the direction of the roll axis in order to specify plane-strain conditions.

Fig. 7.41 Frictional boundary conditions during finite-element simulation of rolling.

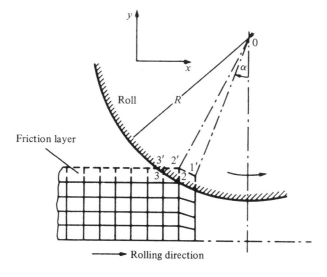

Figure 7.41 illustrates how the boundary conditions were imposed on the FE mesh. Any nodes that came into contact with the roll were constrained to move tangentially to it. The adjacent friction-layer nodes were rotated by a small angular amount about the roll axis each increment, which tended to pull the upper layer of the mesh into the roll gap. Note that the velocity of the top of the strip was not prescribed beforehand but was a function of the stiffness of the friction layer and hence of the frictional conditions at the roll surface. On exit from the roll of course, the surface of the strip was moving faster than the roll, and the effect of the friction layer was to retard this movement. The FE program was therefore able to predict the position of the neutral point on the roll.

Figure 7.42 shows the deformed FE meshes from just after the strip has entered the roll gap until the steady state. This figure shows clearly that plane sections through the workpiece do not remain plane as they pass under the roll (an assumption of simpler theories [7.15]). Also the curvature of these sections reverses during the rolling, an effect that is more pronounced for higher levels of friction.

The velocity vectors, relative to the horizontal velocity of the neutral point of the roll, are shown at various stages in figure 7.43. By subtracting the horizontal

Fig. 7.42 Predicted finite-element deformed meshes at stages during the rolling of wide copper strip. (d) is for steady state conditions.

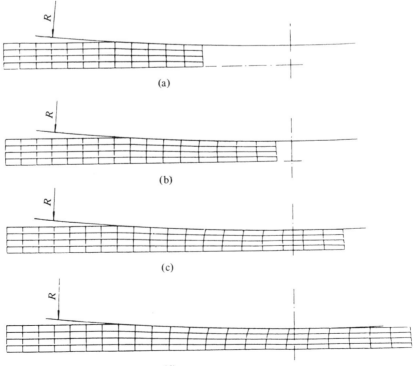

component of the neutral-point velocity in this way, the pattern of flow and the change in position of the neutral surface in the strip can be seen more clearly.

Figure 7.44 compares experimental [7.16] and FE results for the distribution of pressure under the roll during steady-state rolling. Two FE curves are shown, one for $m = 0.2$ and the other for $m = 0.4$. The lower value of m corresponds to the level of friction estimated for the experimental trials, but it can be seen that the FE results for this level of friction are much lower than the experimental measurements. FE predictions for a friction factor of $m = 0.4$ agree very well with experiment. Estimating friction factors from experimental tests using pressure pins is very difficult. The results here suggest that the higher value may be more appropriate.

7.4.2   Slab rolling

In slab rolling, the width of the workpiece is of the same order as its height. In such cases, the bulging and spread of material as the workpiece passes through

Fig. 7.43 Finite-element flow velocity vectors in rolling of copper strip relative to the horizontal velocity of the neutral point on the roll.

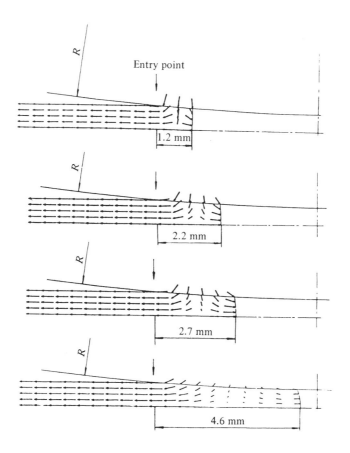

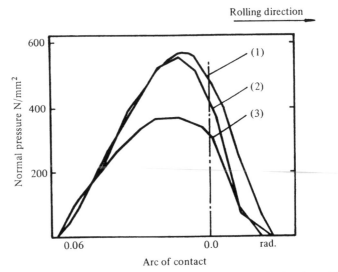

Fig. 7.44 Distribution of pressure across the contact region during the rolling of wide copper strip: (1) experimental measurements; (2) FE predictions for $m = 0.4$; (3) FE predictions for $m = 0.2$.

the roll gap can be quite large and it is very useful to be able to predict spread and the profile of the surface in order to reduce material wastage. The FE treatment of this type of rolling requires a full 3-D approach.

The large-displacement 3-D elastic–plastic program described in this monograph was used by Liu *et al.* [7.17] to examine slab rolling. The deformation was assumed to be carried out at room temperature using mild steel (AISI 1080). Several rectangular sections of slab were studied. The example examined here had a height and width of 25.4 mm ($W/H = 1$). This was subjected to a reduction in thickness of 20% in a single pass between rolls 604.6 mm in diameter. To model one quarter of the slab, 300 eight-noded iso-parametric hexahedral elements were used, arranged as in figure 7.45. The length of the FE mesh was

Fig. 7.45 Initial finite-element mesh for slab rolling.

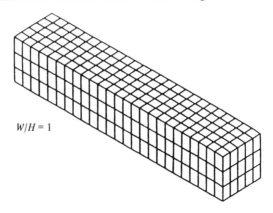

$W/H = 1$

estimated at about three times the nominal contact length, which ensured that steady-state conditions could be attained during the analysis. The beta-stiffness technique was used to model the frictional conditions at the roll surface, the friction factor *m* having a value of 0.5. The slab was drawn into the roll gap by imposing a small rotation to the roll surface each increment.

The shapes of various vertical sections through the steady-state FE mesh are shown in figure 7.46. These indicate that the bulge does not develop until the

Fig. 7.46 Transverse cross-sections through deformed finite-element mesh for slab rolling.

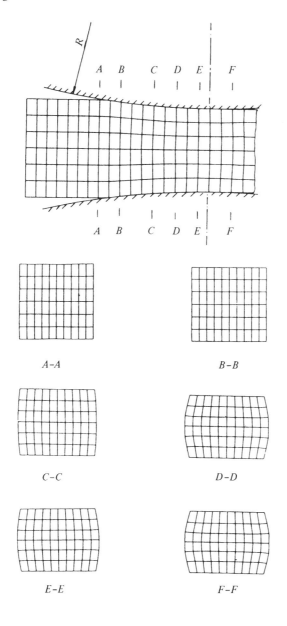

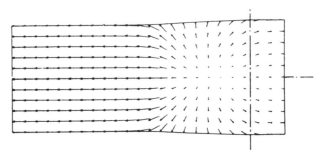

Fig. 7.47 Finite-element flow velocity vectors on horizontal mid-plane section through workpiece relative to the horizontal component of velocity of neutral point of roll surface.

material is well into the roll gap $(C–C)$, and that before this the outer surface of the slab may even be slightly concave.

The velocities of points on the mid-plane horizontal section of the workpiece relative to the horizontal component of velocity of the neutral point are illustrated in figure 7.47. This shows that the transverse flow of material is not uniform throughout the deformation zone, but increases from the centre to the edges of the slab.

Figure 7.48 compares the FE predictions for the overall spread of the material,

Fig. 7.48 Comparison of finite-element predicted values of spread in slab rolling for various reductions with experimental and theoretical results.

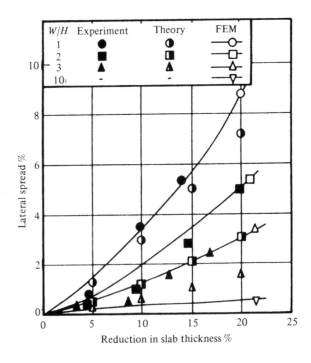

for a range of reductions in thickness from 5% to 20%, with experimental results [7.16] and previous empirical theory [7.18]. The FE results agree very well with experiment except for small reductions in thickness.

7.5   MULTI-STAGE PROCESSES – FORGING SEQUENCE DESIGN

7.5.1   Automobile spigot

For economic reasons, spigots, gear-wheel blanks and other such components are frequently forged from comparitively thin bar stock. Because of the limitations of press capacity and the risk of buckling, this cannot be accomplished in a single operation. Figure 7.49 shows the sequence of operations used to forge a typical automobile spigot [7.19].

The cropped or sawn bar is first forced into a set of conical-ended dies to forge a chamfer on the ends (stage 1). This has the effect of tidying-up the cropped surfaces of the bar, as well as allowing the workpiece to be located easily in the subsequent dies.

The next operation, stage 2, is an upsetting process with the ends of the bar contained within conical dies. The billet spreads to fill these dies and also increases in thickness in the central region.

The third stage is similar to the previous one, but in this operation the ends of the workpiece are also extruded into a narrower cavity, causing an area reduction of 28%.

In the fourth and final stage, the ends of the workpiece are held in a container, and the central portion is upset into a die cavity to form the required shape of the disc.

A small-strain axi-symmetric (2-D) elastic–plastic formulation was used by Al-Sened et al. [7.19] for the FE analysis of this process. The forging was assumed to be done slowly at room temperature on 0.1% C steel. The beta-stiffness

Fig. 7.49 Stages in the production of an automobile spigot.

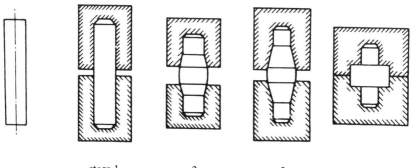

stage 1   2   3   4

Fig. 7.50 Distorted reference grids predicted by finite-element analysis for the four stages in forging of spigot (super-grid technique).

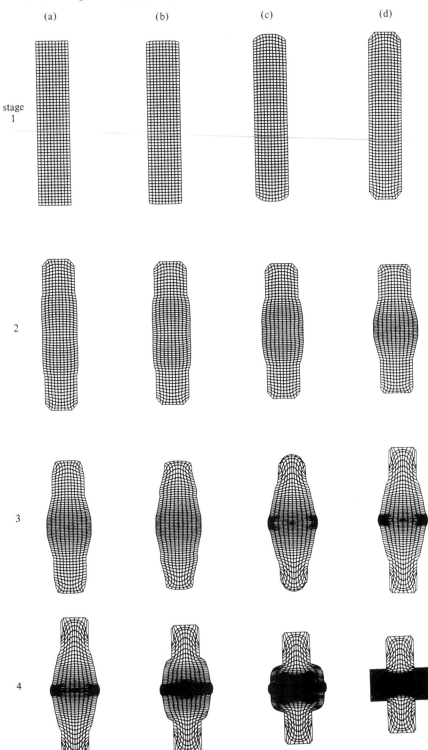

technique was used to model the frictional conditions at the interfaces between the workpiece and the dies. A value of friction factor $m = 0.07$ was selected to generate a frictional restraint similar to the phosphate and soap lubricant used in the experimental trials.

Due to the severe deformation produced in this process, the FE mesh needed to be re-formed at intervals during the analysis. Eight-noded iso-parametric quadrilateral elements were used throughout. Figure 7.50 shows how an initially-rectangular grid inscribed on the cross-section of the bar would have appeared at stages throughout the process. These results were obtained by referring information about FE nodal displacement back to the reference grid (super-grid technique). Throughout the analysis, the length of of the workpiece was reduced by 1% of its original value during each increment.

Stage 1 was performed with a mesh of 125 elements representing one quarter of the original bar cross-section. This stage ended when the required degree of chamfering had been obtained on the ends. No re-meshing was required during this stage and the same mesh was carried over into the FE analysis of the second stage of the process. This time, the ends of the mesh were contained within 15° conical dies. This stage required 24 increments.

The third stage started with a reformed regular mesh of 95 elements constructed within the profile obtained from the previous stage. Values of the important state variables were also obtained from the previous analysis and interpolated to the new nodal points. The third stage was carried out in 36 increments, with another re-meshing after 30 increments.

The recently-reformed mesh from stage 3 was then used to start stage 4. The deformation was complete after another 30 increments. The whole analysis required 180 seconds of CPU time on a Honeywell DPS8 mainframe computer.

The distorted grids in figure 7.50 clearly show the intense band of deformation in the central part of the disc of the spigot at the end of the process.

In figure 7.51, experimental and FE values of VPN hardness are compared at the end of each of the four stages of the process. The FE predictions agree very well with the experimental results at the end of stages 1 and 3, but the agreement is not so close at the end of the other two stages. However, this may be due to variations in the material used in the experiments, since the hardness measurements were carried out on different production components, sectioned at different stages of the deformation. Comparison of figures 7.51c and d show that there appears to be a significant increase in the hardness in the shank of the experimental specimens during stage 4. Since the shank is held within a closely fitting container during this part of the process, there ought to be very little deformation, and very little change in strain distribution as a result. It should be noted that the FE program correctly predicted very similar distributions of hardness in the shank at the ends of stages 3 and 4.

Figure 7.52 shows the FE predictions of forming force for each of the stages as a function of machine stroke. Also shown is the total force for a 4-stage heading machine. Information of this type is obviously very useful in optimising the use of forging machinery.

Fig. 7.51 Comparison of finite-element predicted and experimentally measured distributions of VPN hardness on cross-section of spigot during forging: (a) end of stage 1; (b) end of stage 2; (c) end of stage 3; (d) end of stage 4.

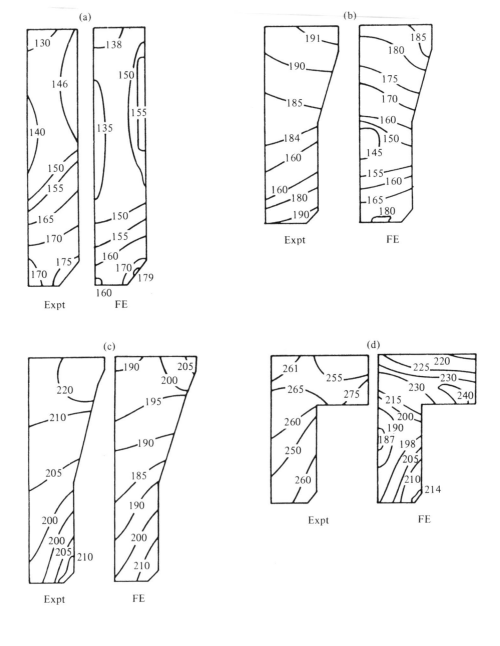

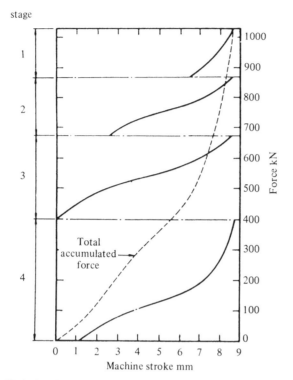

Fig. 7.52 Variation in forging force predicted by finite-element analysis for the four stages of spigot forging and total force in a four-stage heading machine.

7.5.2   Short hollow tube (gudgeon pin)

A four-stage forging process is typical for the production of 'gudgeon pins' (small lengths of tube). The first stage is cropping, and is not included here. The three stages modelled are as follows:

Fig. 7.53   Stages in the forming of a gudgeon pin: upsetting (a), indenting (b) and backward extrusion (c). Initial dimensions of billet and container: billet diameter $D_0 = 22.8$ mm; billet height $H_0 = 34.2$ mm; container diameter $D_c = 28.575$ mm. Indentation with punch of diameter $D_p = 19.05$ mm.

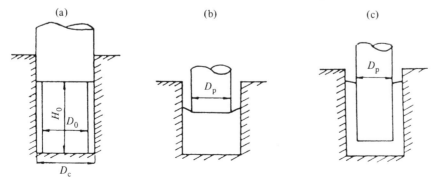

Stage 1: Simple upsetting in a closed container (figure 7.53a). The billet length/diameter ratio was 1.5 after cropping. The die was filled after a reduction in height of the billet by 37%, reducing the length/diameter ratio to just below unity.

Stage 2: Indentation of the dumped billet by 1 mm to guide the punch in the backward extrusion operation (figure 7.53b).

Stage 3: Backward extrusion of the indented slug, with an area reduction of 1·8:1 (figure 7.53c).

The finishing operation involves punching out the base of the cup. This controlled-fracture process is not modelled here.

The model material used for the analysis carried out by Al-Sened *et al.* [7.20] was commercially-pure aluminium. The experimental processes were lubricated with lanolin, $m = 0.1$, and friction was incorporated in the finite-element analysis using the beta-stiffness technique.

An axi-symmetric elastic–plastic, small displacement, isothermal FE program was used. Eight-node iso-parametric quadrilateral elements were used and the initial FE mesh is shown in figure 7.54. Extensive re-meshing was necessary due to the extreme non-linearities associated with the corners of the backward extrusion punch for stage 3. No remeshing was required for the dumping and indentation phases (figures 7.55a–c) but the backward extrusion sequence (figures 7.55d–g) required a re-meshing operation every 5% penetration. This was an extremely laborious process at the time the analysis was conducted as no

Fig. 7.54 Initial FE mesh for forming of a gudgeon pin.

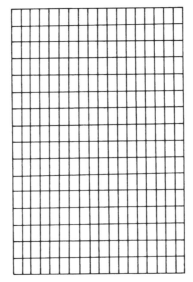

Fig. 7.55 Predicted grid distortions for the dumping, indentation and back-ward extrusion stages.

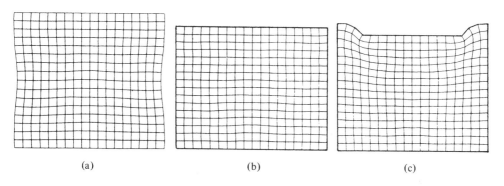

(a)   (b)   (c)

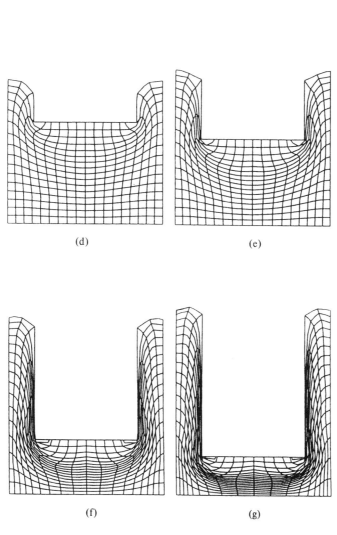

(d)   (e)

(f)   (g)

automatic re-meshing programs were available. For practical use of FE simulations of large deformations re-meshing programs are essential.

The internal flow during the dumping operation could not be determined
experimentally but the changes in the external geometry of the billets was consistent with observations. A double-barrelled shaped billet was produced by the
deformation prior to the billet contacting the walls of the die – this was due to
the height to diameter ratio of the initial billet. When the billet did contact the
die wall, at the top and bottom, these parts were retarded due to die friction.
The last part of the billet to contact the die was in the 'valley' between the ends
of the billet. This is reflected in the predicted strain distribution figure 7.56,
where the maximum strain (0.57) is seen to occur at the outer radius of the billet
a little above the centre line.

The predicted grid distortion obtained from the super-grid technique was in
excellent agreement with experimentally observed distorted grids, figure 7.57.
The 'rigid' zone with a curved high-deformation boundary, the flow of material
around the punch corner, and the spread of the deformation zone throughout
the base of the can during the post steady-state phase are all clearly visible in
both the simulation and the experimental trials.

Fig. 7.56 Generalised strain distribution at the end of the dumping stage.

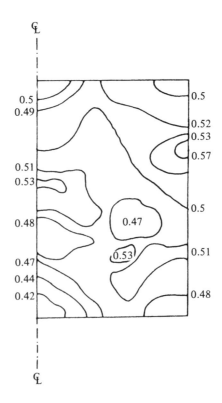

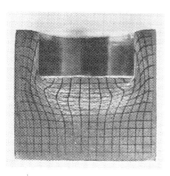
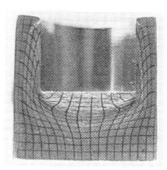
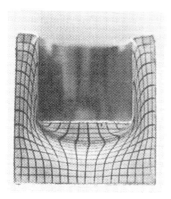
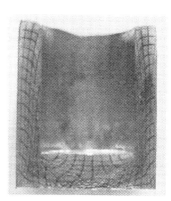

Fig. 7.57 Distorted grids from experimental studies of the backward extrusion stage.

References

[7.1] Pillinger, I., Hartley, P., Sturgess, C.E.N. and Rowe, G.W. An elastic–plastic three-dimensional finite-element analysis of the upsetting of rectangular blocks and experimental comparison. *Int. J. Machine Tool Des. Res.* **25**, 229–43 (1985).

[7.2] Nagamatsu, A. and Takuma, M. Experimental study of pressure and deformation of rectangular blocks in compression. *J. Jap. Soc. for Techn. of Plasticity* **14**, no. 144, 49–57 (in Japanese) (1973).

[7.3] Eames, A.J., Dean, T.A., Hartley, P. and Sturgess, C.E.N. An integrated computer system for forging die design and flow simulation. *Proc. Int. Conf. on Computer-Aided Production Engineering*, ed. J.A. McGeough, Mechanical Engineering Pubs., pp. 231–6 (1986).

[7.4] Hussin, A.A.M., Hartley, P., Sturgess, C.E.N. and Rowe, G.W. Simulation of industrial cold forming processes. *Comm. App. Num. Methods* **3**, 415–26 (1987).

[7.5] Clift, S.E., Hartley, P., Sturgess, C.E.N. and Rowe, G.W. Fracture initiation in plane-strain forging. *Proc. 25th Int. Machine Tool Des. Res. Conf., ed. S.A. Tobias, Macmillan, pp. 413–19 (1985).*

[7.6] Clift, S.E. Identification of defect locations in forged products using the finite-element method. Ph. D. thesis, University of Birmingham, UK (1986) (unpublished).

[7.7] Hartley, P., Clift, S.E., Salimi-Namin, J., Sturgess, C.E.N. and Pillinger, I. The prediction of ductile fracture initiation in metalforming using a finite element model and various fracture criteria. Special issue of *Res. Mechanica*, **28**, 269–93 (1989).

[7.8] Hartley, P., Sturgess, C.E.N. and Rowe, G.W. Prediction of deformation and homogeneity in rim-disc forging. *J. Mech. Wkg. Techn.* **4**, 145–54 (1980).

[7.9] Hussin, A.A.M., Hartley, P., Sturgess, C.E.N. and Rowe, G.W. Elastic–plastic finite-element modelling of a cold-extrusion process using a micro-computer-based system. *J. Mech. Wkg. Techn.* **16**, 7–19 (1988).

[7.10] Kato, K., Rowe, G.W., Sturgess, C.E.N., Hartley, P. and Pillinger, I. Fundamental deformation modes in open die forging – finite element analysis of open die forging I. *J. Jap. Soc. for Techn. of Plasticity* **277**, no. 311, 1383–9 (in Japanese) (1986).

[7.11] Kato, K., Rowe, G.W., Sturgess, C.E.N., Hartley, P. and Pillinger, I. Classification of deformation modes and deformation property maps in open die forging – finite element analysis of open die forging II. *J. Jap. Soc. for Techn. of Plasticity* **277**, no. 312, 67–74 (in Japanese) (1987).

[7.12] Pillinger, I., Hartley, P., Sturgess, C.E.N. and Rowe, G.W. A three-dimensional finite-element analysis of the cold forging of a model aluminium connecting rod. *Proc. Instn. Mech. Engrs.* **199**, no. C4, 319–24 (1985).

[7.13] Liu, C., Hartley, P., Sturgess, C.E.N. and Rowe, G.W. Elastic–plastic finite-element modelling of cold rolling of strip. *Int. J. Mech. Sci.* **27**, 531–41 (1985).

[7.14] Liu, C., Hartley, P., Sturgess, C.E.N. and Rowe, G.W. Simulation of the cold rolling of strip using an elastic–plastic finite-element technique. *Int. J. Mech. Sci.* **27**, 829–39 (1985).

[7.15] Hartley, P., Sturgess, C.E.N., Liu, C. and Rowe, G.W. Experimental and theoretical studies of workpiece deformation, stress and strain during flat rolling. *Int. Mater. Rev.* **34**, 19–34 (1989).

[7.16] Lahoti, G.D., Akgerman, N., Oh, S.I. and Altan, T. Computer-aided analysis of metal flow and stresses in plate rolling. *J. Mech. Wkg. Techn.* **4**, 105–19 (1980).

[7.17] Liu, C., Hartley, P., Sturgess, C.E.N. and Rowe, G.W. Finite-element modelling of deformation and spread in slab rolling. *Int. J. Mech. Sci.* **29**, 271–83 (1987).

[7.18] Sparling, L.G.M. Formula for spread in hot flat rolling (quoting results based on unpublished work of R. Hill) *Proc. Instn. Mech. Engrs.* **175**, 616–40 (1961).

[7.19] Al-Sened, A.A.K., Hartley, P., Sturgess, C.E.N. and Rowe, G.W. Finite-element analysis of a five-stage cold heading process. *J. Mech. Wkg. Techn.* **14**, 225–34 (1987).

[7.20] Al–Sened, A.A.K., Hartley, P., Sturgess, C.E.N. and Rowe, G.W. Forming sequences in axi-symmetric cold-forging. *Proc. 12th Nth. American Manufacturing Res. Conf.*, SME, pp. 151–8 (1984).

8 Future developments

8.1 INTRODUCTION

The programs described in this monograph have been developed specifically for application to metalforming, with the associated large plastic deformation and rotations produced by these processes. The emphasis throughout is on the flow of metal and the quality of the product in terms of homogeneity, residual stress and defect formation.

Various other types of program are available commercially and in laboratories. The earliest ones were extensions of linear elastic problems into small plastic deformation. As such, they are useful in structural engineering design and failure analysis. They may however lead to serious errors in the large plastic deformation range. Others concentrate on plastic deformation to the exclusion of elastic, simulating the classical rigid-plastic material of slip-line field theory. The plastic flow may be represented as a form of viscous motion, either with a constant coefficient of viscosity or of a non-Newtonian type. These are probably most suited to modelling high-temperature deformation. It is then, of course, important to include a thermal analysis which may be independent or preferably fully coupled to the incremental plasticity analysis. Elastic stresses can be determined in such solutions by a subsequent elastic unloading analysis. It is also possible to combine elastic–plastic analysis of the workpiece with an elastic analysis of the tools.

A survey of many of the programs commercially available in 1986 has been published [8.1]. Some very helpful direct comparisons are provided in that book.

Taking a general view of the whole situation, it appears that there is likely to be a convergence of objectives if not of particular methods during the next decade. All formulations intended for use in metalforming are likely to include elastic and plastic deformation, with temperature, speed and strain-hardening influences.

In this chapter we shall consider the main advances that still need to be made in modelling and software, followed by a brief survey of the main areas of application of the large-strain FE programs. The very rapid growth of the use of finite-element methods in many diverse areas of engineering, greatly aided by the enormous increase in power and reduction in cost of computers, suggests that these will soon be achieved and superseded by more ambitious objectives.

8.2   DEVELOPMENTS IN SOFTWARE

Highly complex design problems in structural, aeronautical, electronic and hydraulic engineering can now be solved by commercially-available linear finite-element packages. The main area for development may be in improving the efficiency of programs handling very large matrices, for example by greater use of vectorisation. Pre- and post-processing techniques and the presentation of information have been advancing rapidly during the past few years. One of the problems now is the effective assimilation and use of the mass of information that can be generated.

The problems of computer memory-space and operating time become more serious limitations in non-linear analysis. The iterative types of solution are clearly inefficient, but so far it has always proved necessary to use small increments for non-linear problems such as are inevitable in plasticity. Attention to second-order and perhaps third-order corrections has permitted larger increments without serious error, thus substantially reducing CPU time.

All FE analysis requires a mesh to be generated at the outset. Automatic 2-D mesh generation has considerably reduced the manual labour of initial setting up, and improved the efficiency of the computation. Mesh generation in 3-D involves much greater problems. Some advance has been made with space filling and contour matching using tetrahedral elements but difficulties remain with the more conventional hexahedral or brick elements. The Delaunay method may be mentioned as a recent improvement [8.2]. In many metalforming analyses the overall deformation is large and the meshes may be very distorted locally. Remeshing is therefore required.

8.3   ADVANCES IN MODELLING

Apart from the need for automatic mesh generation, it is important to have a fuller understanding of the requirements of the mesh and the influence of the mesh itself on the final result. The effects of mesh size are obvious. It is desirable to have a fine enough mesh to reveal all the necessary detail of the deformation or other features, but this must be balanced against CPU time and costs. The aspect ratio and type of element can influence the result. As has been seen in earlier chapters, simple triangular elements tend to be overstiff in large plastic deformation. Some intelligent routines for element selection would enhance the general utility of FE analysis packages. A book dealing with error estimation and mesh refinement has been published recently [8.3].

Modelling in 3-D produces particular difficulties, which are enhanced by the need to reduce the total number of elements involved.

Boundary conditions also cause difficulties. It is necessary to match the initial geometric shape as well as possible, but also to introduce tests to determine when a node comes into contact with a rigid or elastic tool. This can be done geometrically, and the node then restored to the surface if it has apparently crossed the boundary. The problem is further complicated when nodes may leave a boundary surface. It is then necessary to determine whether the normal force has become tensile, before re-siting the node. The general problem of contact between two deformable bodies has been approached in a recent publication [8.4].

There is still little understanding of boundary frictional conditions. In civil engineering it is usual to assume some coefficient of friction relating the normal and tangential forces or stresses. For most metalforming operations this is not valid, since the tangential stress is limited to the shear yield stress, which is independent of the stress acting normal to the surface. It is convenient to assume that the frictional resistance is expressible as some fraction m of the yield stress k, such that $0<m<1$. This can be incorporated in the FE formulation as a factor $m/(1-m)$ in the stiffness of a fictitious surface layer. This appears to give useful results, but is inherently unsatisfactory as a physical model. This deficiency is seen when the method is applied to 3-D deformation. A more realistic method of modelling friction is desirable. It has been suggested that a composite model with shear limitation in some zones and Coulomb friction in others may be appropriate to some situations.

Other modelling problems arise in several operations where a shear or tensile crack is produced. This was encountered long ago in a study of pile-driving, where the solution was to provide double nodes along a line. These were considered to be separate and could move apart under the influence of tensile or shear stress, but again the model is analytically convenient rather than realistic. Other studies, in metalforming and in civil engineering, have utilised weak elements in the critical regions by introducing an artificially low modulus. The disadvantage of such methods is that the location of the separable nodes or the weak elements tends to predetermine the deformation mode.

Any studies of crack propagation or cavity formation, especially in composites, require better modelling of this feature. It is also important in all cutting type operations where a new surface is created, for example at the nose of a lathe tool. Very accurate analysis of the active and residual elastic stresses is also required in such problems.

The modelling of inhomogeneous, anisotropic and composite materials is still at a very early stage, but clearly has many practical applications.

There is still doubt about the accuracy of calculation of hydrostatic stress in elastic–plastic problems, because of the large elastic stresses implied by small errors in volume calculation.

8.4   MATERIAL PROPERTIES

Until quite recently, the analytical techniques available for metalforming could deal only with highly idealised material properties. It was common to assume an isotropic, homogeneous, non-hardening rigid-plastic workpiece in slip-line field analysis. This produced surprisingly accurate general pictures of the deformation pattern and the forces, at least for 2-D problems. Since little use could be made of detailed material properties there was no incentive to measure more than fairly crude values of yield stress and, more recently for elastic analyses, the fracture toughness.

FE programs can now, however, produce very detailed information about the distributions of strain, stress, strain-rate and temperature. Indeed, a major problem is to find experimental methods that are sufficiently detailed to check the analyses. To make full use of the available FE packages, it is highly desirable, if not essential, to obtain much more refined and accurate materials property data.

Such data do not exist in the majority of instances, and they are very expensive to obtain. Even the test methods themselves are open to question. A simple upsetting test involves significant inhomogeneity of strain, apart from the dubious nature of the frictional contributions, so FE analyses of the tests are also required. The common hardness test, when subjected to FE analysis, is found to be highly complex, involving steep strain gradients. Its interpretation is by no means clear.

The establishment of a reliable data bank is clearly desirable but equally clearly beyond the resources of a single university department or a single company. This would be a worthy task for a central European institute.

The problem is exacerbated by the well-known variability of material within a standard specification. Very large differences in cutting tool life can, for example, be encountered even in the machining of free-cutting mild steel.

A lack of data is even more apparent when composites are considered. These are of course of great importance not only in the reinforced polymer field but in aerospace technology with metal/ceramic materials. Extreme forms of composite are found in low-density materials with closed or open pores, which incidentally also pose very interesting questions for FE analysis.

A collaborative study of materials properties and the FE analysis of composites may offer for the first time a proper understanding of the role of macroscopic inclusions of hard carbides on the one hand and soft manganese sulphide on the other. These two have been of great importance in metal cutting for many years.

8.5   POST-PROCESSING

The typical product of a finite-element analysis of metalforming is a matrix of strain distribution, possibly also with the local temperatures. The deviatoric and hydrostatic stress components are also available.

It is then necessary to present these results in a form that can be readily assimilated by the investigator, or possibly used in a control system. The most easily understood form for metalforming is a distorted grid based upon a regular initial pattern. Since the distortions themselves produce errors in the FE analysis, it is not possible to work progressively through the whole deformation, up to 80% or more, using the original elements. The technique used is to re-mesh at suitable intervals, as has been described in Chapter 6. To retain the visual image, the original grid is stored in its distorted form, updated at each re-meshing, so that the final display corresponds to the distortion that would actually be seen on a gridded specimen [8.5].

These techniques are useful but the re-meshing in particular takes time, because of the need to average and transfer all the nodal and element centroid data. Improved re-meshing techniques would be valuable.

For more detailed numerical consideration, it is useful to display the strain and other results in the form of contour maps. Improvements in these are no doubt possible, especially for 3-D examination. The FE results can be linked with conventional CAD graphics.

Apart from the visual inspection of the results, from which very useful conclusions can be drawn, there is also a great potential in analysis of the numbers themselves. For example, a predetermined limit may be set to the allowable strain before reheating or inter-annealing the workpiece. The programs can then determine how many reheating operations are needed, and at what stage they should be inserted in the schedule. Taking this a little further, the shapes of preforms can be modified, perhaps eliminating one of the inter-anneals, with attendant cost saving.

The possibility of cracks occurring can also be examined. At the present stage of knowledge, the onset of shear cracking can be predicted from the value of the total plastic work at a particular location. Again, the preform shapes may be modified to avoid this problem, but this type of analysis is still at an early stage and is capable of considerable refinement. It is not, for example, yet known precisely how tensile hydrostatic stress and shear stress interact.

After process cracking has been avoided or overcome, there is still the very important aspect of service properties of the product. The fatigue life can, for example, be greatly influenced by the residual stresses from the working sequence. These depend very strongly on the strain and temperature history of the process, but so far the subject of residual stresses has received little attention in FE analysis.

8.6   EXPERT SYSTEMS

As we have seen, large-strain FEM provides a powerful tool for studying the deformation occurring in a forging operation. As such, it can be used to indicate how the forging operation can be improved, but it cannot design the forging dies, or determine the number of preform operations or choose the forging conditions. These tasks, traditionally the province of craftsmen with many years of experience, are increasingly being undertaken by 'expert systems' [8.6]. Expert systems are computer programs that emulate the cognitive, reasoning and explanatory processes of a human expert within a specialised field of interest, using theoretical or empirical rules and knowledge of the problem domain.

The complementary natures of expert-system design and FE analysis make them eminently suitable for integration with a die design package. In such a package, the expert system would design the dies for each stage of the forging sequence, and specify the initial billet size and the forging conditions. For components that fall within established categories, this would be sufficient. An FE analysis would only be needed if it was uncertain whether the rules built in to the expert system could be applied satisfactorily to an unfamiliar component. The results of the FE analysis would help in the design of the dies in this instance, and they would also provide information about the general applicability of the expert-system rule base. The FE results might even suggest how the rules could be modified in order to deal with future components of a similar type.

In this way, the range of application of the expert-system die-design procedure increases with time, while the need for expensive FE analysis becomes less and less.

This integration of the die-design and FE programs requires the use of an Intelligent Knowledge-Based System (IKBS) to compare the components that are to be manufactured with previous examples and to decide whether FE analysis is required [8.7].

A start has already been made in achieving this kind of integrated facility though much work still needs to be done in this area.

8.7   HARDWARE

It is not the purpose of this monograph to discuss hardware in any detail. The rate of progress is anyway so rapid that such an account would be out of date by the time it was printed.

There seem however to be two general trends as far as FE plasticity analysis is concerned. On the one hand bigger and faster machines of very great cost, such as CRAY-2, are appearing. These clearly extend the capability, especially for 3-D analyses, and they will undoubtedly be used. On the other hand, the advent of personal computers with 20-megabyte memories on hard disc and 0.5 megabytes or more internally has dramatically altered the situation. Whereas a few years ago even linear FE analysis was far beyond the capacity of microcomputers, it has now been demonstrated that plasticity problems can be solved. The accuracy is as good as that of a main-frame computer, and in terms of total turn-round time the 16-bit micros have the advantage of the immediate availability of the solution [8.8]. Their cost is a major factor allowing many organisations to undertake in-house FE analysis if they so decide.

There will certainly be further advances, as evidenced already by the popularity of powerful workstations. These can be used to run the programs to examine 3-D problems, which require large amounts of storage, and so provide a real alternative to mainframe computing for complex and lengthy analyses.

8.8   APPLICATIONS IN THE FUTURE

A number of current applications have been discussed in Chapter 7. These will certainly be multiplied in the next few years, to include many of the variants on common processes.

Here we shall briefly consider a few areas to which relatively little attention has yet been given in non-linear FE analysis.

Possibly the largest is the whole field of polymer processing. Until reliable thermal analysis could be included there was little point in applying FE methods to these materials. There are severe problems, such as the very low conductivity and high temperature sensitivity of polymers. The properties also vary considerably according to structure and chain length, which themselves can be changed by mechanical processing. Strain rate is also an obviously important parameter, but unlike metals, the polymers may change significantly in density and properties by orientational crystallisation.

Sheet forming has been aided by FE analysis, but much remains to be learnt about the analysis of anisotropic materials commonly encountered in the forming of rolled stock.

The effects of anisotropy may also be important in extruded material. Composites have already been mentioned; these usually have a deliberately-produced severe anisotropy.

One of the greater problems in primary metal and alloy forming is the production of homogeneous fine-grained material in a reliably reproducible way. FE analysis of the processes and of the materials themselves may be expected to advance the technology in this respect.

Reproducibility of stock material is of increasing importance as automated or computer control of processes becomes more widespread.

Although in principle it is possible to apply adaptive control in many processes, for much of the large-scale primary processing the individual time cycles do not permit large or even small changes from billet to billet. There may nevertheless be economic advantages and material savings through special processing or recovery of rogue billets detected by prior detailed non-destructive testing. This is of course only to be considered for the very high-cost materials such as are used in aerospace. Rigorous control of these products is clearly essential, including careful attention to residual stresses and the ensuing fatigue properties. In this context, it should be recognised that high-stress low-cycle fatigue is frequently as important as the more common high-cycle failure.

FE analysis probably has a significant role to play in the developments of materials science. One obvious area is in the stress analysis associated with fracture toughness. Beyond this the actual process of tensile and shear cracking may be illuminated by fine-scale FE approaches.

The simulation of friction has been mentioned as a problem in FE modelling. Quite apart from this, it is well known that the friction of metals and polymers itself is closely associated with the elastic, plastic and visco-elastic deformation properties. The stress fields around interacting asperities are little understood, and the interfacial conditions are described only in general terms. There is scope for more detailed analysis.

The problems of wear are universal in machinery but scientifically wear is even less well understood. Wear rates are either measured empirically in each situation or some stochastic model is used. A detailed knowledge of the deformation and fracture processes at local contacts should advance the understanding of this commercially-important topic.

References

[8.1]   Niku-Lari, A. (Ed.) *Structural Analysis Systems: Software, Hardware, Capability, Applications*, Vol. 1, Pergamon (1986).

[8.2]   Cavendish, J.C., Field, D.A. and Frey, W.H. Automatic mesh generation: a finite element/computer aided geometric design interface. *The Mathematics of Finite Elements and Applications V*, Academic Press, pp. 83–96 (1985).

[8.3]   Babuska, I., Zienkiewicz, O.C., Gago, J. and de A. Oliveira, E.R. (Eds.) *Accuracy Estimates and Adaptive Refinements in Finite Element Computations*, Wiley (1986).

[8.4]   Baaijens, F.P.T., Veldpaus, F.E. and Brekelmans, W.A.M. On the numerical simulation of contact problems in forming processes. *Proc. 2nd Int. Conf. on Numerical Methods in Industrial Forming Processes*, ed. K.

Mattiasson, A. Samuelsson, R.D. Wood and O.C. Zienkiewicz, Balkema Press, pp. 85–90 (1986).

[8.5]   Al-Sened, A.A.K., Hartley, P., Sturgess, C.E.N. and Rowe, G.W. Forming sequences in axi-symmetric cold-forging. *Proc. 12th Nth. American Manufacturing Res. Conf.*, SME, pp. 151–8 (1984).

[8.6]   Vemuri, K.R., Raghupathi, P.S. and Altan, T. Automatic design of blocker forging dies. *Proc. 14th Nth. American Manufacturing Res. Conf.*, SME, pp. 372–8 (1986).

[8.7]   Rowe, G.W. An intelligent knowledge-based system to provide design and manufacturing data for forging. *Computer-Aided Eng. J.*, 56–61 (Feb. 1987).

[8.8]   Hussin, A.A.M., Hartley, P., Sturgess, C.E.N. and Rowe, G.W. Non-linear finite-element analysis on microcomputers for metal forging. *J. Strain Analysis* **21**, no. 4, 197–203 (1986).

Appendices
Appendix 1

DERIVATION OF SMALL-STRAIN $[B]$ MATRIX FOR 2-D TRIANGULAR ELEMENT

With a linear interpolation function, the displacement (u_1, u_2) of any point (x_1, x_2) within the element may be expressed as:

$$u_i = a_i + b_{i1}x_1 + b_{i2}x_2, \ i = 1,2 \tag{A1.1}$$

(The bold subscripts that denote values for a particular element have been omitted for the sake of clarity. This will be the case for the rest of this appendix, though it is to be understood that all relationships apply to a given element of the FE mesh.)

If the displacement of node I of the element is (d_{I1}, d_{I2}) and this node has co-ordinates (x_{I1}, x_{I2}) then:

$$d_{11} = a_1 + b_{11}x_{11} + b_{12}x_{12} \tag{A1.2}$$

$$d_{21} = a_1 + b_{11}x_{21} + b_{12}x_{22} \tag{A1.3}$$

$$d_{31} = a_1 + b_{11}x_{31} + b_{12}x_{32} \tag{A1.4}$$

(with similar expressions for the displacement of nodes in the x_2 direction.)

Subtracting equation A1.2 from equation A1.3:

$$d_{21} - d_{11} = b_{11}(x_{21}-x_{11}) + b_{12}(x_{22}-x_{12}) \tag{A1.5}$$

Subtracting equation A1.2 from equation A1.4:

$$d_{31} - d_{11} = b_{11}(x_{31}-x_{11}) + b_{12}(x_{32}-x_{12}) \tag{A1.6}$$

Multiplying equation A1.5 by $(x_{31}-x_{11})/(x_{21}-x_{11})$ and subtracting from equation A1.6:

$$(d_{31}-d_{11}) - \left(\frac{x_{31}-x_{11}}{x_{21}-x_{11}}\right)(d_{21}-d_{11}) = b_{12}\left[(x_{32}-x_{12}) - \left(\frac{x_{31}-x_{11}}{x_{21}-x_{11}}\right)(x_{22}-x_{12})\right]$$

or

$$(x_{21}-x_{11})(d_{31}-d_{11}) - (x_{31}-x_{11})(d_{21}-d_{11})$$
$$= b_{12}\left[(x_{21}-x_{11})(x_{32}-x_{12}) - (x_{31}-x_{11})(x_{22}-x_{12})\right] \tag{A1.7}$$

Thus:

$$b_{12} = \frac{(x_{31}-x_{21})d_{11} + (x_{11}-x_{31})d_{21} + (x_{21}-x_{11})d_{31}}{(x_{21}-x_{11})(x_{32}-x_{12}) - (x_{31}-x_{11})(x_{22}-x_{12})} \tag{A1.8}$$

But, with reference to figure A1.1, the area A of the triangle is given by:

$$A = A^{1453} + A^{3562} - A^{1462}$$

$$= \frac{1}{2}(x_{12}+x_{32})\,(x_{31}-x_{11}) + \frac{1}{2}(x_{32}+x_{22})\,(x_{21}-x_{31}) - \frac{1}{2}(x_{12}+x_{22})\,(x_{21}-x_{11})$$

$$= \frac{1}{2}(x_{21}-x_{11})\,(x_{32}-x_{12}) + \frac{1}{2}(x_{31}-x_{11})\,(x_{22}-x_{12}) \tag{A1.9}$$

So from equation A1.8:

$$b_{12} = \frac{1}{2A}\left((x_{31}-x_{21})d_{11} + (x_{11}-x_{31})d_{21} + (x_{21}-x_{11})d_{31} \right) \tag{A1.10}$$

An analysis similar to the above gives:

$$b_{11} = \frac{1}{2A}\left((x_{22}-x_{32})d_{11} + (x_{32}-x_{12})d_{21} + (x_{12}-x_{22})d_{31} \right) \tag{A1.11}$$

$$b_{21} = \frac{1}{2A}\left((x_{22}-x_{32})d_{12} + (x_{32}-x_{12})d_{22} + (x_{12}-x_{22})d_{32} \right) \tag{A1.12}$$

$$b_{22} = \frac{1}{2A}\left((x_{31}-x_{21})d_{12} + (x_{11}-x_{31})d_{22} + (x_{21}-x_{11})d_{32} \right) \tag{A1.13}$$

$$a_i = \frac{1}{2A}\left((x_{21}x_{32}-x_{22}x_{31})d_{1i} + (x_{12}x_{31}-x_{11}x_{32})d_{2i} + (x_{11}x_{22}-x_{12}x_{21})d_{3i} \right) \tag{A1.14}$$

Now:

$$\epsilon_{11} = \frac{\partial u_1}{\partial x_1} = b_{11} \tag{A1.15}$$

$$\epsilon_{22} = \frac{\partial u_2}{\partial x_2} = b_{22} \tag{A1.16}$$

Fig. A1.1 Calculation of area of triangular element.

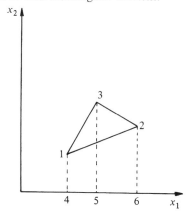

$$\gamma_{12} = 2 \cdot \epsilon_{12} = \frac{\partial u_1}{\partial x_2} + \frac{\partial u_2}{\partial x_1} = b_{12} + b_{21} \qquad (A1.17)$$

so:

$$\boldsymbol{\epsilon} = \begin{pmatrix} \epsilon_{11} \\ \epsilon_{22} \\ \gamma_{12} \end{pmatrix}$$

$$= \frac{1}{2A} \begin{bmatrix} x_{22}-x_{32} & 0 & x_{32}-x_{12} & 0 & x_{12}-x_{22} & 0 \\ 0 & x_{31}-x_{21} & 0 & x_{11}-x_{31} & 0 & x_{21}-x_{11} \\ x_{31}-x_{21} & x_{22}-x_{32} & x_{11}-x_{31} & x_{32}-x_{12} & x_{21}-x_{11} & x_{12}-x_{22} \end{bmatrix} \begin{pmatrix} d_{11} \\ d_{12} \\ d_{21} \\ d_{22} \\ d_{31} \\ d_{32} \end{pmatrix}$$

or

$$\boldsymbol{\epsilon} = [B]\mathbf{d} \qquad (A1.18)$$

Appendix 2

DERIVATION OF ELASTIC [D] MATRIX

In Chapter 2 (equation 2.16) it was shown that:

$$
\begin{pmatrix} \epsilon_{11} \\ \epsilon_{22} \\ \epsilon_{33} \\ \gamma_{12} \\ \gamma_{23} \\ \gamma_{13} \end{pmatrix} = \frac{1}{E}
\begin{bmatrix}
1 & -\nu & -\nu & 0 & 0 & 0 \\
-\nu & 1 & -\nu & 0 & 0 & 0 \\
-\nu & -\nu & 1 & 0 & 0 & 0 \\
0 & 0 & 0 & 2(1+\nu) & 0 & 0 \\
0 & 0 & 0 & 0 & 2(1+\nu) & 0 \\
0 & 0 & 0 & 0 & 0 & 2(1+\nu)
\end{bmatrix}
\begin{pmatrix} \sigma_{11} \\ \sigma_{22} \\ \sigma_{33} \\ \sigma_{12} \\ \sigma_{23} \\ \sigma_{13} \end{pmatrix}
$$

$$(A2.1)$$

Clearly:

$$\sigma_{ij} = \frac{E}{2(1+\nu)}\gamma_{ij}, \ i \neq j \tag{A2.2}$$

so all that remains is to express the three normal components of stress in terms of the strain components.

Multiplying ϵ_{22} by ν and adding to ϵ_{11}:

$$E(\epsilon_{11} + \nu\epsilon_{22}) = (1-\nu^2)\sigma_{11} - \nu(1+\nu)\sigma_{33} \tag{A2.3}$$

Subtracting ϵ_{33} from ϵ_{11}:

$$E(\epsilon_{11} - \epsilon_{33}) = (1+\nu)\sigma_{11} - (1+\nu)\sigma_{33} \tag{A2.4}$$

Multiplying equation A2.4 by ν and subtracting from equation A2.3:

$$E[(1-\nu)\epsilon_{11} + \nu\epsilon_{22} + \nu\epsilon_{33}] = (1 - \nu^2 - \nu - \nu^2)\sigma_{11}$$

$$= (1-2\nu)(1+\nu)\sigma_{11} \tag{A2.5}$$

so:

$$\sigma_{11} = \frac{E}{(1+\nu)(1-2\nu)}[(1-\nu)\epsilon_{11} + \nu\epsilon_{22} + \nu\epsilon_{33}] \tag{A2.6}$$

and similarly:

$$\sigma_{22} = \frac{E}{(1+\nu)(1-2\nu)}[\nu\epsilon_{11} + (1-\nu)\epsilon_{22} + \nu\epsilon_{33}] \tag{A2.7}$$

$$\sigma_{33} = \frac{E}{(1+\nu)(1-2\nu)}[\nu\epsilon_{11} + \nu\epsilon_{22} + (1-\nu)\epsilon_{33}] \tag{A2.8}$$

Hence:

$$\begin{pmatrix} \sigma_{11} \\ \sigma_{22} \\ \sigma_{33} \\ \sigma_{12} \\ \sigma_{23} \\ \sigma_{13} \end{pmatrix} = \frac{E}{1+\nu} \begin{bmatrix} \frac{1-\nu}{1-2\nu} & \frac{\nu}{1-2\nu} & \frac{\nu}{1-2\nu} & 0 & 0 & 0 \\ \frac{\nu}{1-2\nu} & \frac{1-\nu}{1-2\nu} & \frac{\nu}{1-2\nu} & 0 & 0 & 0 \\ \frac{\nu}{1-2\nu} & \frac{\nu}{1-2\nu} & \frac{1-\nu}{1-2\nu} & 0 & 0 & 0 \\ 0 & 0 & 0 & \frac{1}{2} & 0 & 0 \\ 0 & 0 & 0 & 0 & \frac{1}{2} & 0 \\ 0 & 0 & 0 & 0 & 0 & \frac{1}{2} \end{bmatrix} \begin{pmatrix} \epsilon_{11} \\ \epsilon_{22} \\ \epsilon_{33} \\ \gamma_{12} \\ \gamma_{23} \\ \gamma_{13} \end{pmatrix}$$
$$\tag{A2.9}$$

or:

$$\boldsymbol{\sigma} = [D]\boldsymbol{\epsilon} \tag{A2.10}$$

It is useful to be able to express the above relationship using subscript notation. Since $\gamma_{ij} = 2\epsilon_{ij}$:

$$\sigma_{ij} = 2G\epsilon_{ij}, \; i \neq j \tag{A2.11}$$

where G is the Rigidity Modulus. For the normal components of stress, for example $i = j = 1$:

$$\sigma_{11} = \frac{2G}{1-2\nu}[(1-\nu)\epsilon_{11} + \nu\epsilon_{22} + \nu\epsilon_{33}]$$

$$= \frac{2G}{1-2\nu}[(1-2\nu)\epsilon_{11} + \nu\epsilon_{11} + \nu\epsilon_{22} + \nu\epsilon_{33}] \tag{A2.12}$$

$$= 2G(\epsilon_{11} + \frac{\nu}{1-2\nu}\epsilon_{kk})$$

using the usual suffix summation convention. Similar expressions may be obtained for the other two normal components of stress. Equations A2.11 and A2.12 may be combined using the Kronecker delta:

$$\sigma_{ij} = 2G(\epsilon_{ij} + \delta_{ij} \left(\frac{\nu}{1-2\nu}\right) \epsilon_{kk}) \tag{A2.13}$$

Alternatively, we can write:

$$\sigma'_{ij} = 2G\epsilon'_{ij} \text{ and } \sigma^h = \kappa\epsilon_{kk} \tag{A2.14}$$

where the deviatoric components are defined by:

$$\sigma'_{ij} = \sigma_{ij} - \delta_{ij}\,\sigma^h \quad \text{and} \quad \epsilon'_{ij} = \epsilon_{ij} - \frac{1}{3}\delta_{ij}\epsilon_{kk} \tag{A2.15}$$

ϵ_{kk} is the sum of the normal components of strain (the bulk or volume strain), κ is the Bulk Modulus and σ^h is the hydrostatic stress, which is equal to one third of the sum of the three normal components of stress.

Appendix 3

DERIVATION OF ELASTIC–PLASTIC [D] MATRIX

The basic assumption of the elastic–plastic theory is that an increment of strain can be separated into an elastic part, which may be recovered on unloading, and a plastic part, which results in permanent deformation. Since, for all practical purposes, the permanent deformation does not involve any change in volume, the bulk or volume strain occurring during a process must be recoverable and hence elastic. Thus, by equation A2.14, it is simply proportional to the change in the hydrostatic component of stress. It is only necessary, therefore, to consider the deviatoric components of the strain increment:

$$d\epsilon'_{ij} = d\epsilon^{e\prime}_{ij} + d\epsilon^{p\prime}_{ij} \qquad (A3.1)$$

where it is to be noted that $d\epsilon^{p\prime}_{ij} = d\epsilon^{p}_{ij}$, since there is no volume component of plastic strain.

Using the incremental form of equation A2.14, we can write immediately:

$$d\epsilon^{e\prime}_{ij} = \frac{1}{2G}\,d\sigma'_{ij} \qquad (A3.2)$$

With the von Mises yield criterion, the principle of the normality of the plastic strain increment to the yield locus [A3.1] leads to the Lévy–Mises flow rule:

$$d\epsilon^{p\prime}_{ij} = \sigma'_{ij}d\lambda \qquad (A3.3)$$

$d\lambda$ is a proportionality factor. As will be shown later, this depends upon the elastic and plastic moduli, the current state of stress and the components of the strain increment.

Substituting from equations A3.2 and A3.3 into A3.1 gives the Prandtl–Reuss elastic–plastic flow equations:

$$d\epsilon'_{ij} = \frac{1}{2G}\,d\sigma'_{ij} + \sigma'_{ij}d\lambda \qquad (A3.4)$$

We require an expression giving the components of the stress increment in terms of the components of the strain increment and so must determine the proportionality factor $d\lambda$. From equation A3.3 and the definition of an increment of generalised plastic strain given in Chapter 3:

$$d\bar{\epsilon}^{p} = \left(\frac{2}{3}d\epsilon_{ij}^{p\prime}d\epsilon_{ij}^{p\prime}\right)^{1/2}$$

$$= \left(\frac{2}{3}\sigma_{ij}'\sigma_{ij}'\right)^{1/2} d\lambda$$

$$= \frac{2}{3}\left(\frac{3}{2}\sigma_{ij}'\sigma_{ij}'\right)^{1/2} d\lambda$$

$$= \frac{2}{3}\bar{\sigma}\cdot d\lambda \tag{A3.5}$$

$\bar{\sigma}$ is the generalised stress which is also defined in Chapter 3. Define Y' to be the rate of change of Y, the yield stress in simple tension, with respect to plastic strain. This quantity is obtained as required using the empirically-determined relationship between yield stress and plastic strain:

$$Y' = \frac{dY}{d\bar{\epsilon}^{p}} \quad \text{or} \quad Y'd\bar{\epsilon}^{p} = dY \tag{A3.6}$$

But the von Mises yield criterion states that the yield stress equals the generalised stress, so:

$$dY = \frac{\partial\bar{\sigma}}{\partial\sigma_{kl}'}d\sigma_{kl}'$$

$$= \frac{\partial\left(\frac{3}{2}\sigma_{kl}'\sigma_{kl}'\right)^{1/2}}{\partial\sigma_{kl}'}d\sigma_{kl}'$$

$$= \frac{3\sigma_{kl}'d\sigma_{kl}'}{2\bar{\sigma}} \tag{A3.7}$$

Substituting from equations A3.5 and A3.7 into equation A3.6:

$$Y'\left(\frac{2}{3}\bar{\sigma}\cdot d\lambda\right) = \frac{3\sigma_{kl}'d\sigma_{kl}'}{2\bar{\sigma}} \tag{A3.8}$$

or:

$$\frac{4}{9}\bar{\sigma}^2 Y'd\lambda = \sigma_{kl}'d\sigma_{kl}' \tag{A3.9}$$

But from equation A3.4:

$$d\sigma_{kl}' = 2G(d\epsilon_{kl}' - \sigma_{kl}'d\lambda) \tag{A3.10}$$

Hence:

$$\frac{4}{9}\bar{\sigma}^2 Y'd\lambda = \sigma_{kl}'\cdot 2G(d\epsilon_{kl}' - \sigma_{kl}'d\lambda)$$

$$= 2G\left(\sigma_{kl}'d\epsilon_{kl}' - \frac{2}{3}\bar{\sigma}^2 d\lambda\right) \tag{A3.11}$$

Rearranging this equation:

$$\left(\frac{2}{9}\bar{\sigma}^2 Y' + \frac{2}{3}G\cdot\bar{\sigma}^2\right)d\lambda = G\cdot\sigma'_{kl}d\epsilon'_{kl} \qquad (A3.12)$$

or:

$$d\lambda = \frac{\sigma'_{kl}d\epsilon'_{kl}}{\frac{2}{3}\bar{\sigma}^2\left(1 + \frac{Y'}{3G}\right)}$$

$$= \frac{\sigma'_{kl}d\epsilon'_{kl}}{S} \qquad (A3.13)$$

where:

$$S = \frac{2}{3}\bar{\sigma}^2\left(1 + \frac{Y'}{3G}\right) \qquad (A3.14)$$

Substituting equation A3.11 into equation A3.4 and rearranging:

$$d\sigma'_{ij} = 2G\left(d\epsilon'_{ij} - \frac{\sigma'_{ij}\,\sigma'_{kl}d\epsilon'_{kl}}{S}\right) \qquad (A3.15)$$

By definition of the deviatoric components of strain and stress (equation A2.15):

$$d\sigma_{ij} - \delta_{ij}\,d\sigma^h \qquad = 2G\left(d\epsilon_{ij} - \delta_{ij}\frac{d\epsilon_{mm}}{3} - \frac{\sigma'_{ij}\,\sigma'_{kl}d\epsilon'_{kl}}{S}\right) \qquad (A3.16)$$

or:

$$d\sigma_{ij} - \delta_{ij}\left(\frac{E}{3(1-2\nu)}\right)d\epsilon_{mm} = 2G\left(d\epsilon_{ij} - \delta_{ij}\frac{d\epsilon_{mm}}{3} - \frac{\sigma'_{ij}\,\sigma'_{kl}d\epsilon'_{kl}}{S}\right) \qquad (A3.17)$$

Thus the required constitutive relationship between the components of incremental strain and the components of incremental stress is:

$$d\sigma_{ij} = 2G\left[d\epsilon_{ij} + \delta_{ij}\left(\frac{\nu}{1-2\nu}\right)d\epsilon_{mm} - \frac{\sigma'_{ij}\,\sigma'_{kl}d\epsilon'_{kl}}{S}\right] \qquad (A3.18)$$

since $\sigma'_{kl}d\epsilon'_{kl} = \sigma'_{kl}d\epsilon_{kl}$.

In matrix form:

$$d\boldsymbol{\sigma} = ([D^e] - [D^p])d\boldsymbol{\epsilon} \qquad (A3.19)$$

where the elastic matrix is:

$$[D^e] = 2G \begin{bmatrix} \dfrac{1-\nu}{1-2\nu} & \dfrac{\nu}{1-2\nu} & \dfrac{\nu}{1-2\nu} & 0 & 0 & 0 \\[2mm] \dfrac{\nu}{1-2\nu} & \dfrac{1-\nu}{1-2\nu} & \dfrac{\nu}{1-2\nu} & 0 & 0 & 0 \\[2mm] \dfrac{\nu}{1-2\nu} & \dfrac{\nu}{1-2\nu} & \dfrac{1-\nu}{1-2\nu} & 0 & 0 & 0 \\[2mm] 0 & 0 & 0 & \dfrac{1}{2} & 0 & 0 \\[2mm] 0 & 0 & 0 & 0 & \dfrac{1}{2} & 0 \\[2mm] 0 & 0 & 0 & 0 & 0 & \dfrac{1}{2} \end{bmatrix} \tag{A3.20}$$

the plastic matrix is:

$$[D^p] = \frac{2G}{S} \begin{bmatrix} \sigma'_{11}\sigma'_{11} & \sigma'_{11}\sigma'_{22} & \sigma'_{11}\sigma'_{33} & \sigma'_{11}\sigma'_{12} & \sigma'_{11}\sigma'_{23} & \sigma'_{11}\sigma'_{13} \\[2mm] \sigma'_{22}\sigma'_{11} & \sigma'_{22}\sigma'_{22} & \sigma'_{22}\sigma'_{33} & \sigma'_{22}\sigma'_{12} & \sigma'_{22}\sigma'_{23} & \sigma'_{22}\sigma'_{13} \\[2mm] \sigma'_{33}\sigma'_{11} & \sigma'_{33}\sigma'_{22} & \sigma'_{33}\sigma'_{33} & \sigma'_{33}\sigma'_{12} & \sigma'_{33}\sigma'_{23} & \sigma'_{33}\sigma'_{13} \\[2mm] \sigma'_{12}\sigma'_{11} & \sigma'_{12}\sigma'_{22} & \sigma'_{12}\sigma'_{33} & \sigma'_{12}\sigma'_{12} & \sigma'_{12}\sigma'_{23} & \sigma'_{12}\sigma'_{13} \\[2mm] \sigma'_{23}\sigma'_{11} & \sigma'_{23}\sigma'_{22} & \sigma'_{23}\sigma'_{33} & \sigma'_{23}\sigma'_{12} & \sigma'_{23}\sigma'_{23} & \sigma'_{23}\sigma'_{13} \\[2mm] \sigma'_{13}\sigma'_{11} & \sigma'_{13}\sigma'_{22} & \sigma'_{13}\sigma'_{33} & \sigma'_{13}\sigma'_{12} & \sigma'_{13}\sigma'_{23} & \sigma'_{13}\sigma'_{13} \end{bmatrix} \tag{A3.21}$$

and the stress and strain increment vectors are:

$$d\boldsymbol{\sigma} = (d\sigma_{11}, d\sigma_{22}, d\sigma_{33}, d\sigma_{12}, d\sigma_{23}, d\sigma_{13})^T \tag{A3.22}$$

$$d\boldsymbol{\epsilon} = (d\epsilon_{11}, d\epsilon_{22}, d\epsilon_{33}, d\gamma_{12}, d\gamma_{23}, d\gamma_{13})^T \tag{A3.23}$$

References

[A3.1] Drucker, D.C. A more fundamental approach to plastic stress-strain relations. *Proc. 1st Nat. Cong. App. Mech.*, ed. E. Sternberg, ASME, pp. 487–91 (1951).

Appendix 4

DERIVATION OF SMALL-STRAIN STIFFNESS MATRIX [K] FOR PLANE-STRESS TRIANGULAR ELEMENT

To avoid the proliferation of subscripts, assume, without loss of generality, that the body consists of just one element. The element stiffness matrix is then the same as the global stiffness matrix $[K]$ and is defined by:

$$\mathbf{f} = [K]\mathbf{d} \tag{A4.1}$$

in which the global vectors of nodal force $\mathbf{f}$ and nodal displacement $\mathbf{d}$ are identical to the corresponding element vectors.

The element is initially unstressed, with zero nodal forces, and the set of forces $\mathbf{f}$ is then applied to the nodes. A possible final configuration will be one in which the nodes have been displaced by amounts $\mathbf{d}$. The potential energy of this new configuration (relative to the potential energy of the initial configuration) is the difference between the extra strain energy U stored in the element, and the work W done by the applied forces in moving their points of application between the initial and the final configurations. For plane-stress conditions:

$$I = U - W$$

$$= \frac{1}{2} \int (\sigma_{11}\epsilon_{11} + \sigma_{22}\epsilon_{22} + \sigma_{12}\gamma_{12}) \mathrm{d}V$$

$$- f_{11}d_{11} - f_{12}d_{12} - f_{21}d_{21} - f_{22}d_{22} - f_{31}d_{31} - f_{32}d_{32} \tag{A4.2}$$

in which f_{li} denotes the component of force in the x_i direction acting t node I, and d_{li} is the displacement of node I in the x_i direction.

For constant-strain triangular elements, the integral may be evaluated directly, and equation 2.24 may be used to express the stress components in terms of the strain components:

$$I = \frac{V}{2}\left((D_{11}\epsilon_{11} + D_{12}\epsilon_{22})\epsilon_{11} + (D_{21}\epsilon_{11} + D_{22}\epsilon_{22})\epsilon_{22} + (D_{33}\gamma_{12})\gamma_{12} \right)$$

$$- f_{11}d_{11} - f_{12}d_{12} - f_{21}d_{21} - f_{22}d_{22} - f_{31}d_{31} - f_{32}d_{32} \tag{A4.3}$$

or:

$$I = \frac{V}{2}\left(D_{11}(\epsilon_{11})^2 + (D_{12} + D_{21})\epsilon_{11}\epsilon_{22} + D_{33}(\gamma_{12})^2 \right)$$

$$- f_{11}d_{11} - f_{12}d_{12} - f_{21}d_{21} - f_{22}d_{22} - f_{31}d_{31} - f_{32}d_{32} \tag{A4.4}$$

where D_{ij} is the entry in row i, column j of the $[D]$ matrix in equation 2.24. The *actual* configuration that the element takes up will be the one that minimises the potential energy of the system, i.e. the actual set of d_{Ii} will be such that:

$$\frac{\partial I}{\partial d_{Ii}} = 0, I = 1,3 \text{ and } i = 1,2 \tag{A4.5}$$

For example, if $I = 1$ and $i = 1$:

$$\frac{V}{2}\left(D_{11}\frac{\partial(\epsilon_{11})^2}{\partial d_{11}} + (D_{12}+D_{21})\epsilon_{22}\cdot\frac{\partial\epsilon_{11}}{\partial d_{11}} + D_{33}\frac{\partial(\gamma_{12})^2}{\partial d_{11}}\right) - f_{11} = 0 \tag{A4.6}$$

since ϵ_{22} does not depend upon d_{11}. Substituting for the strain components from equation A1.18, and peforming the differentiation gives the result:

$$\frac{V}{2(2A)^2}\left[2D_{11}(x_{22}-x_{32})\Big((x_{22}-x_{32})d_{11} + (x_{32}-x_{12})d_{21} + (x_{12}-x_{22})d_{31} \Big) \right.$$

$$+ \quad 2D_{12}(x_{22}-x_{32})\Big((x_{31}-x_{21})d_{12} + (x_{11}-x_{31})d_{22} + (x_{21}-x_{11})d_{32} \Big)$$

$$+ \quad 2D_{33}(x_{31}-x_{21})\Big((x_{31}-x_{21})d_{11} + (x_{11}-x_{31})d_{21} + (x_{21}-x_{11})d_{31}$$

$$+ \quad \left. (x_{22}-x_{32})d_{12} + (x_{32}-x_{12})d_{22} + (x_{12}-x_{22})d_{32} \Big)\right]$$

$$- f_{11} = 0 \tag{A4.7}$$

in which use has been made of the fact that $D_{12} = D_{21}$. Rewriting equation A4.7 in vector form:

$$f_{11} = \frac{V}{(2A)^2}\begin{pmatrix} D_{11}(x_{22}-x_{32})(x_{22}-x_{32}) + D_{33}(x_{31}-x_{21})(x_{31}-x_{21}) \\ \\ D_{12}(x_{22}-x_{32})(x_{31}-x_{21}) + D_{33}(x_{31}-x_{21})(x_{22}-x_{32}) \\ \\ D_{11}(x_{22}-x_{32})(x_{32}-x_{12}) + D_{33}(x_{31}-x_{21})(x_{11}-x_{31}) \\ \\ D_{12}(x_{22}-x_{32})(x_{11}-x_{31}) + D_{33}(x_{31}-x_{21})(x_{32}-x_{12}) \\ \\ D_{11}(x_{22}-x_{32})(x_{12}-x_{22}) + D_{33}(x_{31}-x_{21})(x_{21}-x_{11}) \\ \\ D_{12}(x_{22}-x_{32})(x_{21}-x_{11}) + D_{33}(x_{31}-x_{21})(x_{12}-x_{22}) \end{pmatrix}^{\text{T}} \begin{pmatrix} d_{11} \\ \\ d_{12} \\ \\ d_{21} \\ \\ d_{22} \\ \\ d_{31} \\ \\ d_{32} \end{pmatrix} \tag{A4.8}$$

in which the transposed vector is the first row of the required element stiffness matrix $[K]$. The other rows may be found in a similar way by considering the other five expressions defined by equation A4.5. In matrix terms:

$$I = \frac{1}{2} \int \boldsymbol{\sigma}^{\mathrm{T}} \boldsymbol{\epsilon} \, \mathrm{d}V - \mathbf{d}^{\mathrm{T}} \mathbf{f}$$

$$= \frac{1}{2} \int \boldsymbol{\epsilon}^{\mathrm{T}} [D] \boldsymbol{\epsilon} \, \mathrm{d}V - \mathbf{d}^{\mathrm{T}} \mathbf{f}$$

$$= \frac{1}{2} \mathbf{d}^{\mathrm{T}} \left(\int [B]^{\mathrm{T}} [D] [B] \mathrm{d}V \right) \mathbf{d} - \mathbf{d}^{\mathrm{T}} \mathbf{f} \qquad (A4.9)$$

Interpreting equation A4.5 as a vector expression, and differentiating equation A4.9 with respect to the vector $\mathbf{d}$:

$$\mathbf{0} = \int [B]^{\mathrm{T}} [D] [B] \mathrm{d}V \cdot \mathbf{d} - \mathbf{f} \qquad (A4.10)$$

so that:

$$[K] = \int [B]^{\mathrm{T}} [D] [B] \mathrm{d}V \qquad (A4.11)$$

For constant-strain triangular elements deformed under plane-stress conditions, it can readily be seen that equation A4.11 leads to the expression given in equation A4.8, and all the similar expressions for the other components of the force vector. However, equation A4.11 is quite generally true for all small-strain formulations, using the appropriate definitions for $[B]$ and $[D]$.

Appendix 5

SOLUTION OF STIFFNESS EQUATIONS BY GAUSSIAN ELIMINATION AND BACK-SUBSTITUTION

The assembled global stiffness relationship:

$$[K]\mathbf{d} = \mathbf{f} \tag{A5.1}$$

must be solved for the displacement vector $\mathbf{d}$, given the applied force vector $\mathbf{f}$. This relationship consists of a set of n linear simultaneous equations in n unknowns, where n is the number of degrees of freedom in the FE discretisation. Thus equation A5.1 may be written as:

$$
\begin{bmatrix}
K_{11} & K_{12} & \cdot & \cdot & \cdot & K_{1n} \\
K_{21} & K_{22} & \cdot & \cdot & \cdot & K_{2n} \\
\cdot & \cdot & & & & \cdot \\
K_{\alpha 1} & K_{\alpha 2} & \cdot & \cdot & \cdot & K_{\alpha n} \\
\cdot & \cdot & & & & \cdot \\
K_{n1} & K_{n2} & \cdot & \cdot & \cdot & K_{nn}
\end{bmatrix}
\begin{Bmatrix}
d_1 \\ d_2 \\ \cdot \\ \cdot \\ \cdot \\ d_n
\end{Bmatrix}
=
\begin{Bmatrix}
f_1 \\ f_2 \\ \cdot \\ f_\alpha \\ \cdot \\ f_n
\end{Bmatrix}
\tag{A5.2}
$$

The solution of this set of equations by Gaussian elimination and back-substitution is a two-stage process. The first stage, the elimination, converts the set of equations into an equivalent ordered set in which each equation involves one less variable than the preceding one.

To do this, use is made of the fact that the solution of a set of equations is unchanged if a multiple of one equation is added, coefficient by coefficient, to another, as long as the same multiple of the right-hand side of this equation is added to the right-hand side of the other. We do not prohibit adding a multiple of one equation to the same equation, that is multiplying this equation by some number, providing this number is not zero. The proof of the above result is straightforward, for if d_β, $(\beta = 1,n)$ is a solution for the set of equations, it is certainly a solution for equations α and γ:

$$\sum_{\beta=1}^{n} K_{\alpha\beta} d_\beta = f_\alpha \tag{A5.3}$$

$$\sum_{\beta=1}^{n} K_{\gamma\beta} d_\beta = f_\gamma \tag{A5.4}$$

It w times equation α is now added to equation γ:

$$\sum_{\beta=1}^{n} (K_{\gamma\beta} + wK_{\alpha\beta})d_\beta = \sum_{\beta=1}^{n} (K_{\gamma\beta} d_\beta + wK_{\alpha\beta} d_\beta)$$

$$= \sum_{\beta=1}^{n} K_{\gamma\beta} d_\beta + w\left(\sum_{\beta=1}^{n} K_{\alpha\beta} d_\beta\right)$$

$$= f_\gamma + w \cdot f_\alpha \tag{A5.5}$$

so the set of d_β is still a solution of the new equation γ.

Returning to equation A5.2, a multiple of $w = K_{21}/K_{11}$ times the first equation can be subtracted from the second:

$$
\begin{bmatrix}
K_{11} & K_{12} & \cdot & K_{1n} \\
K_{21}-wK_{11} & K_{22}-wK_{12} & \cdot & K_{2n}-wK_{1n} \\
\cdot & \cdot & & \cdot \\
K_{\alpha 1} & K_{\alpha 2} & \cdot & K_{\alpha n} \\
\cdot & \cdot & & \cdot \\
K_{n1} & K_{n2} & \cdot & K_{nn}
\end{bmatrix}
\begin{pmatrix}
d_1 \\ d_2 \\ \cdot \\ \cdot \\ d_n
\end{pmatrix}
=
\begin{pmatrix}
f_1 \\ f_2 - wf_1 \\ \cdot \\ f_\alpha \\ \cdot \\ f_n
\end{pmatrix}
\tag{A5.6}
$$

or:

$$
\begin{bmatrix}
K_{11} & K_{12} & \cdot & \cdot & \cdot & K_{1n} \\
0 & K_{22}^* & \cdot & \cdot & \cdot & K_{2n}^* \\
\cdot & \cdot & & & & \cdot \\
K_{\alpha 1} & K_{\alpha 2} & \cdot & \cdot & \cdot & K_{\alpha n} \\
\cdot & \cdot & & & & \cdot \\
K_{n1} & K_{n2} & \cdot & \cdot & \cdot & K_{nn}
\end{bmatrix}
\begin{pmatrix}
d_1 \\ d_2 \\ \cdot \\ \cdot \\ d_n
\end{pmatrix}
=
\begin{pmatrix}
f_1 \\ f_2^* \\ \cdot \\ f_\alpha \\ \cdot \\ f_n
\end{pmatrix}
\tag{A5.7}
$$

where the asterisks denote that the quantities no longer have their initial values. Continuing this process, by subtracting a multiple of $K_{\alpha 1}/K_{11}$ times the first equation from each equation α, $\alpha = 3,n$ gives:

$$
\begin{bmatrix}
K_{11} & K_{12} & \cdot & \cdot & \cdot & K_{1n} \\
0 & K_{22}^* & \cdot & \cdot & \cdot & K_{2n}^* \\
\cdot & \cdot & & & & \cdot \\
0 & K_{\alpha 2}^* & \cdot & \cdot & \cdot & K_{\alpha n}^* \\
\cdot & \cdot & & & & \cdot \\
0 & K_{n2}^* & \cdot & \cdot & \cdot & K_{nn}^*
\end{bmatrix}
\begin{pmatrix}
d_1 \\ d_2 \\ \cdot \\ \cdot \\ \cdot \\ d_n
\end{pmatrix}
=
\begin{pmatrix}
f_1 \\ f_2^* \\ \cdot \\ f_\alpha^* \\ \cdot \\ f_n^*
\end{pmatrix}
\qquad (A5.8)
$$

This process is then repeated, subtracting appropriate multiples of the second equation from all succeeding equations to introduce zeros into the second column below the diagonal, and so on. In general, this can be codified into the following algorithm:

repeat for $\beta = 1,n-1$
    repeat for $\alpha = \beta+1,n$
        repeat for $\gamma = \beta+1,n$
            subtract $K_{\alpha\beta}^* K_{\beta\gamma}^*/K_{\beta\beta}^*$ from $K_{\alpha\gamma}^*$
        end repeat
        subtract $K_{\alpha\beta}^* f_\beta^*/K_{\beta\beta}^*$ from f_α^*
    end repeat
end repeat

Note that in the inner loops, the repeated subscript β does *not* imply a summation. It should also be noted that the above piece of pseudo-code is several removes from an actual computer program. For instance, in practice the computationally time-consuming division would be removed from the innermost loop and instead a multiplication by a previously-calculated inverse would be used. With this algorithm, no entry in the matrix is actually set to zero, it is just ignored in all future calculations.

At the end of the elimination stage, the equations look like:

$$
\begin{bmatrix}
K_{11}^* & K_{12}^* & \cdot & K_{1\alpha}^* & \cdot & K_{1n}^* \\
0 & K_{22}^* & \cdot & K_{2\alpha}^* & \cdot & K_{2n}^* \\
\cdot & \cdot & & \cdot & & \cdot \\
0 & 0 & \cdot & K_{\alpha\alpha}^* & \cdot & K_{\alpha n}^* \\
\cdot & \cdot & & \cdot & & \cdot \\
0 & 0 & \cdot & 0 & \cdot & K_{nn}^*
\end{bmatrix}
\begin{pmatrix}
d_1 \\ d_2 \\ \cdot \\ d_\alpha \\ \cdot \\ d_n
\end{pmatrix}
=
\begin{pmatrix}
f_1^* \\ f_2^* \\ \cdot \\ f_\alpha^* \\ \cdot \\ f_n^*
\end{pmatrix}
\qquad (A5.9)
$$

where asterisks have been added to the first row of coefficients for consistency, even though these values have not been altered.

The second or back-substitution stage of the solution procedure calculates the values of the components of the solution vector **d** by considering, in reverse order, the equations obtained in the first part:

repeat for $\alpha = n,1$

$$
\text{displacement } d_\alpha = \left(f_\alpha^* - \sum_{\beta=\alpha+1}^{n} K_{\alpha\beta}^* d_\beta \right) / K_{\alpha\alpha}^*
$$

end repeat

in which the summation is taken to be zero if $\alpha = n$. It can be seen that when the time comes to calculate any particular d_α, the d_β values that occur on the right-hand side of the expression in the above loop are already known.

In theory, the process of Gaussian elimination and back-substitution will solve any set of simultaneous equations providing the matrix of coefficients is non-singular, that is, has an inverse. (If the matrix is singular, a zero diagonal term will be encountered during the elimination phase, and an attempt will be made to divide by zero. This is unlikely to happen if the stiffness equations are based upon proper physical principles.)

In practice, a computer implementation of the above solution procedure will be subject to accumulated round-off error, particularly as the method involves repeated subtraction of one quantity from another and these quantities may quite easily have approximately the same value. (Round-off error is a consequence of the fact that computers can only represent real numbers to a finite number of significant figures. The effective number of significant figures is decreased even further when a number is subtracted from one of about the same value because the higher-order figures cancel.)

In the worst cases, when the matrix equations are ill-conditioned (roughly speaking, when the matrix is almost singular) accumulated round-off error may cause very small or zero diagonal coefficients to be calculated, leading to floating-point overflow, or division by zero. FE stiffness equations are usually well-conditioned, but even so the effect of round-off error will be to reduce the accuracy of the calculated solution of the equations.

The standard way to reduce the effects of round-off error is to re-order a calculation so as to avoid, as far as possible, the subtraction of quantities of a similar size. In the Gaussian elimination procedure, this may be done by re-ordering the equations, each time a column of zeros is to be introduced, in order to place the largest coefficient in that column on the diagonal. This is called partial pivotting. Full pivotting, in which the unknowns of the equations are also re-ordered, may be carried out, but at the expense of having to keep track of the new orderings.

The solution procedure outlined in this appendix has made no special assumptions about the matrix $[K]$, except that it must be non-singular. For isotropic flow however, $[K]$ will always be symmetric. Examination of the elimination algorithm stated above shows that the symmetry of the submatrix to the right and below the current diagonal coefficient is preserved. This has two consequences. Firstly, only about half of $[K]$ need be assembled and stored, and secondly only those calculations affecting the coefficients of the stored half of the matrix need be performed. Modification of the elimination algorithm to take advantage of these two facts is straightforward and is left to the reader.

Modifications of the basic technique can also be made to take advantage of the sparse (banded) nature of the stiffness matrix. This is discussed in Chapter 6.

Appendix 6

IMPOSITION OF BOUNDARY CONDITIONS

A6.1 Components of displacement prescribed in global axis system

The basic Gaussian elimination and back-substitution procedure solves the stiffness equations to obtain a nodal displacement vector $\mathbf{d}$, given an applied nodal force vector $\mathbf{f}$. In metalforming problems, the effective force acting at all nodes is zero, except for those nodes presumed to be in contact with the dies, or for those nodes otherwise constrained – on a plane of symmetry for example. In general, the value of the force acting at such a boundary node is not known beforehand, though the value of at least some of the components of its displacement will be. Any solution procedure adopted in an FE metalforming program must therefore be capable of imposing prescribed values of certain components of displacement upon the solution, and calculating, as a result, the components of the reactive force.

In the simplest case, one or more of the global Cartesian components of displacement at a node may be prescribed by the boundary conditions of the FE model of the metalforming process. A prescribed value may, of course, be zero. Suppose component d_α of the global nodal displacement vector is known. Consider the stiffness matrix equations at the stage in the Gaussian elimination procedure when zeros are to be introduced into column α below the diagonal (Appendix 5):

$$\begin{bmatrix} K_{11}^* & K_{12}^* & \cdot & K_{1\alpha}^* & \cdot & K_{1n}^* \\ 0 & K_{22}^* & \cdot & K_{2\alpha}^* & \cdot & K_{2n}^* \\ \cdot & \cdot & & \cdot & & \cdot \\ 0 & 0 & \cdot & K_{\alpha\alpha}^* & \cdot & K_{\alpha n}^* \\ \cdot & \cdot & & \cdot & & \cdot \\ 0 & 0 & \cdot & K_{n\alpha}^* & \cdot & K_{nn}^* \end{bmatrix} \begin{pmatrix} d_1 \\ d_2 \\ \cdot \\ d_\alpha \\ \cdot \\ d_n \end{pmatrix} = \begin{pmatrix} f_1^* \\ f_2^* \\ \cdot \\ f_\alpha^* \\ \cdot \\ f_n^* \end{pmatrix} \quad (A6.1)$$

Since the value of the reaction f_α is, as yet, undetermined, so is the modified value f_α^*. Thus it is not possible to subtract multiples of equation α from all the succeeding equations as would normally be the case. However, since d_α is known, each of the coefficients in column α below the diagonal can be replaced by zero, providing d_α times the coefficient is subtracted from each of the right-hand sides:

$$\begin{bmatrix} K_{11}^* & K_{12}^* & \cdot & K_{1\alpha}^* & \cdot & K_{1n}^* \\ 0 & K_{22}^* & \cdot & K_{2\alpha}^* & \cdot & K_{2n}^* \\ \cdot & \cdot & & \cdot & & \cdot \\ 0 & 0 & \cdot & K_{\alpha\alpha}^* & \cdot & K_{\alpha n}^* \\ \cdot & \cdot & & \cdot & & \cdot \\ 0 & 0 & \cdot & 0 & \cdot & K_{nn}^* \end{bmatrix} \begin{pmatrix} d_1 \\ d_2 \\ \cdot \\ d_\alpha \\ \cdot \\ d_n \end{pmatrix} = \begin{pmatrix} f_1^* \\ f_2^* \\ \cdot \\ f_\alpha^* \\ \cdot \\ f_n^* - K_{n\alpha}^* d_\alpha \end{pmatrix} \qquad (A6.2)$$

The Gaussian elimination algorithm given in Appendix 5 may therefore be modified:

```
repeat for α = 1,n
    if dα prescribed then
        fα = 0
    end if
end repeat
repeat for β = 1,n–1
    repeat for α = β+1,n
        if dβ prescribed then
            subtract K*αβ dβ from f*α
        otherwise
            repeat for γ = β+1,n
                subtract K*αβ K*βγ / K*ββ from K*αγ
            end repeat
            subtract K*αβ f*β / K*ββ from f*α
        end if
    end repeat
end repeat
```

and similarly, the back-substitution algorithm becomes:

repeat for $\alpha = n,1$

if d_α prescribed then

$$\text{reaction } f_\alpha = \sum_{\beta=\alpha}^{n} K^*_{\alpha\beta} d_\beta - f^*_\alpha$$

otherwise

$$\text{displacement } d_\alpha = \left(f^*_\alpha - \sum_{\beta=\alpha+1}^{n} K^*_{\alpha\beta} d_\beta \right) / K^*_{\alpha\alpha}$$

end if

end repeat

As in the previous appendix, a summation from $n+1$ to n is assumed to be zero.

A6.2   Components of displacement prescribed in rotated axis system

In general, die surfaces and planes of symmetry may not be aligned with the global axes. If the displacement of node I has some prescribed value in a given arbitrary direction, but unknown values in directions perpendicular to this, then all the global components of displacement will be initially unknown, as will all the global components of force. The method cannot therefore be used. The same thing applies if the displacement is prescribed in two orthogonal directions and unknown in a third. (If the displacement is prescribed in three orthogonal directions, the components in the global axis system can be found immediately, and the previous technique can be used.)

Consider, as before the stiffness equations during the elimination stage, but this time at the point when zeros are to be introduced below the diagonal into the column corresponding to the first degree of freedom at node I. Rewrite the matrix equations in terms of $N \times N$ nodal submatrices $[K_{IJ}]$ containing the four (2-D) or nine (3-D) coefficients relating $\mathbf{f}_I$, the global components of force at node I to $\mathbf{d}_J$, the global components of displacement at node J:

$$
\begin{bmatrix}
[K^*_{11}] & [K^*_{12}] & \cdot & [K^*_{1I}] & \cdot & [K^*_{1N}] \\
[0] & [K^*_{22}] & \cdot & [K^*_{2I}] & \cdot & [K^*_{2N}] \\
\cdot & \cdot & & \cdot & & \cdot \\
[0] & [0] & \cdot & [K^*_{II}] & \cdot & [K^*_{IN}] \\
\cdot & \cdot & & \cdot & & \cdot \\
[0] & [0] & \cdot & [K^*_{NI}] & \cdot & [K^*_{NN}]
\end{bmatrix}
\begin{pmatrix}
\mathbf{d}_1 \\ \mathbf{d}_2 \\ \cdot \\ \mathbf{d}_I \\ \cdot \\ \mathbf{d}_N
\end{pmatrix}
=
\begin{pmatrix}
\mathbf{f}^*_1 \\ \mathbf{f}^*_2 \\ \cdot \\ \mathbf{f}^*_I \\ \cdot \\ \mathbf{f}^*_N
\end{pmatrix}
\qquad \text{(A6.3)}
$$

in which, as before, asterisks denote that the quantities no longer have their original values. It should be noted that the diagonal submatrices above row I are themselves in upper-triangular form.

Define a new set of orthogonal axes X_i such that one of these axes is parallel to the prescribed displacement of node I (or two of the axes are so aligned if the displacement is prescribed in two perpendicular directions). Figure A6.1 shows the situation in 3-D with one prescribed component. Let $[R]$ be the rotational transformation matrix containing the direction cosines of the three new axes with respect to the global Cartesian axes x_i:

$$[R] = \frac{\partial X_j}{\partial x_i} \tag{A6.4}$$

where i denotes one of the rows of $[R]$, and j one of the columns. Denote the components of displacement of node I in the rotated axis system by $\mathbf{d}'_I$. By definition, $[R]$ is an orthonormal matrix. The expressions for the change of basis (equation A8.21, Appendix 8) then give that:

$$\mathbf{d}_I = [R]\mathbf{d}'_I \tag{A6.5}$$

and similarly, if the components of $\mathbf{f}^*_I$ are $\mathbf{f}^{*\prime}_I$:

$$\mathbf{f}^*_I = [R]\mathbf{f}^{*\prime}_I \tag{A6.6}$$

Fig. A6.1 Local rotated axis system at node.

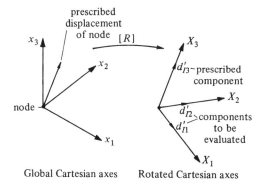

Global Cartesian axes   Rotated Cartesian axes

Substituting equations A6.5 and A6.6 into A6.3:

$$\begin{bmatrix} [K^*_{11}] & [K^*_{12}] & \cdot & [K^*_{1I}] & \cdot & [K^*_{1N}] \\ [0] & [K^*_{22}] & \cdot & [K^*_{2I}] & \cdot & [K^*_{2N}] \\ \cdot & \cdot & & \cdot & & \cdot \\ [0] & [0] & \cdot & [K^*_{II}] & \cdot & [K^*_{IN}] \\ \cdot & \cdot & & \cdot & & \cdot \\ [0] & [0] & \cdot & [K^*_{NI}] & \cdot & [K^*_{NN}] \end{bmatrix} \begin{pmatrix} \mathbf{d}_1 \\ \mathbf{d}_2 \\ \cdot \\ [R]\mathbf{d}'_I \\ \cdot \\ \mathbf{d}_N \end{pmatrix} = \begin{pmatrix} \mathbf{f}^*_1 \\ \mathbf{f}^*_2 \\ \cdot \\ [R]\mathbf{f}^{*\prime}_I \\ \cdot \\ \mathbf{f}^*_N \end{pmatrix} \quad \text{(A6.7)}$$

which is equivalent to the two sets of equations:

$$\begin{bmatrix} [K^*_{11}] & [K^*_{12}] & \cdot & [K^*_{1I}] & \cdot & [K^*_{1N}] \\ [0] & [K^*_{22}] & \cdot & [K^*_{2I}] & \cdot & [K^*_{2N}] \\ \cdot & \cdot & & \cdot & & \cdot \\ [0] & [0] & \cdot & [K^*_{(I-1)I}] & \cdot & [K^*_{(I-1)N}] \end{bmatrix} \begin{pmatrix} \mathbf{d}_1 \\ \mathbf{d}_2 \\ \cdot \\ \mathbf{d}_I \\ \cdot \\ \mathbf{d}_N \end{pmatrix} = \begin{pmatrix} \mathbf{f}^*_1 \\ \mathbf{f}^*_2 \\ \cdot \\ \mathbf{f}^*_{I-1} \end{pmatrix} \quad \text{(A6.8a)}$$

and:

$$\begin{bmatrix} [0] & [0] & \cdot & [R]^T[K^*_{II}][R] & \cdot & [R]^T[K^*_{IN}] \\ \cdot & \cdot & & \cdot & & \cdot \\ [0] & [0] & \cdot & [K^*_{NI}][R] & \cdot & [K^*_{NN}] \end{bmatrix} \begin{pmatrix} \mathbf{d}_1 \\ \mathbf{d}_2 \\ \cdot \\ \mathbf{d}'_I \\ \cdot \\ \mathbf{d}_N \end{pmatrix} = \begin{pmatrix} \mathbf{f}^{*\prime}_I \\ \cdot \\ \mathbf{f}^*_N \end{pmatrix} \quad \text{(A6.8b)}$$

since the inverse of an orthonormal matrix is its transpose. The Gaussian elimination may now proceed as before, with zeros being introduced into the two or three columns of the second matrix corresponding to the components of displacement of node I. This may be accomplished either by subtracting multiples of one of the two or three equations expressing the force at node I, or by modification of the right-hand sides, according to whether the particular component of displacement at node I in the rotated axis system is an unknown or is prescribed.

Back-substitution is also carried out as described earlier except that when the rotated components of displacement and force have been evaluated at node I, they must be pre-multiplied by $[R]$ in order to obtain the components of these nodal vectors in the global axis system. The global components of displacement at node I are then used in the back-substitution process to obtain the displacement at nodes 1 to $I-1$.

It should be noted that, providing the original stiffness matrix was symmetric, the part of the matrix to the right of and including the submatrix $[R]^T[K^*_{II}][R]$ in equation A6.8b is still symmetric, so only the two or three rows of this matrix corresponding to node I actually need to be modified (assuming the upper triangle of the matrix is being stored).

Appendix 7

RELATIONSHIP BETWEEN ELASTIC MODULI E, G AND κ

Rigidity Modulus G

Consider a unit cube of material subjected to pure shear, with engineering shear strain γ_{12} in the x_1x_2 plane (figure A7.1). Since there is no stress or strain in the x_3 direction, this axis has been omitted from the diagram. As a result of the application of the complementary shear stresses σ_{12} and σ_{21}, the square cross-section $ABCD$ is deformed into the rhombus $AB'C'D'$. The engineering shear strain and shear stress are related by the Rigidity or shear Modulus G:

$$\gamma_{12} = \frac{\sigma_{12}}{G} \tag{A7.1}$$

By constructing Mohr's circles for this stress system, or by considering, for example, the equilibrium of forces acting upon triangle $B'C'D'$, it can be shown that the normal stress acting in tension along the diagonal AC' is:

$$\sigma^{AC'} = \sigma_{12} \tag{A7.2}$$

and similarly, the normal stress acting in compression along the diagonal $B'D'$ is:

$$\sigma^{B'D'} = -\sigma_{12} \tag{A7.3}$$

Fig. A7.1 Unit cube subjected to pure shear.

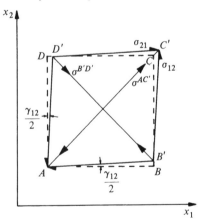

The normal strain in the original diagonal BD, ϵ^{BD}, resulting from these two normal stresses (cf. equation 2.14) is therefore:

$$\epsilon^{BD} = \frac{1}{E}(\sigma^{B'D'} - \nu\sigma^{AC'})$$

$$= -\frac{1+\nu}{E}\sigma_{12} \tag{A7.4}$$

Applying the cosine rule to the triangle $AB'D'$:

$$(B'D')^2 = (AB')^2 + (AD')^2 - 2 \cdot AB' \cdot AD' \cdot \cos\left(\frac{\pi}{2} - \gamma_{12}\right)$$

$$= 2 - 2\gamma_{12} \tag{A7.5}$$

for small angle γ_{12}. But:

$$B'D' = BD \cdot (1 + \epsilon^{BD})$$
$$= \sqrt{2}(1 + \epsilon^{BD}) \tag{A7.6}$$

so:

$$2(1 + \epsilon^{BD})^2 = 2(1 - \gamma_{12}) \tag{A7.7}$$

and ignoring the second-order term involving the square of the strain:

$$\epsilon^{BD} = -\frac{\gamma_{12}}{2} \tag{A7.8}$$

Thus, from equations A7.4 and A7.8:

$$\epsilon^{BD} = -\frac{\gamma_{12}}{2} = -\frac{1+\nu}{E}\sigma_{12} \tag{A7.9}$$

and comparing with equation A7.1:

$$G = \frac{E}{2(1+\nu)} \tag{A7.10}$$

Bulk Modulus κ

The elastic Bulk Modulus determines the relationship between the bulk strain, the sum of the normal strain components ϵ_{ii}, and the hydrostatic stress $\sigma^h = \sigma_{ii}/3$:

$$\sigma^h = \kappa\epsilon_{ii} \tag{A7.11}$$

But as was shown in Chapter 2:

$$\epsilon_{11} = \frac{1}{E}(\sigma_{11} - \nu\sigma_{22} - \nu\sigma_{33})$$

$$\epsilon_{22} = \frac{1}{E}(-\nu\sigma_{11} + \sigma_{22} - \nu\sigma_{33}) \tag{A7.12}$$

$$\epsilon_{33} = \frac{1}{E}(-\nu\sigma_{11} - \nu\sigma_{22} + \sigma_{33})$$

so summing these equations:

$$\epsilon_{11} + \epsilon_{22} + \epsilon_{33} = \frac{1}{E}(1-2\nu)(\sigma_{11} + \sigma_{22} + \sigma_{33}) \tag{A7.13}$$

Hence:

$$\epsilon_{ii} = \frac{(1-2\nu)}{E}\sigma_{jj} = \frac{3(1-2\nu)}{E}\sigma^{h} \tag{A7.14}$$

and so:

$$\kappa = \frac{E}{3(1-2\nu)} \tag{A7.15}$$

Appendix 8

VECTORS AND TENSORS

Let $\{\mathbf{v}\}$ be the set of all vectors in 3-D geometric space. This set forms a vector space over the scalar field of real numbers, in which vector addition and scalar multiplication have their usual geometric interpretations. It is therefore possible to choose a set of three linearly-independent vectors $\mathbf{g}_i$ to form a *basis* of $\{\mathbf{v}\}$. (A set of three vectors is linearly independent if none of them may be expressed as a linear combination of the other two – in geometric terms this means that the three vectors do not lie in the same plane.) Any member of $\{\mathbf{v}\}$ may then be written in the form:

$$\mathbf{v} = v^i \mathbf{g}_i \qquad\qquad (A8.1)$$

in which the scalar quantities v^i are called the *contravariant components* of $\mathbf{v}$ with respect to the basis $\mathbf{g}_i$ (figure A8.1).

Equation A8.1 illustrates the *summation convention*. This states that whenever an index occurs exactly twice in a term of an expression, that term is summed over all three values of the repeated index. The repeated index is called the *dummy index* and may be chosen arbitrarily, except that it may not be the same as any other index occurring in the same term. In addition, if the expression is referred to a basis of $\{\mathbf{v}\}$, one of each pair of dummy indices must be a subscript and the other must be a superscript. This last requirement is relaxed for the special case of an orthonormal basis (see below).

Any index which occurs only once in a term is called a *free* index. The range

Fig. A8.1 Contravariant components of a vector.

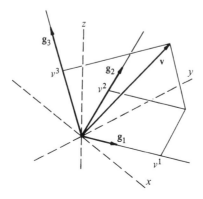

convention states that an expression refers, in turn, to all three values of all the free indices. If indexed expressions are equated, the equality is taken to be true for the complete range of the free indices and so any free index must be present on both sides of the equality sign and all occurrences must either be as subscripts or as superscripts. Again, the last restriction does not apply when an orthonormal basis is being used.

Define the *scalar product* ($\cdot$) of two vectors to be:

$$\mathbf{u} \cdot \mathbf{v} = u^i v^j (\mathbf{g}_i \cdot \mathbf{g}_j) \qquad (A8.2)$$

for:

$$\begin{aligned}\mathbf{g}_i \cdot \mathbf{g}_j &= \mathbf{g}_j \cdot \mathbf{g}_i \\ &= |\mathbf{g}_i| \cdot |\mathbf{g}_j| \cos(\alpha)\end{aligned} \qquad (A8.3)$$

where $|\ |$ represents the magnitude of the enclosed vector, and α is the angle between $\mathbf{g}_i$ and $\mathbf{g}_j$ in 3-D space.

Let δ_{ij} be the *Kronecker delta* in which it is understood that either of the indices may be a subscript or a superscript. This quantity is defined to be equal to one if the values of the indices are the same, and zero otherwise.

For the basis $\mathbf{g}_i$, define the *dual* basis $\mathbf{g}^j$ such that:

$$\mathbf{g}_i \cdot \mathbf{g}^j = \delta_i{}^j \qquad (A8.4)$$

Using the dual basis, an alternative expression for the vector $\mathbf{v}$ may be obtained:

$$\mathbf{v} = v_i \mathbf{g}^i \qquad (A8.5)$$

in which v_i are called the *covariant components* of $\mathbf{v}$ (figure A8.2).

A basis is said to be orthonormal if it is equal to its dual. Under these circumstances, the contravariant and covariant components of a vector are the same, and by convention subscripts are used for all indices.

Suppose a different basis $\mathbf{G}_j$ of $\{\mathbf{v}\}$ is chosen. The position vector $\mathbf{x}$ of an arbitrary point in 3-D space may be written as:

$$\mathbf{x} = x^i \mathbf{g}_i = X^j \mathbf{G}_j \qquad (A8.6)$$

Fig. A8.2 Dual basis.

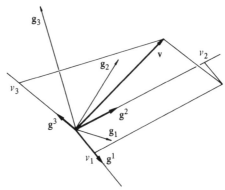

and:

$$\mathbf{x} = x_i\mathbf{g}^i = X_j\mathbf{G}^j \qquad (A8.7)$$

Taking the scalar product of the vectors in equation A8.6 with $\mathbf{g}^k$ gives:

$$x^i\mathbf{g}^k\cdot\mathbf{g}_i = X^j\mathbf{g}^k\cdot\mathbf{G}_j \qquad (A8.8)$$

or, using equation A8.4:

$$x^k = X^j\mathbf{g}^k\cdot\mathbf{G}_j \qquad (A8.9)$$

Similarly:

$$x_k = X_j\mathbf{g}_k\cdot\mathbf{G}^j \qquad (A8.10)$$

$$X^k = x^i\mathbf{g}_j\cdot\mathbf{G}^k \qquad (A8.11)$$

$$X_k = x_j\mathbf{g}^j\cdot\mathbf{G}_k \qquad (A8.12)$$

Differentiating these expressions with respect to the three co-ordinates gives:

$$\frac{\partial x^i}{\partial X^j} = \frac{\partial X_j}{\partial x_i} = \mathbf{g}^i\cdot\mathbf{G}_j \qquad (A8.13)$$

and:

$$\frac{\partial x_i}{\partial X_j} = \frac{\partial X^j}{\partial x^i} = \mathbf{g}_i\cdot\mathbf{G}^j \qquad (A8.14)$$

Substituting equation A8.9 into equation A8.6:

$$X^i\mathbf{G}_i = X^i(\mathbf{g}^j\cdot\mathbf{G}_i)\mathbf{g}_j \qquad (A8.15)$$

or, using equation A8.13:

$$X^i\mathbf{G}_i = X^i\frac{\partial x^j}{\partial X^i}\,\mathbf{g}_j \qquad (A8.16)$$

But the choice of $\mathbf{x}$ was arbitrary, so:

$$\mathbf{G}_i = \frac{\partial x^j}{\partial X^i}\,\mathbf{g}_j \qquad (A8.17)$$

If V^j and V_j are the contravariant and covariant components of $\mathbf{v}$ with respect to the new basis $\mathbf{G}_j$, then by analogy with equations A8.11 and A8.12:

$$V^j = v^i\mathbf{g}_i\cdot\mathbf{G}^j \qquad (A8.18)$$

$$V_j = v_i\mathbf{g}^i\cdot\mathbf{G}_j \qquad (A8.19)$$

which, using equations A8.13 and A8.14 may be written as:

$$V^j = v^i\,\frac{\partial X^j}{\partial x^i} \qquad (A8.20)$$

$$V_j = v_i\,\frac{\partial x^i}{\partial X^j} \qquad (A8.21)$$

These two equations lead to the alternative definition of a vector: a set of three scalar values obtained in co-ordinate system x^i are the contravariant components of a vector if the corresponding values calculated in co-ordinate system X^j with the same origin may be expressed in the form of equation A8.20. If the values may be related by an expression of the form of equation A8.21, they are covariant components.

For example, the three values defining the force acting at a point are contravariant components, whereas it can be easily shown that the three quantities obtained by differentiating the equation of a plane with respect to the three co-ordinates are the covariant components of a normal vector to that plane.

Note that although the theory so far has only been concerned with geometric space, the above definition is quite general, since any 3-D vector (such as force) may be mapped onto an appropriate geometric vector.

Vectors are a special case of *tensors* i.e. first-order tensors. Conversely, higher-order tensors may be expressed in terms of products or quotients of vector components. The definition of a tensor is simply an extension of that just stated for a vector. For example, the nine scalar values t^{ij} obtained in co-ordinate system x^i are the contravariant components of a tensor if the corresponding values T^{kl} obtained in co-ordinate system X^k are related to these by:

$$T^{kl} = t^{ij} \frac{\partial X^k}{\partial x^i} \frac{\partial X^l}{\partial x^j}$$

Similar definitions exist for covariant and mixed tensors.

One important example of a tensor quantity is, of course, the stress in a deforming body.

Appendix 9

STRESS IN A DEFORMING BODY

Consider an infinitesimally-small region of the body located at point P at time t. At some later time, $t+dt$, this region has deformed and is situated at point P' (figure A9.1). During this increment of deformation, an infinitesimally-small plane surface at P, with area da and unit normal $\mathbf{n}$ is deformed into a plane surface with area da' and unit normal $\mathbf{n}'$. Let $\mathbf{g}_i$ be the basis of a stationary *reference* co-ordinate system.

At time $t+dt$, choose a new co-ordinate origin O' and a new basis $\mathbf{G}_i$ so that the co-ordinates of P' with respect to $\mathbf{G}_i$ are the same as the reference co-ordinates of P. The new basis and origin define the *convected* co-ordinate system. Upper-case letters will be used for convected components and lower-case letters will be used for reference components. A prime will be used to indicate that a particular quantity refers to the deformed state at time $t+dt$.

Suppose the reference co-ordinates of P and P' are x^i and x''^i respectively, so that $x''^i = x''^i(x^j,dt)$ is the function defining the deformation of the infinitesimal region originally situated at P. Then:

$$\mathbf{O'P'} = X''^j\mathbf{G}_j = x''^i\mathbf{g}_i - \mathbf{OO'} \qquad (A9.1)$$

or, since X''^j are convected co-ordinates:

$$x^j\mathbf{G}_j = x''^i\mathbf{g}_i - \mathbf{OO'} \qquad (A9.2)$$

so that:

$$x''^i = x^j\mathbf{G}_j\cdot\mathbf{g}^i + \mathbf{OO'}\cdot\mathbf{g}^i \qquad (A9.3)$$

Fig. A9.1 Deformation of infinitesimal region.

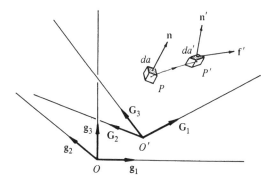

Differentiating and comparing with equation A8.13 (Appendix 8), noting that this expression is independent of the position of the origin of the two co-ordinate systems, results in the relationship:

$$\frac{\partial x'^i}{\partial x^j} = \mathbf{g}^i \cdot \mathbf{G}_j = \frac{\partial x^i}{\partial X^j} \tag{A9.4}$$

or:

$$\frac{\partial x^i}{\partial X^j} = x'^{i,j} \tag{A9.5}$$

in which the superscript ',j' denotes the derivative with respect to the jth reference co-ordinate.

Another important result concerns the deformation of the infinitesimally-small plane originally situated at P (figure A9.2). Since $\mathbf{n}$ is a unit vector, the definition of vector product ($*$) gives that:

$$\mathbf{n}da = \mathbf{v}^1 * \mathbf{v}^2 \tag{A9.6}$$

where $\mathbf{v}^1$ and $\mathbf{v}^2$ are infinitesimally-small vectors bounding area da at time t. At time $t+dt$, these are deformed into $\mathbf{v}'^1$ and $\mathbf{v}'^2$, so:

$$\mathbf{n}'da' = \mathbf{v}'^1 * \mathbf{v}'^2 \tag{A9.7}$$

Choose an arbitrary small vector $\mathbf{v}^3$ in the undeformed body and write:

$$\mathbf{v}^i = v^{ij}\mathbf{g}_j \tag{A9.8}$$

Thus:

$$\mathbf{v}'^i = V'^{ij}\mathbf{G}_j = v^{ij}\mathbf{G}_j \tag{A9.9}$$

by the property of convected co-ordinates, since $\mathbf{v}^i$ are infinitesimally-small vectors. Taking the scalar product of $\mathbf{v}'^3$ with the deformed area normal having covariant components $N'_j da'$:

$$(\mathbf{v}'^3 \cdot \mathbf{n}')da' = V'^{3i}N'_j da'(\mathbf{G}_i \cdot \mathbf{G}^j)$$
$$= v^{3i}N'_i da' \tag{A9.10}$$

Fig. A9.2 Undeformed and deformed area elements.

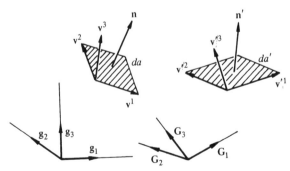

from the definition of dual basis given in equation A8.4 (Appendix 8). Alternatively:

$$(\mathbf{v}'^3 \cdot \mathbf{n}') da' = (\mathbf{v}'^1 * \mathbf{v}'^2) \cdot \mathbf{v}'^3$$
$$= (V'^{1k} \mathbf{G}_k * V'^{2m} \mathbf{G}_m) \cdot V'^{3n} \mathbf{G}_n$$
$$= \det(V'^{ij})(\mathbf{G}_1 * \mathbf{G}_2) \cdot \mathbf{G}_3 \tag{A9.11}$$

by the definition of triple product, where det ( ) denotes the determinant of the enclosed matrix. Substitution of equations A8.17 (Appendix 8) and A9.5 then gives that:

$$\mathbf{v}'^3 \cdot \mathbf{n}' da' = \det(v^{ij})(x'^{k,1} \mathbf{g}_k * x'^{m,2} \mathbf{g}_m) \cdot x'^{n,3} \mathbf{g}_n$$
$$= \det(v^{ij}) \det(x'^{i,j})(\mathbf{g}_1 * \mathbf{g}_2) \cdot \mathbf{g}_3$$
$$= \det(x'^{i,j})(v^{1k} \mathbf{g}_k * v^{2m} \mathbf{g}_m) \cdot v^{3n} \mathbf{g}_n$$
$$= \det(x'^{i,j})(\mathbf{v}^1 * \mathbf{v}^2) \cdot \mathbf{v}^3 \tag{A9.12}$$

Thus:

$$v'^{3i} N_i' da' = \mathbf{v}'^3 \cdot \mathbf{n}' da' = J \mathbf{v}^3 \cdot \mathbf{n} da$$
$$= J v^{3i} n_i \, da \tag{A9.13}$$

or since the choice of $\mathbf{v}^3$ is arbitrary:

$$N_i' da' = J n_i \, da \tag{A9.14}$$

in which J is the Jacobian of the deformation i.e. the determinant of the matrix of the transformation of the reference co-ordinates of particle P during the time interval dt (equivalent to the ratio of deformed to undeformed infinitesimal volumes).

The stress acting at a point is a tensor describing the force per unit area acting upon any infinitesimal plane situated at that point. The *reference* components of True or *Cauchy* stress are denoted by σ^{ij} and are defined by:

$$f'^i = \sigma^{ij} n_i' \, da' \tag{A9.15}$$

Where f'^i are the contravariant reference components of the force vector $\mathbf{f}'$ acting upon the deformed infinitesimal plane, which has area da' and a unit normal with covariant reference components n_i'. It can be easily shown from equation A9.15 that σ^{ij} obey the co-ordinate transformation rule stated in equation A8.22 (Appendix 8), so these are contravariant components.

The variational principle used in Chapter 5 is expressed in terms of the rate of *nominal* stress (also called the Lagrange or Piola–Kirchhoff I stress). The nominal stress s^{ij} is calculated assuming that the forces in the deformed configuration act on the body in its undeformed state. Thus:

$$f'^i = s^{ij} n_i \, da \tag{A9.16}$$

The rules for change of basis state that:

$$n_l' = N_k' \frac{\partial X^k}{\partial x^l} \tag{A9.17}$$

so that equation A9.15 may be written as:

$$f'^j = \sigma'^{lj} \frac{\partial X^k}{\partial x^l} N'_k da' = s^{kj} n_k da \tag{A9.18}$$

Using equation A9.14, this gives the result:

$$J\sigma'^{lj} \frac{\partial X^k}{\partial x^l} = s^{kj} \tag{A9.19}$$

or:

$$J\sigma^{ij} = s^{kj} \frac{\partial x^i}{\partial X^k} = s^{kj} x'^{i,k} \tag{A9.20}$$

The conditions for force equilibrium require Cauchy stress to be symmetric, but equation A9.20 shows that, in general, the same is not true for nominal stress.

A third types of stress, the *Kirchhoff* or Piola–Kirchhoff II stress, may be defined in the following manner.

Consider the undeformed infinitesimal surface with area da. This has a *unit* normal with covariant components n_i in the reference system. As mentioned above, the covariant components of the normal to a plane in a co-ordinate system are obtained by differentiating the equation of that plane with respect to the three co-ordinates in turn. But by the definition of convected co-ordinates, the equation of the undeformed plane in the reference system is the same as the equation of the deformed plane in the convected system. Hence n_i are proportional to N'_i (though not necessarily equal).

Define $\hat{\mathbf{f}}$ to be the force acting upon a plane, in the deformed body, which has area da' and a normal with reference components N'_i. That is:

$$\hat{f}^j = \sigma^{ij} N'_i da' \tag{A9.21}$$

The Kirchhoff stress τ^{ij} is then defined so that this is the force which would be calculated to act upon the undeformed infinitesimal plane:

$$\hat{f}^j = \tau^{ij} n_i da \tag{A9.22}$$

Using equation A9.14 Kirchhoff stress may be written in terms of Cauchy stress and nominal stress:

$$\tau^{ij} = J\sigma^{ij} = s^{kj} x'^{i,k} \tag{A9.23}$$

The following example will illustrate the difference between the three definitions of stress.

Figure A9.3 shows a cubic element of unit dimension. The orthonormal basis $\mathbf{g}_i$ define a Cartesian reference co-ordinate system x^i. The element is first extended in simple tension by amount e in the x^1 direction (with Poisson contraction ve in the other two directions) and then rotated by an angle α about the x^3 axis. The edges of the deformed element define the convected basis $\mathbf{G}_i$. Note that although $\mathbf{G}_i$ is an orthogonal basis, it is not orthonormal.

The deformation matrix for this example is:

$$x'^{i,j} = \frac{\partial x^i}{\partial X^j} = \begin{bmatrix} (1+e)\cos(\alpha) & -(1-\nu e)\sin(\alpha) & 0 \\ (1+e)\sin(\alpha) & (1-\nu e)\cos(\alpha) & 0 \\ 0 & 0 & 1-\nu e \end{bmatrix} \quad (A9.24)$$

and so:

$$J = (1+e)(1-\nu e)^2 \quad (A9.25)$$

Suppose that at the end of this simple deformation there is a tensile stress of $T\ N/m^2$ acting in the X^1 direction. Equilibrium of force (figure A9.4) means that the reference components of Cauchy stress are:

$$\sigma^{ij} = \begin{bmatrix} T\cos^2(\alpha) & T\sin(\alpha)\cos(\alpha) & 0 \\ T\sin(\alpha)\cos(\alpha) & T\sin^2(\alpha) & 0 \\ 0 & 0 & 0 \end{bmatrix} \quad (A9.26)$$

Fig. A9.3 Deformation of unit element.

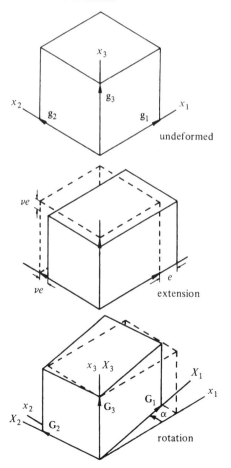

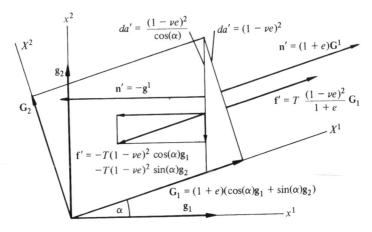

Fig. A9.4 Cauchy stress.

The nominal stress s^{ij} in this example (figure A9.5) is:

$$s^{ij} = \begin{bmatrix} T(1-\nu e)^2\cos(\alpha) & T(1-\nu e)^2\sin(\alpha) & 0 \\ 0 & 0 & 0 \\ 0 & 0 & 0 \end{bmatrix} \quad (A9.27)$$

The Kirchhoff stress τ^{ij} (figure A9.6) is:

$$\tau^{ij} = \begin{bmatrix} T(1+e)(1-\nu e)^2\cos^2(\alpha) & T(1+e)(1-\nu e)^2\sin(\alpha)\cos(\alpha) & 0 \\ T(1+e)(1-\nu e)^2\sin(\alpha)\cos(\alpha) & T(1+e)(1-\nu e)^2\sin^2(\alpha) & 0 \\ 0 & 0 & 0 \end{bmatrix} \quad (A9.28)$$

Fig. A9.5 Nominal stress.

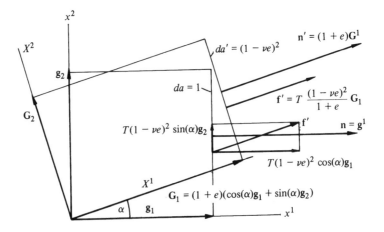

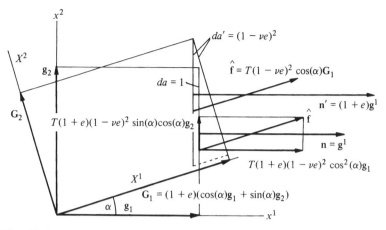

Fig. A9.6 Kirchhoff stress.

Thus if $e = 0.01$, a is $5°$ and ν is 0.5 (for a plastically-deforming body):

$$\sigma^{ij} = T \begin{bmatrix} 0.992404 & 0.086824 & 0 \\ 0.086824 & 0.007596 & 0 \\ 0 & 0 & 0 \end{bmatrix} \quad \text{(A9.29)}$$

$$s^{ij} = T \begin{bmatrix} 0.986258 & 0.086286 & 0 \\ 0 & 0 & 0 \\ 0 & 0 & 0 \end{bmatrix} \quad \text{(A9.30)}$$

$$\tau^{ij} = T \begin{bmatrix} 0.992330 & 0.086818 & 0 \\ 0.086818 & 0.007596 & 0 \\ 0 & 0 & 0 \end{bmatrix} \quad \text{(A9.31)}$$

from which it can be seen that the Cauchy and Kirchhoff stress measures are, to a good approximation, the same for conditions of near incompressibility.

Appendix 10

STRESS RATES

The FE governing equations are expressed in terms of stress rates. These are evaluated at the start of each increment (time t) when the reference and convected co-ordinate systems coincide. Therefore, at this time, all three stresses (Cauchy, nominal, Kirchhoff) have the same value. However, their time derivatives are, in general, not equal.

The variational principle involves the rate of nominal stress. Therefore, differentiating equation A9.23 (Appendix 9) with respect to time gives:

$$\dot{\tau}^{ij} = s^{kj}\dot{x}^{\prime i,k} + s^{kj}\dot{u}^{i,k}$$
$$= \dot{s}^{ij} + \sigma^{kj}\dot{u}^{i,k} \tag{A10.1}$$

since $x^{\prime i,j} = \delta^{ij}$ at the start of the increment. $\dot{u}^{i,j}$ is the rate of deformation tensor, the time derivative of $x^{\prime i,j}$.

In the constitutive relation, both the stress rate and the strain rate must be independent of any rigid-body component of the deformation. As explained in Chapter 5, Kirchhoff stress is the appropriate measure to use, and so the constitutive law is expressed in terms of the Jaumann rate of this stress. This is defined to be the rate of change of the components of Kirchhoff stress calculated in a co-ordinate system which rotates with the material.

The deformation tensor is clearly real, and is non-singular since the inverse transformation may easily be defined. It may therefore be uniquely expressed as a product of a rotational transformation r^{ij} and a symmetric deformation q^{ij}:

$$x^{\prime i,j} = r^{ik}q^{kj} \tag{A10.2}$$

Differentiating this expression with respect to time:

$$\dot{u}^{i,j} = \dot{r}^{ik}q^{kj} + r^{ik}\dot{q}^{kj}$$
$$= \dot{r}^{ij} + \dot{q}^{ij} \tag{A10.3}$$

since $r^{ij} = q^{ij} = \delta^{ij}$ at the start of the increment. Therefore:

$$\dot{r}^{ij} = \frac{1}{2}(\dot{u}^{i,j} - \dot{u}^{j,i}) = <\dot{u}^{i,j}> \tag{A10.4}$$

where the angle brackets denote the skew-symmetric part of the enclosed matrix. Define a transformation which is the rotational part of $x^{\prime i,j}$, that is:

$$x^{*i,j} = r^{ij} \tag{A10.5}$$

then the rotationally-invariant Kirchhoff stress, τ^{*ij} is given by:

$$\tau^{ij} = \tau^{*mn}x^{*i,m}x^{*j,n}$$
$$= \tau^{*mn}r^{im}r^{jn} \qquad (A10.6)$$

using equation A8.22 (Appendix 8) for the transformation of the components of a tensor. Differentiating this expression with respect to time and noting again that $r^{ij} = \delta^{ij}$ at the start of the increment:

$$\dot{\tau}^{ij} = \dot{\tau}^{*ij} + \tau^{*mj}\dot{r}^{im} + \tau^{*in}\dot{r}^{jn}$$
$$= \dot{\tau}^{*ij} + \sigma^{kj}<\dot{u}^{i,k}> + \sigma^{ik}<\dot{u}^{j,k}> \qquad (A10.7)$$

from equation A10.4 and using the fact that rotationally-invariant Kirchhoff stress is the same as the reference components of Cauchy stress at the start of the increment. $\dot{\tau}^{*ij}$ is the Jaumann rate of Kirchhoff stress.

Combining equations A10.1 and A10.7:

$$\dot{s}^{ij} = \dot{\tau}^{*ij} + \sigma^{kj}<\dot{u}^{i,k}> + \sigma^{ik}<\dot{u}^{j,k}> - \sigma^{kj}\dot{u}^{i,k}$$
$$= \dot{\tau}^{*ij} - \sigma^{kj}\dot{\epsilon}^{ik} - \sigma^{ik}\dot{\epsilon}^{kj} + \sigma^{ik}\dot{u}^{j,k} \qquad (A10.8)$$

where $\dot{\epsilon}^{ij}$, the strain-rate tensor is defined to be the symmetric part of the deformation-rate tensor:

$$\dot{\epsilon}^{ij} = \frac{1}{2}(\dot{u}^{i,j} + \dot{u}^{j,i}) \qquad (A10.9)$$

Finally, since the reference co-ordinate system is chosen to be the orthonormal Cartesian system, equation A10.8 may be rewritten as:

$$\dot{s}_{ij} = \dot{\tau}^{*}_{ij} - \sigma_{kj}\dot{\epsilon}_{ik} - \sigma_{ik}\dot{\epsilon}_{kj} + \sigma_{ik}\dot{u}_{j,k} \qquad (A10.10)$$

Appendix 11

LISTING OF BASIC PROGRAM FOR SMALL-DEFORMATION ELASTIC-PLASTIC FE ANALYSIS

The program listed in this appendix was designed principally for demonstration purposes and to form an introduction to non-linear FE plasticity analyses. In order to simplify the approach, only constant-strain triangular elements have been used. The program is sufficiently flexible, however, to be able to deal with plane-stress, plane-strain and axi-symmetric examples.

The data contained within the sub-program FILDATA are specifically for sticking-friction axi-symmetric upsetting. It should be a simple matter to change the appropriate entries to tackle different problems. Comment statements are used liberally throughout the program in order to try and explain what is happening at each stage. The programs could have been written in a more condensed and efficient form, but the format given here was chosen deliberately to allow the reader and potential user to follow the program structure and operation as easily as possible.

```
1000 REM========================================================================
1010 REM==           This is file FILT used  to  display the  on-screen      ==
1020 REM==           titles and call the first of the processing programs.   ==
1030 REM==                                                                   ==
1040 REM==           This is a BASIC finite element program that can be      ==
1050 REM==           used either for elasticity problems or for elastic      ==
1060 REM==           plastic problems based  on the  small  deformation      ==
1070 REM==           formulation described in chapter 3 and applied  in      ==
1080 REM==           chapter 4 of the book. The program is intended for      ==
1090 REM==           educational or demonstration purposes only.            ==
1100 REM==           The program has  been developed on an Olivetti M24      ==
1110 REM==           personal computer. It can be used for plane stress,     ==
1120 REM==           plane strain or axi-symmetric analysis.                 ==
1130 REM==           Remember that as a small deformation formulation is     ==
1140 REM==           used, the stress  components  may  be unreliable in     ==
1150 REM==           the plastic range.                                      ==
1160 REM========================================================================
1170 REM
1180 CLS
1190 PRINT "    A Finite Element program for Elasticity or for Elastic-Plastic"
1200 PRINT "          problems using a small-deformation formulation"
1210 PRINT
1220 PRINT "                          From the book"
1230 PRINT
1240 PRINT "         Finite Element Plasticity and Metalforming Analysis"
1250 PRINT
1260 PRINT "                              by"
1270 PRINT
1280 PRINT "        G.W.Rowe, C.E.N.Sturgess, P.Hartley and I.Pillinger"
1290 PRINT
1300 PRINT "              Cambridge University Press, 1990"
1310 PRINT
1320 PRINT
1330 PRINT
1340 PRINT
1350 PRINT
1360 INPUT "                              press RETURN to continue",A$
1370 REM==        Load and run next program                                  ==
1380 REM
1390 CHAIN "FILDATA"
1400 REM========================================================================
1410 REM== FILT version 1.0 completed  16-2-89          Peter Hartley        ==
1420 REM========================================================================
1430 END
```

```
1000 REM=====================================================================
1010 REM==     This is file DISPLAY used to present a graphic display      ==
1020 REM==     of the finite-element mesh.                                 ==
1030 REM=====================================================================
1040 REM
1050 REM==     Set array dimensions and define variables                  ==
1060 REM
1070 DEFSTR A-B
1080 DEFINT E-N
1090 DIM CO(25,2),EC(32,3),PO(5)
1100 REM
1110 REM==     Retrieve data from disc file INTLDATA                      ==
1120 REM
1130 OPEN "INTLDATA" FOR INPUT AS 1
1140 INPUT#1,IEP,IGE,NP,NE,NB,NF,INF,NW,INO,INL,IFLAG,IVER
1150 INPUT#1,PO(1),PO(2),PO(3)
1160 IF IEP=2 THEN INPUT#1,PO(4),PO(5)
1170 FOR I=1 TO NE
1180 INPUT#1,EC(I,1),EC(I,2),EC(I,3)
1190 NEXT I
1200 FOR I=1 TO NP
1210 INPUT#1,CO(I,1),CO(I,2)
1220 NEXT I
1230 CLOSE#1
1240 REM
1250 REM==     Shift X and Y coordinates to suit screen display           ==
1260 REM
1270 FOR I=1 TO NP
1280 CO(I,1)=CO(I,1)+50
1290 CO(I,2)=-CO(I,2)+350
1300 NEXT I
1310 REM
1320 REM==     Set screen mode                                            ==
1330 REM
1340 CLS
1350 SCREEN 3
1360 LINE (0,0)-(635,0)
1370 LINE (635,0)-(635,370)
1380 LINE (635,370)-(0,370)
1390 LINE (0,370)-(0,0)
1400 REM
1410 REM==     Identify the connecting nodes of each element and          ==
1420 REM==     plot the element.                                          ==
1430 REM
1440 FOR I=1 TO NE
1450 J=EC(I,1)
1460 K=EC(I,2)
1470 L=EC(I,3)
1480 LINE (CO(J,1),CO(J,2))-(CO(K,1),CO(K,2))
1490 LINE (CO(K,1),CO(K,2))-(CO(L,1),CO(L,2))
1500 LINE (CO(L,1),CO(L,2))-(CO(J,1),CO(J,2))
1510 NEXT I
1520 REM
1530 IF IFLAG=2 THEN GOTO 1690
1540 PRINT
1550 PRINT
1560 PRINT "               Initial finite element mesh"
1570 PRINT
1580 INPUT "               Do you wish to run the FE program?    ",A
1590 REM
1600 REM==     return to normal screen mode                              ==
1610 REM
1620 SCREEN 0
1630 IF A="N" THEN GOTO 1810 ELSE IF A="n" THEN GOTO 1810
```

```
1640 REM
1650 REM==      Load next program                                        ==
1660 REM
1670 IF IEP=1 THEN CHAIN "FILBDME"
1680 IF INO>1 THEN CHAIN "FILBDMP" ELSE CHAIN "FILBDME"
1690 PRINT
1700 PRINT
1710 INO=INO-1
1720 PRINT "               Finite element mesh at end of increment",INO
1730 PRINT
1740 PRINT
1750 IF INO>=INL THEN PRINT "               Increment limit has been reached"
1760 PRINT
1770 INPUT "          Do you wish to continue with another increment  ",A
1780 IF A="N" THEN GOTO 1810 ELSE IF A="n" THEN GOTO 1810
1790 INO=INO+1
1800 GOTO 1610
1810 PRINT
1820 PRINT
1830 PRINT "       ***************** PROGRAM ENDED *****************"
1840 REM
1850 REM=======================================================================
1860 REM== DISPLAY version 1.0  completed 20-2-89          Peter Hartley   ==
1870 REM=======================================================================
1880 END
```

```
1000 REM=====================================================================
1010 REM==      This is file FILDATA used to create the initial data file.    ==
1020 REM==                                                                    ==
1030 REM==      The file has been arranged as a demonstration of the  type    ==
1040 REM==      of structure required by the finite element program.    As    ==
1050 REM==      an example the following program has been set up to create    ==
1060 REM==      a datafile specifically for simple upsetting with sticking    ==
1070 REM==      friction. The data are stored in file INTLDATA.              ==
1080 REM==      To use this program for  other  problems  it  is necessary    ==
1090 REM==      only to change the appropriate entries.If several problems    ==
1100 REM==      are  to  be  analysed it would be more  useful  to use  an    ==
1110 REM==      interactive version of this program and  store the data in    ==
1120 REM==      different files. It is advisable  to have the mesh details    ==
1130 REM==      carefully prepared prior to using either type of program.     ==
1140 REM=====================================================================
1150 REM
1160 REM
1170 REM==      Define integer and string variables                          ==
1180 REM
1190 DEFSTR A-B
1200 DEFINT E-N
1210 REM
1220 REM==      Set array dimensions                                         ==
1230 REM
1240 OPTION BASE 1
1250 DIM CO(25,2),LN(10),EC(32,3),NC(15),NT(15),PO(5),R(20),RD(10,2)
1260 REM
1270 REM
1280 REM==      Specify whether the problem is elastic or elastic-plastic    ==
1290 REM==      IEP=1 for elastic,  IEP=2 for elastic-plastic               ==
1300 REM
1310 IEP=2
1320 REM
1330 REM==      Specify the geometry of the problem                         ==
1340 REM==      IGE=1 for axial symmetry                                    ==
1350 REM==      IGE=2 for plane stress                                      ==
1360 REM==      IGE=3 for plane strain                                      ==
1370 REM
1380 IGE=1
1390 REM
1400 REM==      Specify the material properties. The elastic modulus  in    ==
1410 REM==      Newtons per square mm, Poissons ratio  and  the  elastic    ==
1420 REM==      yield stress are  required  for  elastic  analyses.  The    ==
1430 REM==      plastic modulus and the plastic Poissons ratio are  also    ==
1440 REM==      required for elastic-plastic analyses.   The values for    ==
1450 REM==      each property are stored in array PO respectively.          ==
1460 REM==      Linear work hardening is assumed.                           ==
1470 REM
1480 I=3
1490 IF IEP=2 THEN I=5
1500 FOR J=1 TO I
1510 READ PO(J)
1520 NEXT J
1530 DATA 50000,0.33,55,100,0.499
1540 REM
1550 REM==      Specify the number of  nodal  points  in  the  mesh  (NP),   ==
1560 REM==      the number of elements (NE), the  number  of  constrained   ==
1570 REM==      boundary nodes (NB), the number of  boundary  nodes  with   ==
1580 REM==      prescribed finite displacements or with prescribed forces   ==
1590 REM==      (NF), and the bandwidth of the stiffness matrix (NW).        ==
1600 REM
1610 READ NP,NE,NB,NF,NW
1620 DATA 9,8,7,3,10
1630 REM
```

```
1640 REM==      Specify the nodal connections of each element in  an anti   ==
1650 REM==      clockwise rotation.                                          ==
1660 REM==      EC(I) is the list of three nodes defining the element.       ==
1670 REM
1680 FOR I=1 TO NE
1690 READ EC(I,1),EC(I,2),EC(I,3)
1700 NEXT I
1710 DATA 1,5,4
1720 DATA 1,2,5
1730 DATA 2,6,5
1740 DATA 2,3,6
1750 DATA 4,8,7
1760 DATA 4,5,8
1770 DATA 5,9,8
1780 DATA 5,6,9
1790 REM
1800 REM==      Specify the nodal point co-ordinates(mm). x positive is      ==
1810 REM==      horizontal to the right, y positive is vertical upwards      ==
1820 REM==      Co-ordinates are stored in the array CO(I,J),  I is the      ==
1830 REM==      nodal point, J=1 indicates x, J=2 indicates y.               ==
1840 REM
1850 FOR I=1 TO NP
1860 READ CO(I,1),CO(I,2)
1870 NEXT I
1880 DATA 0,0
1890 DATA 50,0
1900 DATA 100,0
1910 DATA 0,50
1920 DATA 50,50
1930 DATA 100,50
1940 DATA 0,100
1950 DATA 50,100
1960 DATA 100,100
1970 REM
1980 REM==      Specify the constrained boundary nodes and the type of       ==
1990 REM==      constraint. Nodes are stored in NC(I).                       ==
2000 REM==      Type of constraint is stored in NT(I),                       ==
2010 REM==      NT=1    indicates zero displacement in the y direction,      ==
2020 REM==      NT=10   indicates zero displacement in the x direction.      ==
2030 REM==      NT=11   indicates zero displacement in both directions.      ==
2040 REM
2050 FOR I=1 TO NB
2060 READ NC(I),NT(I)
2070 NEXT I
2080 DATA 1,11
2090 DATA 2,1
2100 DATA 3,1
2110 DATA 4,10
2120 DATA 7,10
2130 DATA 8,10
2140 DATA 9,10
2150 REM
2160 REM==      Specify nodes with prescribed displacements or forces        ==
2170 REM==      and store in LN(I). Specify the x and y components of        ==
2180 REM==      the displacement(mm) or force(N) and store in RD(I,J)        ==
2190 REM==      J=1 indicates x component, J=2 indicates y component.        ==
2200 REM==      If displacements are being specified set  INF  to 2,         ==
2210 REM==      if forces are being specified set INF to 1.   If only        ==
2220 REM==      one component  is  being specified then set the other        ==
2230 REM==      to zero.                                                     ==
2240 REM
2250 INF=2
2260 REM
2270 FOR I=1 TO NF
```

```
2280 READ LN(I),RD(I,1),RD(I,2)
2290 NEXT I
2300 DATA 7,0,-1
2310 DATA 8,0,-1
2320 DATA 9,0,-1
2330 REM
2340 REM==     Specify the total number of increments for this analysis    ==
2350 REM==     If an elastic treatment has been selected a warning will     ==
2360 REM==     be printed if any element will exceed the yield stress.      ==
2370 REM==     INO is the increment counter initially set at 1.             ==
2380 REM==     INL is the increment limit.                                  ==
2390 REM
2400 INO=1
2410 INL=20
2420 REM
2430 REM==     All the initial data have now been specified. This will     ==
2440 REM==     be displayed on the screen prior to storing on disc.        ==
2450 REM
2460 CLS
2470 IF IEP=1 THEN PRINT "Elastic";
2480 IF IEP=2 THEN PRINT "Elastic-Plastic";
2490 PRINT " Finite Element Analysis in ";
2500 IF IGE=1 THEN PRINT "Axial Symmetry"
2510 IF IGE=2 THEN PRINT "Plane-Stress"
2520 IF IGE=3 THEN PRINT "Plane-Strain"
2530 PRINT
2540 PRINT "Material Properties"
2550 PRINT
2560 PRINT "        Elastic modulus = ",PO(1)
2570 PRINT "        Poissons ratio  = ",PO(2)
2580 PRINT "        Yield stress    = ",PO(3)
2590 IF IEP=2 THEN PRINT "        Plastic modulus = ",PO(4)
2600 IF IEP=2 THEN PRINT "        Plastic ratio   = ",PO(5)
2610 PRINT
2620 PRINT
2630 PRINT "The finite element mesh  contains "NE" elements, "NP" nodal points,"
2640 PRINT NB" nodes constrained on the boundaries, and "NF" nodes subject to"
2650 PRINT "specified displacements or forces."
2660 PRINT "The stiffness matrix has a bandwidth of "NW
2670 PRINT INL" Increments have been specified for this analysis."
2680 PRINT
2690 PRINT
2700 PRINT
2710 PRINT
2720 INPUT "                                Press RETURN to continue",A
2730 CLS
2740 PRINT "     Element connectivity"
2750 PRINT
2760 PRINT "           Element        Nodes defining given element"
2770 PRINT
2780 FOR I=1 TO NE
2790 PRINT ,I,EC(I,1),EC(I,2),EC(I,3)
2800 NEXT I
2810 INPUT "                                Press RETURN to continue",A
2820 CLS
2830 PRINT "     Nodal point co-ordinates"
2840 PRINT
2850 PRINT "           Node       co-ordinates(mm)"
2860 PRINT "                       x          y"
2870 FOR I=1 TO NP
2880 PRINT ,I;
2890 PRINT USING "   ####.##";CO(I,1),CO(I,2)
2900 NEXT I
2910 INPUT "                                Press RETURN to continue",A
```

```
2920 CLS
2930 PRINT "          Constraints on boundary nodes"
2940 PRINT
2950 FOR I=1 TO NB
2960 IF NT(I)=10 THEN A="the x direction"
2970 IF NT(I)=1 THEN A="the y direction"
2980 IF NT(I)=11 THEN A="both x and y"
2990 PRINT "               Node "NC(I)" is constrained in "A
3000 NEXT I
3010 INPUT "                         Press RETURN to continue",A
3020 CLS
3030 IF INF=2 THEN A="displacements (mm)"
3040 IF INF=1 THEN A="forces (N)"
3050 PRINT "        Prescribed nodal point "A
3060 PRINT
3070 PRINT "            Node",A
3080 PRINT "                       x           y"
3090 FOR I=1 TO NF
3100 PRINT ,LN(I),RD(I,1),RD(I,2)
3110 NEXT I
3120 INPUT "                         Press RETURN to continue",A
3130 PRINT
3140 PRINT "    If you wish to see stresses, strains and other results"
3150 PRINT "    displayed on the screen at the end  of each  increment"
3160 PRINT "    type Y to the following prompt. If not type N and  the"
3170 INPUT "    results will be displayed for the final increment only ",A
3180 IVER=1
3190 IF A="Y" THEN IVER=2
3200 PRINT
3210 PRINT
3220 PRINT "Initial data are now being transferred to disc in INTLDATA"
3230 REM
3240 REM==     Open a disc file called INTLDATA in which the initial data    ==
3250 REM==     can be stored                                                 ==
3260 REM
3270 OPEN "INTLDATA" FOR OUTPUT AS 1
3280 REM
3290 REM==     Store initial data  (IFLAG is used in DISPLAY)                 ==
3300 REM
3310 IFLAG=1
3320 PRINT#1,IEP;IGE;NP;NE;NB;NF;INF;NW;INO;INL;IFLAG;IVER
3330 PRINT#1,PO(1);PO(2);PO(3)
3340 IF IEP=2 THEN PRINT#1,PO(4);PO(5)
3350 FOR I=1 TO NE
3360 PRINT#1,EC(I,1);EC(I,2);EC(I,3)
3370 NEXT I
3380 FOR I=1 TO NP
3390 PRINT#1,CO(I,1);CO(I,2)
3400 NEXT I
3410 FOR I=1 TO NB
3420 PRINT#1,NC(I);NT(I)
3430 NEXT I
3440 FOR I=1 TO NF
3450 PRINT#1,LN(I);RD(I,1);RD(I,2)
3460 NEXT I
3470 CLOSE#1
3480 REM
3490 REM==     Open a disc file called FORCEV to contain vector of           ==
3500 REM==     specified forces or displacements.                            ==
3510 REM
3520 OPEN "FORCEV" FOR OUTPUT AS 2
3530 REM
3540 REM==     Assemble vector. Displacements  are  multiplied  by           ==
3550 REM==     1E+26 to eliminate any influence from  other  nodes.          ==
```

```
3560 REM==      The vector is stored in R(I).      See also the       ==
3570 REM==      formulation  of  the  stiffness  matrix  in FILSTFF.  ==
3580 REM
3590 FOR J=1 TO NP
3600 L1=(J-1)*2+1
3610 L2=L1+1
3620 R(L1)=0
3630 R(L2)=0
3640 FOR I=1 TO NF
3650 IF J=LN(I) THEN 3660 ELSE 3680
3660 R(L1)=RD(I,1)*(INF-1)*1E+26
3670 R(L2)=RD(I,2)*(INF-1)*1E+26
3680 NEXT I
3690 PRINT#2,R(L1),R(L2)
3700 NEXT J
3710 CLOSE#2
3720 REM
3730 REM==                Disc storage complete                       ==
3740 REM
3750 PRINT
3760 PRINT "Data storage is now complete"
3770 REM
3780 REM==      Load and run next program                             ==
3790 REM
3800 PRINT
3810 PRINT
3820 PRINT
3830 INPUT "              Do you want a graphical display of the mesh?",A
3840 IF A="Y" THEN CHAIN "DISPLAY"
3850 IF A="N" GOTO 3870 ELSE INPUT "              Please type Y or N",A
3860 GOTO 3840
3870 IF IEP=1 THEN CHAIN "FILBDME"
3880 IF INO>1 THEN CHAIN "FILBDMP" ELSE CHAIN "FILBDME"
3890 REM===============================================================
3900 REM== FILDATA version 1.0 completed 23-1-89      Peter Hartley ==
3910 REM===============================================================
3920 END
```

```
1000 REM=====================================================================
1010 REM==    This is file FILBDME used to set up the [B] and [D]        *=
1020 REM==    matrices for elasticity problems.                          ==
1030 REM=====================================================================
1040 REM
1050 REM==    Set array dimensions and define variables.                 ==
1060 REM
1070 DEFSTR A
1080 DEFINT E-N
1090 DEFDBL B-D,O-Z
1100 DIM B(4,6),CO(25,2),D(4,4),EC(32,3),PO(5)
1110 REM
1120 REM==    Retrieve data from disc file INTLDATA                      ==
1130 REM
1140 OPEN "INTLDATA" FOR INPUT AS 1
1150 INPUT#1,IEP,IGE,NP,NE,NB,NF,INF,NW,INO,INL,IFLAG,IVER
1160 INPUT#1,PO(1),PO(2),PO(3)
1170 IF IEP=2 THEN INPUT #1,PO(4),PO(5)
1180 FOR I=1 TO NE
1190 INPUT#1,EC(I,1),EC(I,2),EC(I,3)
1200 NEXT I
1210 FOR I=1 TO NP
1220 INPUT#1,CO(I,1),CO(I,2)
1230 NEXT I
1240 CLOSE#1
1250 CLS
1260 REM
1270 PRINT
1280 PRINT
1290 PRINT "          Program FILBDME has been loaded"
1300 PRINT
1310 PRINT "          Data are being retrieved from INTLDATA"
1320 REM
1330 REM==    Open files for storing matrices on disc                    ==
1340 REM
1350 OPEN "MATDATB" FOR OUTPUT AS 2
1360 OPEN "MATDATD" FOR OUTPUT AS 3
1370 REM
1380 REM
1390 REM==    Set elastic coefficients for [D] matrix                    ==
1400 REM
1410 T1=(1-PO(2))/(1-2*PO(2))
1420 T2=PO(2)/(1-2*PO(2))
1430 T3=1/(1-PO(2))
1440 T4=PO(2)/(1-PO(2))
1450 TM=PO(1)/(1+PO(2))
1460 REM
1470 REM==    Call subroutine to set up [B] and [D] matrices            ==
1480 REM
1490 IF IGE=1 THEN GOSUB 1520        :REM Axial symmetry
1500 IF IGE=2 THEN GOSUB 2060        :REM Plane stress
1510 IF IGE=3 THEN GOSUB 2500        :REM Plane strain
1520 REM
1530 REM==    Subroutine to set up [D] and [B] matrices for axial symmetry  ==
1540 REM
1550 REM==        [D] Matrix                                             ==
1560 D(1,1)=T1: D(1,2)=T2: D(1,3)=T2: D(1,4)=0
1570 D(2,1)=T2: D(2,2)=T1: D(2,3)=T2: D(2,4)=0
1580 D(3,1)=T2: D(3,2)=T2: D(3,3)=T1: D(3,4)=0
1590 D(4,1)=0:  D(4,2)=0:  D(4,3)=0:  D(4,4)=.5
1600 REM
1610 REM==        Store matrix in MATDATD                                ==
1620 FOR I=1 TO 4
1630 FOR J=1 TO 4
```

```
1640 D(I,J)=D(I,J)*TM
1650 PRINT#3,D(I,J)
1660 NEXT J
1670 NEXT I
1680 REM==      [B] Matrix                                                    ==
1690 REM==      Set coefficients and assemble matrix for each element         ==
1700 REM
1710 FOR N=1 TO NE
1720 I=EC(N,1)
1730 J=EC(N,2)
1740 M=EC(N,3)
1750 C1=CO(J,1)-CO(I,1)
1760 C2=CO(M,1)-CO(I,1)
1770 C3=CO(J,2)-CO(I,2)
1780 C4=CO(M,2)-CO(I,2)
1790 C5=(CO(I,1)+CO(J,1)+CO(M,1))/3
1800 C6=(CO(I,2)+CO(J,2)+CO(M,2))/3
1810 C7=CO(J,1)*CO(M,2)-CO(M,1)*CO(J,2)
1820 C8=CO(M,1)*CO(I,2)-CO(I,1)*CO(M,2)
1830 C9=CO(I,1)*CO(J,2)-CO(J,1)*CO(I,2)
1840 C10=C7/C5+C3-C4+(C2-C1)*C6/C5
1850 C11=C8/C5+C4-C2*C6/C5
1860 C12=C9/C5-C3+C1*C6/C5
1870 C34=C3-C4
1880 C21=C2-C1
1890 CAREA=(C1*C4-C2*C3)/2
1900 B(1,1)=C34: B(1,2)=0:   B(1,3)=C4: B(1,4)=0:   B(1,5)=-C3: B(1,6)=0
1910 B(2,1)=0:   B(2,2)=C21: B(2,3)=0:   B(2,4)=-C2: B(2,5)=0:   B(2,6)=C1
1920 B(3,1)=C10: B(3,2)=0:   B(3,3)=C11: B(3,4)=0:   B(3,5)=C12: B(3,6)=0
1930 B(4,1)=C21: B(4,2)=C34: B(4,3)=-C2: B(4,4)=C4:  B(4,5)=C1:  B(4,6)=-C3
1940 REM
1950 REM==      Store matrix in MATDATB                                       ==
1960 FOR G=1 TO 4
1970 FOR H=1 TO 6
1980 B(G,H)=B(G,H)/(CAREA*2)
1990 PRINT#2,B(G,H)
2000 NEXT H
2010 NEXT G
2020 VOL=CAREA*2*3.14158*C5
2030 PRINT#2,VOL
2040 NEXT N
2050 RETURN 2940
2060 REM
2070 REM==      Subroutine to set up [D] and [B] matrices for plane stress     ==
2080 REM
2090 REM==      [D] Matrix                                                    ==
2100 D(1,1)=T3: D(1,2)=T4: D(1,3)=0
2110 D(2,1)=T4: D(2,2)=T3: D(2,3)=0
2120 D(3,1)=0:   D(3,2)=0:   D(3,3)=.5
2130 REM
2140 REM==      Store matrix in MATDATD                                       ==
2150 FOR I=1 TO 3
2160 FOR J=1 TO 3
2170 D(I,J)=D(I,J)*TM
2180 PRINT#3,D(I,J)
2190 NEXT J
2200 NEXT I
2210 REM==      [B] Matrix                                                    ==
2220 REM==      Set coefficients and assemble matrix for each element         ==
2230 REM
2240 FOR N=1 TO NE
2250 I=EC(N,1)
2260 J=EC(N,2)
2270 M=EC(N,3)
```

```
2280 C1=CO(J,1)-CO(I,1)
2290 C2=CO(M,1)-CO(I,1)
2300 C3=CO(J,2)-CO(I,2)
2310 C4=CO(M,2)-CO(I,2)
2320 C34=C3-C4
2330 C21=C2-C1
2340 CAREA=(C1*C4-C2*C3)/2
2350 B(1,1)=C34: B(1,2)=0:   B(1,3)=C4:  B(1,4)=0:   B(1,5)=-C3: B(1,6)=0
2360 B(2,1)=0:   B(2,2)=C21: B(2,3)=0:   B(2,4)=-C2: B(2,5)=0:   B(2,6)=C1
2370 B(3,1)=C21: B(3,2)=C34: B(3,3)=-C2: B(3,4)=C4:  B(3,5)=C1:  B(3,6)=-C3
2380 REM
2390 REM==      Store matrix in MATDATB                                       ==
2400 FOR G=1 TO 3
2410 FOR H=1 TO 6
2420 B(G,H)=B(G,H)/(CAREA*2)
2430 PRINT#2,B(G,H)
2440 NEXT H
2450 NEXT G
2460 VOL=CAREA
2470 PRINT#2,VOL
2480 NEXT N
2490 RETURN 2940
2500 REM
2510 REM==      Subroutine to set up [D] and [B] matrices for plane strain    ==
2520 REM
2530 REM==      [D] Matrix                                                    ==
2540 D(1,1)=T1: D(1,2)=T2: D(1,3)=0
2550 D(2,1)=T2: D(2,2)=T1: D(2,3)=0
2560 D(3,1)=0:  D(3,2)=0:  D(3,3)=.5
2570 REM
2580 REM==      Store matrix in MATDATD                                       ==
2590 FOR I=1 TO 3
2600 FOR J=1 TO 3
2610 D(I,J)=D(I,J)*TM
2620 PRINT#3,D(I,J)
2630 NEXT J
2640 NEXT I
2650 REM==      [B] Matrix                                                    ==
2660 REM==      Set coefficients and assemble matrix for each element         ==
2670 REM
2680 FOR N=1 TO NE
2690 I=EC(N,1)
2700 J=EC(N,2)
2710 M=EC(N,3)
2720 C1=CO(J,1)-CO(I,1)
2730 C2=CO(M,1)-CO(I,1)
2740 C3=CO(J,2)-CO(I,2)
2750 C4=CO(M,2)-CO(I,2)
2760 C34=C3-C4
2770 C21=C2-C1
2780 CAREA=(C1*C4-C2*C3)/2
2790 B(1,1)=C34: B(1,2)=0:   B(1,3)=C4:  B(1,4)=0:   B(1,5)=-C3: B(1,6)=0
2800 B(2,1)=0:   B(2,2)=C21: B(2,3)=0:   B(2,4)=-C2: B(2,5)=0:   B(2,6)=C1
2810 B(3,1)=C21: B(3,2)=C34: B(3,3)=-C2: B(3,4)=C4:  B(3,5)=C1:  B(3,6)=-C3
2820 REM
2830 REM==      Store matrix in MATDATB                                       ==
2840 FOR G=1 TO 3
2850 FOR H=1 TO 6
2860 B(G,H)=B(G,H)/(CAREA*2)
2870 PRINT#2,B(G,H)
2880 NEXT H
2890 NEXT G
2900 VOL=CAREA
2910 PRINT#2,VOL
```

```
2920 NEXT N
2930 RETURN 2940
2940 REM
2950 REM==     Close all disc files                              ==
2960 REM
2970 CLOSE#2
2980 CLOSE#3
2990 REM
3000 PRINT
3010 PRINT
3020 PRINT "          [B] and [D] matrices have been set up and"
3030 PRINT "          stored in files MATDATB and MATDATD respectively"
3040 PRINT
3050 PRINT
3060 PRINT "          The next program is now being loaded"
3070 REM
3080 REM==     Load next program                                 ==
3090 REM
3100 CHAIN "FILSTFF"
3110 REM
3120 REM======================================================================
3130 REM==     FILBDME version 1.0 completed 20-2-89     Peter Hartley    ==
3140 REM======================================================================
3150 END
```

```
1000 REM=========================================================================
1010 REM==       This is file FILBDMP used to set up the [B] and [D]        ==
1020 REM==       matrices for elastic-plastic problems.                      ==
1030 REM=========================================================================
1040 REM
1050 REM==       Set array dimensions and define variables.                  ==
1060 REM
1070 DEFSTR A
1080 DEFINT F-N
1090 DEFDBL B-E,O-Z
1100 DIM B(4,6),CO(25,2),D(4,4),EC(32,3),PO(5)
1110 DIM SG(32),SNC(32,3),SNG(32),SNS(32),ST(32,3),STD(32,3),SS(32)
1120 REM
1130 REM==       Retrieve data from disc files INTLDATA and STSDATA          ==
1140 REM
1150 OPEN "INTLDATA" FOR INPUT AS 1
1160 INPUT#1,IEP,IGE,NP,NE,NB,NF,INF,NW,INO,INL,IFLAG,IVER
1170 INPUT#1,PO(1),PO(2),PO(3),PO(4),PO(5)
1180 FOR I=1 TO NE
1190 INPUT#1,EC(I,1),EC(I,2),EC(I,3)
1200 NEXT I
1210 FOR I=1 TO NP
1220 INPUT#1,CO(I,1),CO(I,2)
1230 NEXT I
1240 CLOSE#1
1250 OPEN "STSDATA" FOR INPUT AS 4
1260 FOR I=1 TO NE
1270 INPUT#4,STD(I,1),STD(I,2),STD(I,3)
1280 INPUT#4,SG(I),ST(I,1),ST(I,2),ST(I,3),SS(I)
1290 INPUT#4,SNG(I),SNC(I,1),SNC(I,2),SNC(I,3),SNS(I)
1300 NEXT I
1310 CLOSE#4
1320 CLS
1330 REM
1340 PRINT
1350 PRINT
1360 PRINT "            Program FILBDMP has been loaded"
1370 PRINT
1380 PRINT "            Data are being retrieved from INTLDATA and STSDATA"
1390 REM
1400 REM==       Open files for storing matrices on disc                     ==
1410 REM
1420 OPEN "MATDATB" FOR OUTPUT AS 2
1430 OPEN "MATDATD" FOR OUTPUT AS 3
1440 REM
1450 REM==       Call subroutine to set up [B] and [D] matrices             ==
1460 REM
1470 IF IGE=1 THEN GOSUB 1500          :REM Axial symmetry
1480 IF IGE=2 THEN GOSUB 2290          :REM Plane stress
1490 IF IGE=3 THEN GOSUB 2890          :REM Plane strain
1500 REM==       Subroutine to set up [D] and [B] matrices for axial symmetry ==
1510 REM==       [D] Matrix                                                  ==
1520 REM==       Set coefficients and assemble matrix for each element       ==
1530 REM
1540 FOR N=1 TO NE
1550 REM==       Set deviatoric stresses                                     ==
1560 SDR=STD(N,1)
1570 SDZ=STD(N,2)
1580 SDT=STD(N,3)
1590 REM==       Set flags related to current level of yield stress          ==
1600 IF SG(N)>PO(3) THEN I=5 ELSE I=2
1610 IF SG(N)>PO(3) THEN IM=1 ELSE IM=0
1620 TM=PO(1)/(1+PO(2))
1630 REM==       Evaluate multiplying term for plastic part                  ==
```

```
1640 TG=PO(1)/(2*(1+PO(2)))          :REM Torsion modulus
1650 SD=SG(N)^2*(1+PO(4)/(3*TG))*2/3
1660 SD=1/SD*IM
1670 REM==       Evaluate coefficients                                    ==
1680 T1=(1-PO(I))/(1-2*PO(I))-SDR^2*SD
1690 T2=PO(I)/(1-2*PO(I))-SDR*SDZ*SD
1700 T3=PO(I)/(1-2*PO(I))-SDR*SDT*SD
1710 T4=-SDR*SS(N)*SD
1720 T5=(1-PO(I))/(1-2*PO(I))-SDZ^2*SD
1730 T6=PO(I)/(1-2*PO(I))-SDZ*SDT*SD
1740 T7=-SDZ*SS(N)*SD
1750 T8=(1-PO(I))/(1-2*PO(I))-SDT^2*SD
1760 T9=-SDT*SS(N)*SD
1770 T10=.5-SS(N)^2*SD
1780 REM==       Assemble matrix                                          ==
1790 D(1,1)=T1: D(1,2)=T2: D(1,3)=T3: D(1,4)=T4
1800 D(2,1)=T2: D(2,2)=T5: D(2,3)=T6: D(2,4)=T7
1810 D(3,1)=T3: D(3,2)=T6: D(3,3)=T8: D(3,4)=T9
1820 D(4,1)=T4: D(4,2)=T7: D(4,3)=T9: D(4,4)=T10
1830 REM==       Store matrix in MATDATD                                  ==
1840 FOR I=1 TO 4
1850 FOR J=1 TO 4
1860 D(I,J)=D(I,J)*TM
1870 PRINT#3,D(I,J)
1880 NEXT J
1890 NEXT I
1900 NEXT N
1910 REM==       [B] Matrix                                               ==
1920 REM==       Set coefficients and assemble matrix for each element    ==
1930 REM
1940 FOR N=1 TO NE
1950 I=EC(N,1)
1960 J=EC(N,2)
1970 M=EC(N,3)
1980 C1=CO(J,1)-CO(I,1)
1990 C2=CO(M,1)-CO(I,1)
2000 C3=CO(J,2)-CO(I,2)
2010 C4=CO(M,2)-CO(I,2)
2020 C5=(CO(I,1)+CO(J,1)+CO(M,1))/3
2030 C6=(CO(I,2)+CO(J,2)+CO(M,2))/3
2040 C7=CO(J,1)*CO(M,2)-CO(M,1)*CO(J,2)
2050 C8=CO(M,1)*CO(I,2)-CO(I,1)*CO(M,2)
2060 C9=CO(I,1)*CO(J,2)-CO(J,1)*CO(I,2)
2070 C10=C7/C5+C3-C4+(C2-C1)*C6/C5
2080 C11=C8/C5+C4-C2*C6/C5
2090 C12=C9/C5-C3+C1*C6/C5
2100 C34=C3-C4
2110 C21=C2-C1
2120 CAREA=(C1*C4-C2*C3)/2
2130 B(1,1)=C34: B(1,2)=0:   B(1,3)=C4:  B(1,4)=0:   B(1,5)=-C3: B(1,6)=0
2140 B(2,1)=0:   B(2,2)=C21: B(2,3)=0:   B(2,4)=-C2: B(2,5)=0:   B(2,6)=C1
2150 B(3,1)=C10: B(3,2)=0:   B(3,3)=C11: B(3,4)=0:   B(3,5)=C12: B(3,6)=0
2160 B(4,1)=C21: B(4,2)=C34: B(4,3)=-C2: B(4,4)=C4:  B(4,5)=C1:  B(4,6)=-C3
2170 REM
2180 REM==       Store matrix in MATDATB                                  ==
2190 FOR G=1 TO 4
2200 FOR H=1 TO 6
2210 B(G,H)=B(G,H)/(CAREA*2)
2220 PRINT#2,B(G,H)
2230 NEXT H
2240 NEXT G
2250 VOL=CAREA*2*3.14158*C5
2260 PRINT#2,VOL
2270 NEXT N
```

```
2280 RETURN 3540
2290 REM==     Subroutine to set up [D] and [B] matrices for plane stress    ==
2300 REM==     [D] Matrix                                                     ==
2310 REM==     Set coefficients and assemble matrix each element             ==
2320 REM
2330 FOR N=1 TO NE
2340 REM==     Set deviatoric stresses                                       ==
2350 SDX=STD(N,1)
2360 SDY=STD(N,1)
2370 REM==     Set coefficients for elastic-plastic element                  ==
2380 TP=PO(4)/PO(1)*SG(N)^2*2/9+SS(N)^2/(1+PO(5))
2390 TR=SDX^2+SDY^2+2*PO(5)*SDX*SDY
2400 TQ=2*(1-PO(5)^2)*TP+TR
2410 TM=PO(1)/TQ
2420 T1=2*TP+SDY^2
2430 T2=2*PO(5)*TP-SDX*SDY
2440 T3=-(SDX+SDY*PO(5))*SS(N)/(1+PO(5))
2450 T4=2*TP+SDX^2
2460 T5=-(SDY+SDX*PO(5))*SS(N)/(1+PO(5))
2470 T6=(1-PO(5))*2/9*PO(4)/PO(1)*SG(N)^2+TR/(2*(1+PO(5)))
2480 REM==     Assemble matrix                                               ==
2490 D(1,1)=T1: D(1,2)=T2: D(1,3)=T3
2500 D(2,1)=T2: D(2,2)=T4: D(2,3)=T5
2510 D(3,1)=T3: D(3,2)=T5: D(3,3)=T6
2520 REM==     Store matrix in MATDATD                                       ==
2530 FOR I=1 TO 3
2540 FOR J=1 TO 3
2550 D(I,J)=D(I,J)*TM
2560 PRINT#3,D(I,J)
2570 NEXT J
2580 NEXT I
2590 NEXT N
2600 REM==     [B] Matrix                                                    ==
2610 REM==     Set coefficients and assemble matrix for each element         ==
2620 REM
2630 FOR N=1 TO NE
2640 I=EC(N,1)
2650 J=EC(N,2)
2660 M=EC(N,3)
2670 C1=CO(J,1)-CO(I,1)
2680 C2=CO(M,1)-CO(I,1)
2690 C3=CO(J,2)-CO(I,2)
2700 C4=CO(M,2)-CO(I,2)
2710 C34=C3-C4
2720 C21=C2-C1
2730 CAREA=(C1*C4-C2*C3)/2
2740 B(1,1)=C34: B(1,2)=0:  B(1,3)=C4: B(1,4)=0:  B(1,5)=-C3: B(1,6)=0
2750 B(2,1)=0:  B(2,2)=C21: B(2,3)=0:  B(2,4)=-C2: B(2,5)=0:  B(2,6)=C1
2760 B(3,1)=C21: B(3,2)=C34: B(3,3)=-C2: B(3,4)=C4: B(3,5)=C1:  B(3,6)=-C3
2770 REM
2780 REM==     Store matrix in MATDATB                                       ==
2790 FOR G=1 TO 3
2800 FOR H=1 TO 6
2810 B(G,H)=B(G,H)/(CAREA*2)
2820 PRINT#2,B(G,H)
2830 NEXT H
2840 NEXT G
2850 VOL=CAREA
2860 PRINT#2,VOL
2870 NEXT N
2880 RETURN 3540
2890 REM==     Subroutine to set up [D] and [B] matrices for plane strain    ==
2900 REM==     [D] Matrix                                                    ==
2910 REM==     Set coefficients and assemble matrix for each element         ==
```

```
2920 REM
2930 FOR N=1 TO NE
2940 REM==      Set deviatoric stresses                                    ==
2950 SDX=STD(N,1)
2960 SDY=STD(N,2)
2970 REM==      Set flags related to current level of yield stress         ==
2980 IF SG(N)>PO(3) THEN I=5 ELSE I=2
2990 IF SG(N)>PO(3) THEN IM=1 ELSE IM=0
3000 TM=PO(1)/(1+PO(2))
3010 REM==      Evaluate multiplying term for plastic part                 ==
3020 TG=PO(1)/(2*(1+PO(2)))          :REM Torsion modulus
3030 SD=SG(N)^2*(1+PO(4)/(3*TG))*2/3
3040 SD=1/SD*IM
3050 REM==      Evaluate coefficients                                      ==
3060 T1=(1-PO(I))/(1-2*PO(I))-SDX^2*SD
3070 T2=PO(I)/(1-2*PO(I))-SDX*SDY*SD
3080 T3=-SDX*SS(N)*SD
3090 T4=(1-PO(I))/(1-2*PO(I))-SDY^2*SD
3100 T5=-SDY*SS(N)*SD
3110 T6=.5-SS(N)^2*SD
3120 REM==      Assemble matrix                                            ==
3130 D(1,1)=T1: D(1,2)=T2: D(1,3)=T3
3140 D(2,1)=T2: D(2,2)=T4: D(2,3)=T5
3150 D(3,1)=T3: D(3,2)=T5: D(3,3)=T6
3160 REM
3170 REM==      Store matrix in MATDATD                                    ==
3180 FOR I=1 TO 3
3190 FOR J=1 TO 3
3200 D(I,J)=D(I,J)*TM
3210 PRINT#3,D(I,J)
3220 NEXT J
3230 NEXT I
3240 NEXT N
3250 REM==      [B] Matrix                                                 ==
3260 REM==      Set coefficients and assemble matrix for each element      ==
3270 REM
3280 FOR N=1 TO NE
3290 I=EC(N,1)
3300 J=EC(N,2)
3310 M=EC(N,3)
3320 C1=CO(J,1)-CO(I,1)
3330 C2=CO(M,1)-CO(I,1)
3340 C3=CO(J,2)-CO(I,2)
3350 C4=CO(M,2)-CO(I,2)
3360 C34=C3-C4
3370 C21=C2-C1
3380 CAREA=(C1*C4-C2*C3)/2
3390 B(1,1)=C34: B(1,2)=0:   B(1,3)=C4:  B(1,4)=0:   B(1,5)=-C3: B(1,6)=0
3400 B(2,1)=0:   B(2,2)=C21: B(2,3)=0:   B(2,4)=-C2: B(2,5)=0:   B(2,6)=C1
3410 B(3,1)=C21: B(3,2)=C34: B(3,3)=-C2: B(3,4)=C4:  B(3,5)=C1:  B(3,6)=-C3
3420 REM
3430 REM==      Store matrix in MATDATB                                    ==
3440 FOR G=1 TO 3
3450 FOR H=1 TO 6
3460 B(G,H)=B(G,H)/(CAREA*2)
3470 PRINT#2,B(G,H)
3480 NEXT H
3490 NEXT G
3500 VOL=CAREA
3510 PRINT#2,VOL
3520 NEXT N
3530 RETURN 3540
3540 REM
3550 REM==      Close all disc files                                       ==
```

```
3560 REM
3570 CLOSE#2
3580 CLOSE#3
3590 REM
3600 PRINT
3610 PRINT
3620 PRINT "          [B] and [D] matrices have been set up and"
3630 PRINT "          stored in files MATDATB and MATDATD respectively"
3640 PRINT
3650 PRINT
3660 PRINT "          The next program is now being loaded"
3670 REM
3680 REM==     Load next program                                    ==
3690 REM
3700 CHAIN "FILSTFF"
3710 REM
3720 REM=================================================================
3730 REM==FILBDMP version 1.0  23-1-89          Peter Hartley & Ian Pillinger ==
3740 REM=================================================================
3750 END
```

```
1000 REM=====================================================================
1010 REM==        This is file FILSTFF used to assemble the element [k]      ==
1020 REM==        and global [K] stiffness matrices                          ==
1030 REM=====================================================================
1040 REM
1050 REM==        Set array dimensions and define variables                  ==
1060 REM
1070 DEFSTR A
1080 DEFINT F-N
1090 DEFDBL B-E,O-Z
1100 DIM B(4,6),CO(25,2),D(4,4),E(4,6),EC(32,3),LN(10),PO(5),RD(10,2)
1110 DIM S(25,14),SK(6,6)
1120 CLS
1130 REM
1140 REM==        Retrieve data from disc file INTLDATA                      ==
1150 REM
1160 OPEN "INTLDATA" FOR INPUT AS 1
1170 INPUT#1,IEP,IGE,NP,NE,NB,NF,INF,NW,INO,INL,IFLAG,IVER
1180 INPUT#1,PO(1),PO(2),PO(3)
1190 IF IEP=2 THEN INPUT#1,PO(4),PO(5)
1200 FOR I=1 TO NE
1210 INPUT#1,EC(I,1),EC(I,2),EC(I,3)
1220 NEXT I
1230 FOR I=1 TO NP
1240 INPUT#1,CO(I,1),CO(I,2)
1250 NEXT I
1260 FOR I=1 TO NB
1270 INPUT#1,NC(I),NT(I)
1280 NEXT I
1290 FOR I=1 TO NF
1300 INPUT#1,LN(I),RD(I,1),RD(I,2)
1310 NEXT I
1320 CLOSE#1
1330 REM
1340 PRINT "               FILSTFF has been loaded and the element"
1350 PRINT "               and nodal point data are being read in "
1360 PRINT "               from INTLDATA"
1370 PRINT
1380 PRINT "               The global stiffness matrix is now being"
1390 PRINT "               determined"
1400 REM
1410 REM==        Zero all terms of the banded stiffness matrix              ==
1420 REM
1430 NEQ=NP*2            :REM 2 is the number of degress of freedom per node
1440 FOR I=1 TO NEQ      :REM NEQ is the number of equations for a given problem
1450 FOR J=1 TO NW       :REM NW is the bandwidth of the global stiffness matrix
1460 S(I,J)=0
1470 NEXT J
1480 NEXT I
1490 REM==        Open data files                                           ==
1500 REM
1510 OPEN "MATDATB" FOR INPUT AS 2
1520 OPEN "MATDATD" FOR INPUT AS 3
1530 REM
1540 REM==        Consider each element in turn and assemble the global     ==
1550 REM==        stiffness matrix                                           ==
1560 REM
1570 FOR N=1 TO NE
1580 REM
1590 REM==        Call subroutine to evaluate element stiffness             ==
1600 REM
1610 GOSUB 2220
1620 REM
1630 REM==        Assemble global matrix                                     ==
```

```
1640 REM
1650 FOR J1=1 TO 3      :REM 3 nodes per element
1660 M1=(EC(N,J1)-1)*2
1670 FOR J2=1 TO 2      :REM 2 degrees of freedom
1680 M1=M1+1            :REM Locates row in global matrix
1690 I=(J1-1)*2+J2      :REM Locates row in element matrix
1700 FOR J3=1 TO 3
1710 M2=(EC(N,J3)-1)*2
1720 FOR J4=1 TO 2
1730 L=(J3-1)*2+J4      :REM Locates column in element matrix
1740 M3=M2+J4+1-M1      :REM Locates column in banded global matrix
1750 IF M3>0 THEN GOTO 1760 ELSE GOTO 1770
1760 S(M1,M3)=S(M1,M3)+SK(I,L)   :REM Adds element coefficient to global matrix
1770 NEXT J4
1780 NEXT J3
1790 NEXT J2
1800 NEXT J1
1810 NEXT N
1820 REM
1830 REM==      Modify [K] to include zero displacement on specified      ==
1840 REM==      boundary nodes                                            ==
1850 REM
1860 FOR N=1 TO NB
1870 M1=10
1880 M2=(NC(N)-1)*2            :REM NC(N) is the constrained node
1890 FOR M3=1 TO 2
1900 M2=M2+1                   :REM Locates row in banded matrix
1910 I=INT(NT(N)/M1)           :REM NT(N) is the type of constraint
1920 IF I<=0 THEN GOTO 2010
1930 S(M2,1)=1                 :REM Set leading term to unity
1940 FOR J=2 TO NW
1950 S(M2,J)=0                 :REM Set all other terms on specified row to zero
1960 M4=M2+1-J
1970 IF M4<=0 THEN GOTO 1990
1980 S(M4,J)=0
1990 NEXT J
2000 NT(N)=NT(N)-M1*I
2010 M1=M1/10
2020 NEXT M3
2030 NEXT N
2040 REM
2050 REM==      Modify [K] to include the influence of specified      ==
2060 REM==      node displacements                                    ==
2070 REM
2080 IF INF=1 THEN GOTO 2170   :REM Only required if displacements are specified
2090 FOR N=1 TO NF
2100 M1=(LN(N)-1)*2
2110 FOR J=1 TO 2
2120 M1=M1+1                   :REM Locates row in banded matrix
2130 IF RD(N,J)=0 THEN GOTO 2150
2140 S(M1,1)=1E+26             :REM Set leading term to large number
2150 NEXT J
2160 NEXT N
2170 REM
2180 REM==      Close disc files                                      ==
2190 CLOSE#2
2200 CLOSE#3
2210 GOTO 2660
2220 REM
2230 REM== Subroutine to evaluate element stiffness                   ==
2240 REM
2250 REM== Input data from disc files                                 ==
2260 REM
2270 IF IGE=1 THEN I=4 ELSE I=3    :REM axial symmetry or plane stress/strain
```

```
2280 FOR J=1 TO I
2290 FOR K=1 TO 6
2300 INPUT#2,B(J,K)
2310 NEXT K
2320 NEXT J
2330 INPUT#2,VOL
2340 IF IEP=1 THEN IF N>1 THEN GOTO 2410 ELSE GOTO 2350
2350 IF IEP=2 THEN IF N>1 THEN IF INO<2 THEN GOTO 2410 ELSE GOTO 2360
2360 FOR J=1 TO I
2370 FOR K=1 TO I
2380 INPUT#3,D(J,K)
2390 NEXT K
2400 NEXT J
2410 REM
2420 REM== Evaluate [D]*[B]                                              ==
2430 REM
2440 FOR J=1 TO I
2450 FOR K=1 TO 6
2460 Z=0
2470 FOR L=1 TO I
2480 Z=Z+D(J,L)*B(L,K)
2490 NEXT L
2500 E(J,K)=Z
2510 NEXT K
2520 NEXT J
2530 REM
2540 REM== Evaluate [B](transpose)*[D]*[B]*element volume               ==
2550 REM
2560 FOR J=1 TO 6
2570 FOR K=1 TO 6
2580 Z=0
2590 FOR L=1 TO I
2600 Z=Z+B(L,J)*E(L,K)
2610 NEXT L
2620 SK(J,K)=Z*VOL
2630 NEXT K
2640 NEXT J
2650 RETURN
2660 REM
2670 REM==     Save [K] on disc in file STFDATA                         ==
2680 REM
2690 OPEN "STFDATA" FOR OUTPUT AS 5
2700 FOR I=1 TO NEQ
2710 FOR J=1 TO NW
2720 PRINT#5,S(I,J)
2730 NEXT J
2740 NEXT I
2750 CLOSE#5
2760 REM
2770 PRINT
2780 PRINT
2790 PRINT "            Evaluation of the stiffness matrix is"
2800 PRINT "            complete"
2810 PRINT
2820 PRINT "            The next program is being loaded"
2830 REM
2840 REM==     Load next program                                        ==
2850 REM
2860 CHAIN "FILDSTS"
2870 REM
2880 REM===============================================================
2890 REM==    FILSTFF version 1.0 completed 20-2-89      Peter Hartley  ==
2900 REM===============================================================
2910 END
```

```
1000 REM=======================================================================
1010 REM==      This is file FILDSTS used to evaluate nodal point      ==
1020 REM==      displacements and element stresses                     ==
1030 REM=======================================================================
1040 REM
1050 REM==      Set array dimensions and define variables              ==
1060 REM
1070 DEFSTR A
1080 DEFINT E-N
1090 DEFDBL B-D,O-Z
1100 DIM B(4,6),CO(25,2),D(4,4),DEL(6),DIS(20),EC(32,3),LN(10)
1110 DIM NC(15),NT(15),PO(5),R(20),RD(10,2)
1120 DIM S(25,14),SG(32),SK(6,6),SNC(32,3),SNG(32)
1130 DIM SNS(32),SS(32),ST(32,3),STD(32,3),TI(4),X(4)
1140 REM
1150 CLS
1160 PRINT "              File FILDSTS has been loaded"
1170 PRINT
1180 PRINT "              Displacements, stresses and strains are"
1190 PRINT "              now being calculated"
1200 REM
1210 REM==      Retrieve data from INTLDATA, FORCEV and STFDATA         ==
1220 OPEN "INTLDATA" FOR INPUT AS 1
1230 INPUT#1,IEP,IGE,NP,NE,NB,NF,INF,NW,INO,INL,IFLAG,IVER
1240 INPUT#1,PO(1),PO(2),PO(3)
1250 IF IEP=2 THEN INPUT#1,PO(4),PO(5)
1260 FOR I=1 TO NE
1270 INPUT#1,EC(I,1),EC(I,2),EC(I,3)
1280 NEXT I
1290 FOR I=1 TO NP
1300 INPUT#1,CO(I,1),CO(I,2)
1310 NEXT I
1320 FOR I=1 TO NB
1330 INPUT#1,NC(I),NT(I)
1340 NEXT I
1350 FOR I=1 TO NF
1360 INPUT#1,LN(I),RD(I,1),RD(I,2)
1370 NEXT I
1380 CLOSE#1
1390 REM
1400 OPEN "FORCEV" FOR INPUT AS 2
1410 FOR J=1 TO NP
1420 L1=(J-1)*2+1
1430 L2=L1+1
1440 INPUT#2,R(L1),R(L2)
1450 NEXT J
1460 CLOSE#2
1470 REM
1480 OPEN "STFDATA" FOR INPUT AS 3
1490 NEQ=2*NP
1500 FOR I=1 TO NEQ          :REM NEQ is the number of equations
1510 FOR J=1 TO NW           :REM NW is the bandwidth
1520 INPUT#3,S(I,J)
1530 NEXT J
1540 NEXT I
1550 CLOSE#3
1560 REM
1570 REM==      Temporarily store forces in vector DIS                 ==
1580 REM
1590 FOR I=1 TO NEQ
1600 DIS(I)=R(I)
1610 NEXT I
1620 REM
1630 REM==      Reduce equations of global stiffness matrix            ==
```

```
1640 REM
1650 FOR N=1 TO NEQ
1660 I=N
1670 FOR L=2 TO NW
1680 I=I+1
1690 IF S(N,L)=0 THEN GOTO 1790
1700 Q=S(N,L)/S(N,1)
1710 J=0
1720 FOR K=L TO NW
1730 J=J+1
1740 IF S(N,K)=0 THEN GOTO 1760
1750 S(I,J)=S(I,J)-Q*S(N,K)
1760 NEXT K
1770 S(N,L)=Q
1780 DIS(I)=DIS(I)-Q*DIS(N)
1790 NEXT L
1800 DIS(N)=DIS(N)/S(N,1)
1810 NEXT N
1820 REM
1830 REM==      Back substitute to evaluate displacements        ==
1840 REM
1850 N=NEQ
1860 N=N-1
1870 IF N<=0 THEN GOTO 1950
1880 L=N
1890 FOR K=2 TO NW
1900 L=L+1
1910 IF S(N,K)=0 THEN GOTO 1930
1920 DIS(N)=DIS(N)-S(N,K)*DIS(L)
1930 NEXT K
1940 GOTO 1860
1950 REM
1960 REM==      Determine new co-ordinates, store in INTLDATA     ==
1970 REM
1980 FOR I=1 TO NP
1990 FOR J=1 TO 2
2000 L=(I-1)*2+J
2010 CO(I,J)=CO(I,J)+DIS(L)
2020 NEXT J
2030 NEXT I
2040 REM==      Set flag for graphical display                   ==
2050 IFLAG=2
2060 REM==      Set counter for next increment                   ==
2070 INO=INO+1
2080 OPEN "INTLDATA" FOR OUTPUT AS 1
2090 PRINT#1,IEP;IGE;NP;NE;NB;NF;INF;NW;INO;INL;IFLAG;IVER
2100 PRINT#1,PO(1);PO(2);PO(3)
2110 IF IEP=2 THEN PRINT#1,PO(4);PO(5)·
2120 FOR I=1 TO NE
2130 PRINT#1,EC(I,1);EC(I,2);EC(I,3)
2140 NEXT I
2150 FOR I=1 TO NP
2160 PRINT#1,CO(I,1);CO(I,2)
2170 NEXT I
2180 FOR I=1 TO NB
2190 PRINT#1,NC(I);NT(I)
2200 NEXT I
2210 FOR I=1 TO NF
2220 PRINT#1,LN(I);RD(I,1);RD(I,2)
2230 NEXT I
2240 CLOSE#1
2250 REM
2260 IF INO>INL THEN GOTO 2280
2270 IF IVER=1 THEN GOTO 2420
```

```
2280 CLS
2290 PRINT
2300 PRINT "                     NODAL DISPLACEMENTS AND NEW CO-ORDINATES"
2310 PRINT
2320 PRINT "             NODE        HORDIS          VERDIS";
2330 PRINT "        HORCO-ORD     VERCO-ORD"
2340 FOR I=1 TO NP
2350 L1=(I-1)*2+1
2360 L2=L1+1
2370 PRINT ,I;
2380 PRINT USING "      ####.####";DIS(L1),DIS(L2),CO(I,1),CO(I,2)
2390 NEXT I
2400 PRINT
2410 INPUT "                                    Press RETURN to continue",A
2420 REM
2430 REM==     Begin calculation of strains and stresses              ==
2440 REM
2450 REM==     Open disc files                                        ==
2460 REM
2470 OPEN "MATDATD" FOR INPUT AS 4
2480 OPEN "MATDATB" FOR INPUT AS 5
2490 ING=INO-1
2500 IF ING>1 THEN OPEN "STSDATA" FOR INPUT AS 6
2510 REM
2520 IF IGE=1 THEN I=4 ELSE I=3    :REM Axial symmetry or plane stress/strain
2530 FOR N=1 TO NE                 :REM NE is the total number of elements
2540 REM==     If this is the first increment zero strain and stress matrices ==
2550 IF ING>1 GOTO 2610
2560 STD(N,1)=0: STD(N,2)=0: STD(N,3)=0
2570 SG(N)=0:  ST(N,1)=0:  ST(N,2)=0:  ST(N,3)=0:  SS(N)=0
2580 SNG(N)=0: SNC(N,1)=0: SNC(N,2)=0: SNC(N,3)=0: SNS(N)=0
2590 GOTO 2640
2600 REM==     Read data from disc                                    ==
2610 INPUT#6,STD(N,1),STD(N,2),STD(N,3)
2620 INPUT#6,SG(N),ST(N,1),ST(N,2),ST(N,3),SS(N)
2630 INPUT#6,SNG(N),SNC(N,1),SNC(N,2),SNC(N,3),SNS(N)
2640 FOR J=1 TO I
2650 FOR K=1 TO 6
2660 INPUT#5,B(J,K)                :REM Read [B] matrix
2670 NEXT K
2680 NEXT J
2690 INPUT#5,VOL
2700 IF IEP=1 THEN IF N>1 THEN GOTO 2780 ELSE GOTO 2710
2710 IF IEP=2 THEN IF N>1 THEN IF ING=1 THEN GOTO 2780 ELSE GOTO 2720
2720 FOR J=1 TO I
2730 FOR K=1 TO I
2740 INPUT#4,D(J,K)                :REM Read [D] matrix
2750 NEXT K
2760 NEXT J
2770 REM
2780 REM==     Set up vector of nodal displacements for given element    ==
2790 REM
2800 FOR M1=1 TO 3                 :REM 3 nodes per element
2810 FOR M2=1 TO 2                 :REM 2 degrees of freedom per nodes
2820 L1=(EC(N,M1)-1)*2+M2          :REM Location in global vector
2830 L2=(M1-1)*2+M2                :REM Location in element vector
2840 DEL(L2)=DIS(L1)
2850 NEXT M2
2860 NEXT M1
2870 REM
2880 REM
2890 REM==     Evaluate strain components                             ==
2900 REM
2910 FOR J=1 TO I
```

```
2920 Z=0
2930 FOR K=1 TO 6
2940 Z=Z+B(J,K)*DEL(K)
2950 NEXT K
2960 X(J)=Z              :REM Temporary storage of incremental strains
2970 IF J=I THEN SNS(N)=Z ELSE SNC(N,J)=Z
2980 NEXT J
2990 REM
3000 REM==   Store stress components and evaluate incremental values       ==
3010 REM
3020 TI(1)=ST(N,1)
3030 TI(2)=ST(N,2)
3040 TI(3)=ST(N,3)
3050 TI(4)=SS(N)
3060 SG1=SG(N)
3070 FOR J=1 TO I
3080 Z=0
3090 FOR K=1 TO I
3100 IF K=I THEN Z=Z+D(J,K)*SNS(N) ELSE Z=Z+D(J,K)*SNC(N,K)
3110 NEXT K
3120 IF J=I THEN SS(N)=Z ELSE ST(N,J)=Z   :REM incremental stresses
3130 NEXT J
3140 REM== Select axial symmetry(1), plane stress(2) or plane strain(3)     ==
3150 ON IGE GOTO 3160,3500,3870
3160 REM==   GENERALISED STRESS AND STRAIN FOR AXIAL SYMMETRY              ==
3170 REM
3180 REM==       Incremental mean stress and deviatorics                    ==
3190 SM=(ST(N,1)+ST(N,2)+ST(N,3))/3
3200 TT1=ST(N,1)-SM
3210 TT2=ST(N,2)-SM
3220 TT3=ST(N,3)-SM
3230 REM==       Incremental generalised stress based on deviatorics         ==
3240 SGEN=SQR(3/2*(TT1^2+TT2^2+TT3^2+2*SS(N)^2))
3250 REM==       Incremental generalised strain                             ==
3260 W=SQR(2/3*(X(1)^2+X(2)^2+X(3)^2+.5*X(4)^2))
3270 REM==       Generalised stress based on total strain                   ==
3280 SNG(N)=SNG(N)+W
3290 YSN=PO(3)/PO(1)         :REM YSN is the yield stress
3300 IF YSN>SNG(N) THEN SG(N)=PO(3)*SNG(N)
3310 IF IEP=2 THEN IF YSN<=SNG(N) THEN SG(N)=PO(3)+PO(4)*(SNG(N)-YSN)
3320 IF IEP=1 THEN IF YSN<=SNG(N) THEN PRINT "*****WARNING******"
3330 IF IEP=1 THEN IF YSN<=SNG(N) THEN PRINT "ELEMENT ",N," HAS YIELDED"
3340 REM==Reduce components                                                 ==
3350 SRATIO=SG(N)/SGEN
3360 IF ING>1 THEN SRATIO=SG(N)/SG1
3370 FOR J=1 TO 3
3380 ST(N,J)=ST(N,J)+TI(J)
3390 ST(N,J)=ST(N,J)*SRATIO
3400 IF ING>1 THEN ST(N,J)=TI(J)*SRATIO
3410 NEXT J
3420 SS(N)=(TI(4)+SS(N))*SRATIO
3430 IF ING>1 THEN SS(N)=TI(4)*SRATIO
3440 REM Deviatoric stresses
3450 HYD=(ST(N,1)+ST(N,2)+ST(N,3))/3
3460 STD(N,1)=ST(N,1)-HYD
3470 STD(N,2)=ST(N,2)-HYD
3480 STD(N,3)=ST(N,3)-HYD
3490 GOTO 4230
3500 REM==   GENERALISED STRESS AND STRAIN FOR PLANE STRESS               ==
3510 REM
3520 REM== Incremental mean stress and deviatorics                          ==
3530 ST(N,3)=0!            :REM Axial stress component
3540 SM=(ST(N,1)+ST(N,2)+ST(N,3))/3!
3550 TT1=ST(N,1)-SM
```

```
3560 TT2=ST(N,2)-SM
3570 TT3=ST(N,3)-SM
3580 REM==   Incremental generalised stress based on deviatorics      ==
3590 SGEN=SQR(3/2*(TT1^2+TT2^2+TT3^2+2*SS(N)^2))
3600 REM==       Generalised strain                                    ==
3610 REM==       W is incremental value                               ==
3620 REM
3630 IF SGEN>PO(3) THEN X(4)=-(X(1)+X(2)) ELSE X(4)=PO(2)/(PO(2)-1)*(X(1)-X(2))
3640 W=SQR(2/3*(X(1)^2+X(2)^2+X(4)^2+.5*X(3)^2))
3650 SNG(N)=SNG(N)+W
3660 YSN=PO(3)/PO(1)          :REM YSN is the yield strain
3670 IF YSN>SNG(N) THEN SG(N)=PO(3)*SNG(N)
3680 IF IEP=2 THEN IF YSN<=SNG(N) THEN SG(N)=PO(3)+PO(4)*(SNG(N)-YSN)
3690 IF IEP=1 THEN IF YSN<=SNG(N) THEN PRINT "*****WARNING******";
3700 PRINT "ELEMENT ",N," HAS PASSED THE PLASTIC LIMIT"
3710 REM== Reduce components                                           ==
3720 SRATIO=SG(N)/SGEN
3730 IF ING>1 THEN SRATIO=SG(N)/SG1
3740 FOR J=1 TO 3
3750 ST(N,J)=ST(N,J)+TI(J)
3760 ST(N,J)=ST(N,J)*SRATIO
3770 IF ING>1 THEN ST(N,J)=TI(J)*SRATIO
3780 NEXT J
3790 SS(N)=(TI(4)+SS(N))*SRATIO
3800 IF ING>1 THEN SS(N)=TI(4)*SRATIO
3810 REM== Deviatoric stresses                                         ==
3820 HYD=(ST(N,1)+ST(N,2)+ST(N,3))/3
3830 STD(N,1)=STD(N,1)-HYD
3840 STD(N,2)=STD(N,2)-HYD
3850 STD(N,3)=STD(N,3)-HYD
3860 GOTO 4230
3870 REM==       GENERALISED STRESS AND STRAIN FOR PLANE STRAIN        ==
3880 REM
3890 REM==       Generalised strain                                    ==
3900 REM==       W is incremental value                               ==
3910 REM
3920 W=SQR(2/3*(X(1)^2+X(2)^2+.5*X(3)^2))
3930 SNG(N)=SNG(N)+W
3940 YSN=PO(3)/PO(1)          :REM YSN is the yield strain
3950 IF YSN>SNG(N) THEN SG(N)=PO(3)*SNG(N)
3960 IF IEP=2 THEN IF YSN<=SNG(N) THEN SG(N)=PO(3)+PO(4)*(SNG(N)-YSN)
3970 IF IEP=1 THEN IF YSN<=SNG(N) THEN PRINT "*****WARNING******";
3980 PRINT "ELEMENT ",N," HAS PASSED THE PLASTIC LIMIT"
3990 IF YSN>SNG(N) THEN ST(N,3)=PO(2)*(ST(N,1)+ST(N,2))
4000 IF YSN<=SNG(N) THEN ST(N,3)=.5*(ST(N,1)+ST(N,2))
4010 REM== Incremental mean stress and deviatorics                    ==
4020 SM=(ST(N,1)+ST(N,2)+ST(N,3))/3!
4030 TT1=ST(N,1)-SM
4040 TT2=ST(N,2)-SM
4050 TT3=ST(N,3)-SM
4060 REM== Incremental generalised stress based on devaitorics        ==
4070 SGEN=SQR(3/2*(TT1^2+TT2^2+TT3^2+2*SS(N)^2))
4080 REM== Reduce components                                           ==
4090 SRATIO=SG(N)/SGEN
4100 IF ING>1 THEN SRATIO=SG(N)/SG1
4110 FOR J=1 TO 3
4120 ST(N,J)=ST(N,J)+TI(J)
4130 ST(N,J)=ST(N,J)*SRATIO
4140 IF ING>1 THEN ST(N,J)=TI(J)*SRATIO
4150 NEXT J
4160 SS(N)=(TI(4)+SS(N))*SRATIO
4170 IF ING>1 THEN SS(N)=TI(4)*SRATIO
4180 REM== Deviatoric stresses
4190 HYD=(ST(N,1)+ST(N,2)+ST(N,3))/3
```

```
4200 STD(N,1)=ST(N,1)-HYD
4210 STD(N,2)=ST(N,2)-HYD
4220 STD(N,3)=ST(N,3)-HYD
4230 NEXT N
4240 REM== Close files                                              ==
4250 CLOSE#4
4260 CLOSE#5
4270 CLOSE#6
4280 IF INO>INL THEN GOTO 4300
4290 IF IVER=1 THEN GOTO 4690
4300 REM==      Display stresses and strains on screen              ==
4310 REM
4320 CLS
4330 PRINT "           ELEMENT GENERALISED STRESS GENERALISED STRAIN"
4340 FOR N=1 TO NE
4350 PRINT ,N;
4360 PRINT USING "      #######.####";SG(N),SNG(N)
4370 NEXT N
4380 PRINT
4390 INPUT "                                    Press RETURN to continue",A
4400 CLS
4410 PRINT "                    STRESS COMPONENTS"
4420 A1="XSTRESS": A2="YSTRESS": A3="ZSTRESS": A4="XYSTRESS"
4430 A5="RSTRESS": A6="ZSTRESS": A7="TSTRESS": A8="RZSTRESS"
4440 A9="ELEMENT"
4450 IF IGE=1 THEN PRINT A9,A5,A6,A7,A8
4460 IF IGE=2 THEN PRINT A9,A1,A2,A3,A4
4470 IF IGE=2 THEN PRINT A9,A1,A2,A3,A4
4480 PRINT
4490 FOR N=1 TO NE
4500 PRINT N;
4510 PRINT USING "      ######.###";ST(N,1),ST(N,2),ST(N,3),SS(N)
4520 NEXT N
4530 PRINT
4540 INPUT "                                    Press RETURN to continue",A
4550 CLS
4560 PRINT "                    STRAIN COMPONENTS"
4570 A1="XSTRAIN": A2="YSTRAIN": A3="ZSTRAIN": A4="XYSTRAIN"
4580 A5="RSTRAIN": A6="ZSTRAIN": A7="TSTRAIN": A8="RZSTRAIN"
4590 IF IGE=1 THEN PRINT A9,A5,A6,A7,A8
4600 IF IGE=2 THEN PRINT A9,A1,A2,A3,A4
4610 IF IGE=2 THEN PRINT A9,A1,A2,A3,A4
4620 PRINT
4630 FOR N=1 TO NE
4640 PRINT N;
4650 PRINT USING "      ######.####";SNC(N,1),SNC(N,2),SNC(N,3),SNS(N)
4660 NEXT N
4670 PRINT
4680 INPUT "                                    Press RETURN to continue",A
4690 REM
4700 REM==    Store stresses and strains in STSDATA                 ==
4710 REM
4720 OPEN "STSDATA" FOR OUTPUT AS 7
4730 FOR N=1 TO NE
4740 PRINT#7,USING"########.####";STD(N,1);STD(N,2);STD(N,3)
4750 PRINT#7,USING"########.####";SG(N);ST(N,1);ST(N,2);ST(N,3);SS(N)
4760 PRINT#7,SNG(N);SNC(N,1);SNC(N,2);SNC(N,3);SNS(N)
4770 NEXT N
4780 CLOSE#7
4790 REM
4800 CLS
4810 PRINT
4820 PRINT "          Increment number ",ING," is complete"
4830 PRINT
```

```
4840 IF INO>INL THEN GOTO 4860
4850 IF IVER=1 THEN IF IEP=1 THEN CHAIN "FILBDME" ELSE CHAIN "FILBDMP"
4860 INPUT "          Do you wish to display the distorted mesh",A
4870 IF A="Y" THEN CHAIN "DISPLAY"
4880 PRINT
4890 IF INO>INL THEN PRINT "          Specified increment limit has been reached"
4900 PRINT
4910 INPUT "          Do you wish to continue to another increment ",A
4920 IF A="N" THEN GOTO 4940
4930 IF IEP=1 THEN CHAIN "FILBDME" ELSE CHAIN "FILBDMP"
4940 PRINT "**********************PROGRAM ENDED**************************"
4950 REM================================================================
4960 REM==    FILDSTS version 1.0 completed 16-2-89      Peter Hartley      ==
4970 REM================================================================
4980 END
```

Bibliography

The bibliography is not set out as a straightforward alphabetical list but has been arranged in groups under the separate headings listed below so that publications on related material are kept together.

BACKGROUND READING

The publications in this section will be particularly useful for anyone approaching the subject of finite-element simulation of metalforming for the first time.

Continuum mechanics – tensor theory

Eringen, A.C. *Nonlinear Theory of Continuous Media*, McGraw-Hill (1962).
Hunter, S.C. *Mechanics of Continuous Media*, Ellis Horwood (1976).
Leigh, D.C. *Nonlinear Continuum Mechanics*, McGraw-Hill (1968).
Spencer, A.J.M. *Continuum Mechanics*, Longman (1980).

Finite-element methods

Cheung, Y.K. and Yeo, M.F. *A Practical Introduction to Finite Element Analysis*, Pitman (1979).

Hinton, E. and Owen, D.R.J. *Finite Element Programming*, Academic Press (1977).
Irons, B. and Ahmad, S. *Techniques of Finite Elements*, Halsted Press (1980).
Livesley, R. *Finite-elements: An Introduction for Engineers*, Cambridge University Press (1984).
Rao, S.S. *The Finite Element Method in Engineering*, Pergamon Press (1982).
Zienkiewicz, O.C. *The Finite-Element Method*, McGraw-Hill (1977).

Metalworking

Avitzur, B. *Metalforming Processes and Analysis*, McGraw-Hill (1968).
Rowe, G.W. *Principles of Industrial Metalworking Processes*, Arnold (1977).

Numerical modelling of metalforming

Boer, C.R., Rebelo, N., Rydstad, H. and Schroder, G. (Eds.) *Process Modelling of Metal Forming and Thermomechanical Treatment*, Springer (1986).
Owen, D.R.J. and Hinton, E. *Finite Elements in Plasticity: Theory and Practice*, Pineridge Press (1980).
Pittman, J.F.T., Zienkiewicz, O.C., Wood, R.D. and Alexander, J.M. (Eds.) *Numerical Analysis of Forming Processes*, Wiley (1984).

Numerical Techniques

Boas, M.L. *Mathematical Methods in the Physical Sciences*, Wiley (1983).
Day, A.C. *Fortran Techniques*, Cambridge University Press (1972).
Press, W.H., Flannery, B.P., Teukolsky, S.A. and Vetterling, W.T. *Numerical Recipes*, Cambridge University Press (1986).

Plasticity

Ford, H. and Alexander, J.M. *Advanced Mechanics of Materials*, Ellis Horwood (1977).
Hill, R. *The Mathematical Theory of Plasticity*, Clarendon Press (1950).
Hoffmann, O. and Sachs, G. *Introduction to the Theory of Plasticity for Engineers*, McGraw-Hill (1953).
Johnson, W. and Mellor, F.B. *Engineering Plasticity*, von Nostrand, Reinhold (1973).
Nadai, A. *Plasticity, a Mechanics of the Plastic State of Matter*, McGraw-Hill (1931).
Nadai, A. *Theory of Flow and Fracture*, McGraw-Hill (1950).
Tabor, D. *The Hardness of Metals*, Oxford University Press (1951).

RESEARCH PUBLICATIONS

This section is intended to represent a comprehensive and up-to-date survey of publications relating to the finite-element simulation of metalforming. The authors apologise in advance for any omissions or errors, and will be happy to receive suggestions for inclusion in any future editions of this monograph.

Computer systems

Integrated systems

Altan, T. and Oh, S.I. CAD/CAM of tooling and process for plastic working. *Proc. 1st Int. Conf. on Technology of Plasticity*, ed. H. Kudo, JSTP/JSPE, pp. 531–44 (1984).

Eames, A.J., Dean, T.A., Hartley, P. and Sturgess, C.E.N. An integrated computer system for forging die design and flow simulation. *Proc. Int. Conf. on Computer Aided Production Engineering*, ed. J.A. McGeough, Mechanical Engineering Pubs., pp. 231–6 (1986).

Eames, A.J., Dean, T.A., Hartley, P., Sturgess, C.E.N. and Rowe, G.W. An IKBS for upset forging die design. *Proc. 2nd Int. Conf. on Computer Aided Production Engineering*, ed. J.A. McGeough, Mechanical Engineering Pubs., pp. 37–41 (1987).

Hartley, P., Sturgess, C.E.N. and Rowe, G.W. Expert systems for design and manufacture. *Proc. 12th All India Machine Tool Des. Res. Conf.*, ed. U.R.K. Rao, P.N. Rao and N.K. Tiwari, Tata McGraw-Hill, pp. xxi–xxvii (1986).

Hartley, P., Sturgess, C.E.N., Dean, T.A. and Rowe, G.W. Forging die design and flow simulation: their integration in intelligent knowledge based systems. *J. Mech. Wkg. Techn.* **15**, 1–13 (1987).

Hartley, P., Sturgess, C.E.N., Dean, T.A., Rowe, G.W. and Eames, A.J. Development of a forging expert system *Expert Systems in Engineering*, ed. D.T. Pham, IFS Pubs., pp. 425–43 (1988).

Kopp, R. and Arfmann, G. The application of CAD/CAE/CAM from the viewpoint of plastic working technology. *Proc. 1st Int. Conf. on Technology of Plasticity*, ed. H. Kudo, JSTP/JSPE, pp. 489–97 (1984).

Pillinger, I., Hartley, P., Sturgess, C.E.N. and Dean, T.A. An intelligent knowledge based system for the design of forging dies. *Artificial Intelligence in Design*, ed. D.T. Pham, IFS Pubs. (1990).

Pillinger, I., Sims, P., Hartley, P., Sturgess, C.E.N. and Dean, T.A. Integrating CAD/CAM of forging dies with finite-element simulation of metal flow in an intelligent knowledge based system. *Integrating Information and Material Flow, Proc. Int. Conf. on Factory 2000*, Institution of Electronic and Radio Engineers, pp. 205–11 (1988).

Microcomputer applications

Dung, N.L. The use of personal computer for FE simulation of forging process. *Proc. 11th CANCAM Conf.*, Edmonton, Alberta (1987).

Fadzil, M., Chuah, K.C., Hartley, P., Sturgess, C.E.N. and Rowe, G.W. Metal-forming analysis on desk-top microcomputers using non-linear elastic–plastic finite element techniques. *Proc. 22nd Int. Machine Tool Des. Res. Conf.*, ed. B.J. Davies, Macmillan, pp. 533–9 (1981).

Hussin, A.A.M., Hartley, P., Sturgess, C.E.N. and Rowe, G.W. Finite element plasticity on microcomputers. *Proc. Stress Analysis and the Micro Conf. (SAM'85)*, City University, London, pp. 106–15 (1985).

Hussin, A.A.M., Hartley, P., Sturgess, C.E.N. and Rowe, G.W. Non-linear finite-element simulation of metalforming processes on a 16 bit microcomputer. *Proc. 2nd Int. Conf. on Microcomputers in Engineering: Development and Application of Software*, ed. B.A. Schrefler and R.W. Lewis, Pineridge Press, pp. 517–28 (1986).

Hussin, A.A.M., Hartley, P., Sturgess, C.E.N. and Rowe, G.W. Non-Linear finite-element analysis on microcomputers for metal forging. *J. Strain Analysis* **21**, 197–203 (1986).

Hussin, A.A.M., Hartley, P., Sturgess, C.E.N. and Rowe, G.W. Simulation of cold industrial forming processes. Special issue of *J. Comm. Appl. Num. Meth.* **3**, 415–26 (1987).

Hussin, A.A.M., Hartley, P., Sturgess, C.E.N. and Rowe, G.W. Elastic-plastic finite-element modelling of cold backward extrusion using a microcomputer based system. *J. Mech. Wkg. Techn.* **16**, 7–20 (1988).

Rowe, G.W. Finite element analysis in metalforming. *III Jugoslav, Simpozijum o Metalurgiji*, Beograd, pp. 53–65 (1984).

Finite-element techniques

Friction and boundary conditions

Baaijens, F.P.T., Veldpaus, F.E. and Brekelmans, W.A.M. On the numerical simulation of contact problems in forming processes. *Proc. 2nd Int. Conf. on Numerical Methods in Industrial Forming Processes*, ed. K. Mattiasson, A. Samuelsson, R.D. Wood and O.C. Zienkiewicz, Balkema Press, pp. 85–90 (1986).

Charlier, R. and Habraken, A.M. On the modelling of three dimensional contact with friction problem in context of large displacement problems. Supplement to *Proc. 2nd Int. Conf. on Numerical Methods in Industrial Forming Processes*, ed. K. Mattiasson, A. Samuelsson, R.D. Wood and O.C. Zienkiewicz, Balkema Press, (1986).

Charlier, R., Godinas, A. and Cescotto, S. On the modelling of contact problems with friction by the finite element method. *Proc. 8th Conf. on Structural Mechanics in Reactor Technology, Brussels*, (1985).

Chen, C.C. and Kobayashi, S. Rigid plastic finite element analysis of ring compression. *Applications of Numerical Methods to Forming Processes, Winter Annual Meeting of ASME*, AMD 28, ed. H. Armen and R.F. Jones Jr., ASME, pp. 163–74 (1978).

Gunasekera, J.S. and Mahadeva, S. An alternative method to predict the effects of friction in metalforming. *Friction and Material Characterizations, Winter Annual Meeting of ASME*, MD 10, ed. I. Haque; J.E. Jackson Jr., A.A. Tseng and J.L. Rose, ASME, pp. 55–62 (1988).

Haber, R.B. and Hariandia, B.H. An Eulerian–Lagrangian finite element approach to large-deformation frictional contact. *Comput. Struct.* **20**, 193–201 (1985).

Hallquist, J.O., Goudreau, G.L. and Benson, D.J. Sliding interfaces with contact-impact in large-scale Lagrangian computations. *Comp. Meth. Appl. Mech. Eng.* **51**, 107–37 (1985).

Hartley, P., Pillinger, I. and Sturgess, C.E.N. Modelling frictional effects in finite-element simulation of metal forming. *Friction and Material Characterizations, Winter Annual Meeting of ASME*, MD 10, ed. I. Haque, J.E. Jackson Jr., A.A. Tseng and J.L. Rose, ASME, pp. 91–6 (1988).

Hartley, P., Sturgess, C.E.N. and Rowe, G.W. An examination of frictional boundary conditions and their effect in an elastic–plastic finite element solution. *Proc. 20th Int. Machine Tool Des. Res. Conf.*, ed. S.A. Tobias, Macmillan, pp. 157–63 (1979).

Hartley, P., Sturgess, C.E.N. and Rowe, G.W. Friction in finite-element analyses of metalforming processes. *Int. J. Mech. Sci.* **21**, 301–11 (1979).

Hartley, P., Sturgess, C.E.N. and Rowe, G.W. A prediction of the influence of friction in the ring test by the finite element method. *Proc. 7th Nth. American Metalworking Res. Conf.*, SME, pp. 151–8 (1979).

Huetink, J., Van der Lugt, J. and Miedama, J.R. A mixed Eulerian–Lagrangian contact element to describe boundary and interface behaviour in forming processes. *Proc. Int. Conf. on Numerical Methods in Engineering, Theory and Applications (NUMETA)*, Martinus-Nijhof (1987).

Hwang, S.M. and Kobayashi, S. A note on evaluation of interface friction in ring tests. *Proc. 11th Nth. American Metalworking Res. Conf.*, SME, pp. 193–6 (1983).

Jackson Jr., J.E. Haque, I., Gangjee, T. and Ramesh, M. Some numerical aspects of frictional modeling in material forming processes. *Friction and Material Characterizations, Winter Annual Meeting of ASME*, MD 10, ed. I. Haque, J.E. Jackson Jr., A.A. Tseng and J.L. Rose, ASME, pp 39–46 (1988).

Kikuchi, N. and Cheng, J-H. Finite element analysis of large deformation problems including unilateral contact and friction. *Comp. Meth. for Non-linear Solids and Structural Mechanics*, AMD **54**, ed. S.N. Alturi, ASME pp. 121–32 (1983).

Kobayashi, S. Thermoviscoplastic analysis of metal forming problems by the finite element method. *Proc. 1st Int. Conf. on Numerical Methods in Industrial Forming Processes*, ed. J.F.T. Pittman, R.D. Wood, J.M. Alexander and O.C. Zienkiewicz, Pineridge Press, pp. 17–25 (1982).

Kopp, R. and Cho, M.L. Influence of the boundary conditions on results of the finite-element simulation. *Proc. 2nd Int. Conf. on Technology of Plasticity*, ed. K. Lange, Springer, pp. 43–50 (1987).

Mahrenholtz, O. Different finite element approaches to large plastic deformations. *Comp. Meth. Appl. Mech. Eng.* **33**, 453–68 (1982).

Makinouchi, A., Ike, H., Murakawa, M., Noga, N. and Ciupik, L.F. Finite element analysis of flattening of surface asperities by rigid dies in metal working processes. *Proc. 2nd Int. Conf. on Technology of Plasticity*, ed. K. Lange, Springer, pp. 59–66 (1987).

Matsumoto, H., Oh, S.I. and Kobayashi, S. A note on the matrix method for rigid-plastic finite element analysis of ring compression. *Proc. 18th Int. Machine Tool Des. Res. Conf.*, ed. J.M. Alexander, Macmillan, pp. 3–9 (1977).

Mori, K., Osakada, K., Nakadoi, K. and Fukuda, M. Coupled analysis of steady state forming process with elastic tools. *Proc. 2nd Int. Conf. on Numerical Methods in Industrial Forming Processes*, ed. K. Mattiasson, A. Samuelsson, R.D. Wood and O.C. Zienkiewicz, Balkema Press, pp. 237–42 (1986).

Oh, S.I. Finite element analysis of metal forming processes with arbitrarily shaped dies. *Int. J. Mech. Sci.* **24**, no. 8, 479–93 (1982).

Oh, S.I., Rebelo, N. and Kobayashi, S. Finite element formulation for the analysis of plastic deformation of rate sensitive materials in metal forming. *Metal Forming Plasticity*, ed. H. Lippmann, Springer, pp. 273–91 (1979).

Pillinger, I., Hartley, P. and Sturgess, C.E.N. Modelling of frictional tool surfaces in finite element metalforming analyses. *Modelling of Metal Forming Processes, Proc. Euromech 233 Colloquium*, ed. J.L. Chenot and E. Onate, Kluwer, pp. 85–92 (1988).

Price, J.W.H. and Alexander, J.M. The finite-element analysis of two high-temperature metal deformation processes. *Proc. 2nd Int. Symp. on Finite Elements in Flow Problems*, Santa Margherita Ligure, Italy, pp. 715–20 (1976).

Price, J.W.H. and Alexander, J.M. Specimen geometries predicted by computer model of high deformation forging. *Int. J. Mech. Sci.* **21**, 417–30 (1979).

Tatenami, T. FEM for rigid plastic material using dependent nodal point displacement with consideration of frictional force. *Proc. 1st Int. Conf. on Technology of Plasticity*, ed. H. Kudo, JSTP/JSPE, pp. 1015–20 (1984).

Webster, W.D. and Davis R.L. Development of a friction element for metal forming analysis. *J. Eng. Ind.* **104**, 253–6 (1982).

Yijian, H. and Guoji, L. Friction problems involved in the rigid-plastic finite-element analysis of metal forming processes. *Proc. 3rd National Conf. on Theory of Plasticity and Technology*, Guilin, China, (1984).

Re-meshing

Al-Sened, A.A.K., Hartley, P., Sturgess, C.E.N. and Rowe, G.W. Forming sequences in axi-symmetric cold forging. *Proc. 12th Nth. American Manufacturing Res. Conf.*, SME, pp. 151–8 (1984).

Badawy, A., Oh, S.I. and Altan, T. A remeshing technique for the FEM simulation of metalforming processes. *Comp. in Eng.* 143–6 (1983).

Badawy, A., Oh, S.I. and Malas, J. FEM simulation of closed die forging of titanium disk using ALPID. *Proc. Int. Conf. Computer Engr.*, ASME (1983).

Cescutti, J.P. and Chenot, J.L. A geometrical continuous procedure for application to finite element calculation of non-steady state forming processes. *Proc. Int. Conf. on Numerical Methods in Engineering, Theory and Applications (NUMETA)*, Martinus-Nijhof, (1987).

Cescutti, J.P., Soyris, N., Surdon, G. and Chenot, J.L. Thermo-mechanical finite element calculation of three dimensional hot forging with re-meshing. *Proc. 2nd Int. Conf. on Technology of Plasticity*, ed. K. Lange, Springer, pp. 1051–8 (1987).

Cesuctti, J.P., Wey, E., Chenot, J.L. and Mosser, P.E. Finite element calculation of hot forging with continuous remeshing. *Modelling of Metal Forming Processes, Proc. Euromech 233 Colloquium*, ed. J.L. Chenot and E. Onate, Kluwer, pp. 207–16 (1988).

Cheng, J.H. Automatic adaptive remeshing for finite element simulation of forming processes. *Int. J. Num. Meth. Eng.* **26**, 1–18 (1988).

Cheng, J.H. and Kikuchi, N. A mesh rezoning technique for finite element simulations of metal forming process. *Int. J. Num. Meth. Eng.* **23**, 219–28 (1986).

Ficke, J.A., Oh, S.I. and Malas, J. FEM simulation of closed die forging of isothermal titanium disk forging using ALPID. *Proc. 12th Nth. American Manufacturing Res. Conf,.* SME, pp. 166–172 (1984).

Gelten, C.J.M. and Konter, A.W.A. Application of mesh-rezoning in the updated Lagrangian method to metal forming analyses. *Proc. 1st Int. Conf. on Numerical Methods in Industrial Forming Processes*, ed. J.F.T. Pittman, R.D. Wood, J.M. Alexander and O.C. Zienkiewicz, Pineridge Press, pp. 511–21 (1982).

Gelten, C.J.M. and de Jong, J.E. A method to redefine a finite element mesh and its application to metal forming and crack growth analysis. *Proc. of the Int. FEM-Congress at Baden-Baden*, ed. IKOSS GmbH, W. Germany (1981).

Huang, G.C., Liu, Y.C. and Zienkiewicz, O.C. Error control, mesh updating schemes and automatic adaptive remeshing for finite element analysis of unsteady extrusion processes. *Modelling of Metal Forming Processes, Proc. Euromech 233 Collolquium*, ed. J.L. Chenot and E. Onate, Kluwer, pp. 75–84 (1988).

Kikuchi, N. Adaptive grid design methods for finite element analysis. *Comp. Meth. in Appl. Mech. and Eng.* **55**, 129–60 (1986).

Mori, K., Osakada, K. and Fukuda, M. Simulation of severe plastic deformation by finite element method with spatially fixed elements. *Int. J. Mech. Sci.* **26**, 515–25 (1984).

Oh, S.I., Tang, J.P. and Badawy, A. Finite element mesh rezoning and its applications to metal forming analysis. *Proc. 1st Int. Conf. on Technology of Plasticity*, ed. H. Kudo, JSTP/JSPE, pp. 1051–8 (1984).

Roll, K. Possibilities for the use of the finite element method for the analysis of bulk metal forming processes. *Annals of CIRP* **31**, 145–54 (1982).

Roll, K. and Neitzert, Th. On the application of different numerical methods to calculate cold forming processes. *Proc. 1st Int. Conf. on Numerical Methods in Industrial Forming Processes*, ed. J.F.T. Pittman, R.D. Wood, J.M. Alexander and O.C. Zienkiewicz, Pineridge Press, pp. 97–107 (1982).

Zienkiewicz, O.C. and Godbole, P.N. A penalty function approach to problems of plastic flow of metals with large surface deformations. *J. Strain Analysis* **10**, 180–5 (1975).

Zienkiewicz, O.C, Toyoshima, S., Liu, Y.C. and Zhu, J.Z. Flow formulation for numerical solution of forming processes II – some new directions. *Proc. 2nd Int. Conf. on Numerical Methods in Industrial Forming Processes*, ed. K. Mattiasson, A. Samuelsson, R.D. Wood and O.C. Zienkiewicz, Balkema Press, pp. 3–10 (1986).

General finite-element techniques

Abo-Elkhier, M., Oravas, G. AE. and Dokainish, M.A. A consistent Eulerian formulation for large deformation with reference to metal-extrusion process. *Int. J. Non-linear Mech.* **23**, 37–52 (1988).

Abouaf, M., Braudel, H.J. and Chenot, J.L. An implicit and incrementally objective finite element formulation for isotropic generalized standard materials at finite strain. Application to metal forming. *Numerical Methods for Non-Linear Problems*, Vol.3, ed. C.Taylor *et al.*, Pineridge Press, (1986).

Anand, S.C., Weisgerber, F.E. and Shaw, H. Direct solution vs quadratic programming technique in elastic–plastic finite element analysis. *Comput. Struct.* **7**, 221–8 (1977).

Argyris, J.H., Dunne, P.C., Angelopoulos, Th. and Bichat, B. Large natural strains and some special difficulties due to non-linearity and incompressibility in finite elements. *Comp. Meth. Appl. Mech. Eng.* **4**, 219–78 (1974).

Argyris, J.H., Dunne, P.C., Haase, M. and Orkisz, J. Higher-order simpler elements for large strain analysis – natural approach. *Comp. Meth. Appl. Mech. Eng.* **16**, 369–403 (1978).

Argyris, J.H., Haase, M. and Mlejnek, H.P. Some considerations on the natural approach. *Comp. Meth. Appl. Mech.* **30**, 335–46 (1982).

Barnard, A.J. and Sharman, P.W. The elasto-plastic analysis of plates using hybrid finite elements. *Int. J. Num. Meth. Eng.* **10**, 1343–56 (1976).

Bathe, K.J., Ramm, E. and Wilson, E.L. Finite element formulations for large deformation dynamic analysis. *Int. J. Num. Meth. Eng.* **9**, 353–86 (1975).

Besdo, D. Numerical applications of a strain-space represented plasticity. *Proc. 2nd Int. Conf. on Technology of Plasticity*, ed. K. Lange, Springer, pp. 51–7 (1987).

Braudel, H.J., Abouaf, M. and Chenot, J.L. An implicit and incremental formulation for the solution of elastoplastic problems by the FEM. *Comput. Struct.* **22**, 801–14 (1986).

Cannizzaro, L., Lo Valvo, E., Micari, F. and Riccobono, R. H and P mesh refinement in the metal forming FEM analysis. *Modelling of Metal Forming Processes, Proc. Euromech 233 Colloquium*, ed. J.L. Chenot and E. Onate, Kluwer, pp. 49–56 (1988).

Cormeau, I. Numerical stability in quasi-static elasto/visco-plasticity. *Int. J. Num. Meth. Eng.* **9**, 104–27 (1975).

Dafalias, Y.F. The plastic spin concept and a simple illustration of its role in finite transformations. *Mech. of Mater.* **3**, 223 (1984).

Derbalian, K.A., Lee, E.H., Mallett, R.L. and McMeeking, R.M. Finite element metal forming analysis with spacially fixed mesh. *Applications of Numerical Methods to Forming Processes, Winter Annual Meeting of ASME*, AMD 28, ed. H. Armen, R.F. Jones Jr., ASME, pp. 39–47 (1978).

El-Magd, E. and Ahmed, A-N. Influence of manufacturing nonaccuracies on the mechanical behaviour of the product. *Proc. 1st Int. Conf. on Numerical Methods in Industrial Forming Processes*, ed. J.F.T. Pittman, R.D. Wood, J.M. Alexander and O.C. Zienkiewicz, Pineridge Press, pp. 363–72 (1982).

Ergatoudis, I., Irons, B.M. and Zienkiewicz, O.C. Curved, isoparametric, 'quadrilateral' elements for finite element analysis. *Int. J. Solids Struct.* **4**, 31–42 (1968).

Frey, F. and Cescotto, S. Some new aspects of the incremental total Lagrangian description in nonlinear analysis. *Proc. Int. Conf. on Finite Elements in Nonlinear Solid and Structural Mechanics*, Geilo, Norway, pp. 322–43 (1977).

Gadala, M.S., Oravas, G.AE. and Dokainish, M.A. A consistent Eulerian formulation of large deformation problems in statics and dynamics. *Int. J. Non-linear Mech.* **18**, 21–35 (1983).

Gelin, J.C. and Picart, P. Some computational aspects on the integration of rate constitutive equations – application to bulk metal forming processes. *Proc. 2nd Int. Conf. on Numerical Methods in Industrial Forming Processes*, ed. K. Mattiasson, A. Samuelsson, R.D. Wood and O.C. Zienkiewicz, Balkema Press, pp. 103–10 (1986).

Gupta, A.K. and Mohraz, B.M. A method of computing numerically integrated stiffness matrices. *Int. J. Num. Meth. Eng.* **5**, 83–9 (1972).

Habraken, A.M. and Charlier, R. A three-dimensional finite element for simulation of metal forming processes. *Proc. 2nd Int. Conf. on Numerical Methods in Industrial Forming Processes*, ed. K. Mattiasson, A. Samuelsson, R.D. Wood and O.C. Zienkiewicz, Balkema Press, pp. 111–16 (1986).

Hinton, E. and Campbell, J.S. Local and global smoothing of discontinuous finite element functions using a least squares method. *Int. J. Num. Meth. Eng.* **8**, pp. 461–80 (1974).

Hughes, T.J.R. Numerical implementation of constitutive models: rate-independent deviatoric plasticity. *Theoretical Foundation for Large-Scale*

Computations for Nonlinear Material Behaviour, ed. S. Nemat-Nasser, Martinus Nijhoff, p. 29 (1984).

Irons, B.M. Engineering applications of numerical integration in stiffness methods. *AIAA J.* **4**, no.11, 2035–7 (1966).

Johnson, C. A mixed finite element method for plasticity problems with hardening. *SIAM J. Num. Anal.* **14**, 575–84 (1977).

Lee, E.H., Mallett, R.L. and Wertheimer, T.B. Stress analysis for anisotropic hardening in finite-deformation plasticity. *J. Appl. Mech.* **50**, 554–60 (1983).

Loret, B. On the effects of plastic rotation in the finite deformation of anisotropic elastoplastic materials. *Mech. Mater.* **2**, 287–304 (1983).

Marcal, P.V. Finite element analysis of combined problems of non-linear material and geometric behaviour. *Proc. Joint Computer Conf. on Comp. Approach to Applied Mechanics*, ASME, (1969).

Nayak, G.C. and Zienkiewicz, O.C. Note on the 'alpha'-constant stiffness method for the analysis of non-linear problems. *Int. J. Num. Meth. Eng.* **4**, 579–82 (1972).

Needleman, A. Finite elements for finite strain plasticity problems. *Plasticity of Metals at Finite Strain: Theory, Computation and Experiment*, ed. E.H. Lee and R.L. Mallett, Stanford University, 387–436 (1982).

Oden, J.T. A general theory of finite elements – I. topological considerations. *Int. J. Num. Meth. Eng.* **1**, 205–21 (1969).

Oden, J.T. A general theory of finite elements – II. applications. *Int. J. Num. Meth. Eng.* **1**, 247–59 (1969).

Oden, J.T. and Key, J.E. On some generalizations of the incremental stiffness relations for finite deformations of compressible and incompressible finite elements. *Nuc. Eng. Des.* **15**, 121–34 (1971).

Ortiz, M. and Popov, E.P. Accuracy and stability of integration algorithms for elastoplastic constitutive relations. *Int. J. Num. Meth. Eng.* **21**, 1561–76 (1985).

Polat, M.U. and Dokainish, M.A. An automatic subincrementation scheme for accurate integration of elasto-plastic constitutive relations. *Comput. Struct.* **31**, 339–47 (1989).

Ponthot, J.P. A method to reduce cost of mesh deformation in Eulerian–Lagrangian formulation. *Modelling of Metal Forming Processes, Proc. Euromech 233 Colloquium*, ed. J.L. Chenot and E. Onate, Kluwer, pp. 65–74 (1988).

Rebelo, N. A comparative study of algorithms applied in finite element analysis of metal forming problems. *Proc. Symp. Manufacturing Simulation and Processes*, ASME, p. 17 (1986).

Rice, J.R., McMeeking, R.M., Parks, D.M. and Sorensen, E.P. Recent finite element studies in plasticity and fracture mechanics. *Comp. Meth. Appl. Mech. Eng.* **17/18**, 411–42 (1979).

Runesson, K., Bernspang, L. and Mattiasson, K. Implicit integration in plasticity. *Numerical Methods for Non-linear Problems*, Vol. 3, ed. C. Taylor *et al.*, Pineridge Press, (1986).

Son, N.Q. On the elastic plastic initial-boundary value problem and its numerical integration. *Int. J. Num. Meth. Eng.* **11**, 817–32 (1977).

Thompson, E.G. Average and complete incompressibility in the finite element method. *Int. J. Num. Meth. Eng.* **9**, 925–32 (1975).

Webster, W.D. An isoparametric finite element with nodal derivatives. *Trans ASME* **48**, 64–8 (1981).

Wifi, A.S. Finite element correction matrices in metal forming analysis with application to hydrostatic bulging of a circular sheet. *Int. J. Mech. Sci* **24**, 393–406 (1982).

Williams, J.R. A method for three-dimensional dynamic contact analysis of numerous deformable discrete bodies including automatic fracturing. *Proc. Int. Conf. on Computational Plasticity*, ed. D.R.J. Owen, E. Hinton and E. Onate, Pineridge Press, pp. 1177–202 (1987).

Yamada, Y. Nonlinear matrices, their implications and applications in inelastic large deformation analysis. *Comp. Meth. Appl. Mech. Eng.* **23**, 417–37 (1982).

Yoder, P.J. and Whirley, R.G. On the numerical implementation of elastoplastic models. *J. Appl. Mech.* **51**, 283–8 (1984).

Zienkiewicz, O.C. and Cormeau, I. Visco-plasticity – plasticity and creep in elastic solids – a unified numerical solution approach. *Int. J. Num. Meth. Eng.* **8**, 821–45 (1974).

Zienkiewicz, O.C. and Hinton, E. Reduced integration, function smoothing and non-conformity in finite element analysis (with special reference to thick plates). *J. Franklin Inst.* **302**, 443–60 (1976).

Zienkiewicz, O.C., Li, X-K. and Nakazawa, S. Iterative solution of mixed problems and stress recovery procedures. *Comm. in Appl. Num. Anal.* **1**, 3–9 (1985).

Zienkiewicz, O.C., Liu, Y.C. and Huang, G.C. Error estimation and adaptivity in flow formulation for forming problems. *Int. J. Num. Meth. Eng.* **25**, *Special Issue on Numerical Methods in Industrial Forming Processes*, 23–42 (1988).

Material Behaviour

Fracture and defect prediction

Ayada, M., Higashino, T. and Mori, K. Central bursting in extrusion of inhomogeneous materials. *Proc. 2nd Int. Conf. on Technology of Plasticity*, ed. K. Lange, Springer, pp. 553–8 (1987).

Branswyck, O., David, C., Levaillant, C., Chenot, J.L., Billard, J.P., Weber, D. and Guerlet, J.P. Surface defects in hot rolling of copper based heavy ingots: three dimensional simulation including failure criterion. *Proc. Int. Conf. on Computational Methods for Predicting Material Processing Defects*, ed. M. Predeleanu, Elsevier, pp. 29–38 (1987).

Clift, S.E., Hartley, P., Sturgess, C.E.N. and Rowe, G.W. Fracture initiation in plane-strain forging. *Proc. 25th Int. Machine Tool Des. Res. Conf.*, ed. S.A. Tobias, Macmillan, pp. 413–19 (1985).

Clift, S.E., Hartley, P., Sturgess, C.E.N. and Rowe, G.W. Further studies on fracture initiation in plane strain forging. *Proc. Applied Mechanics-2 Conf.*, ed. A.S. Tooth and J. Spence, Elsevier, pp. 89–100 (1988).

Dung, N.L. Fracture initiation in upsetting tests. *Proc. 2nd Int. Conf. on Numerical Methods in Industrial Forming Processes*, ed. K. Mattiasson, A. Samuelsson, R.D. Wood and O.C. Zienkiewicz, Balkema Press, pp. 261–70 (1986).

Dung, N.L. Indentification of defect locations in metal forming using a personal computer oriented finite element method. *Modelling of Metal Forming Processes, Proc. Euromech 233 Colloquium*, ed. J.L. Chenot and E. Onate, Kluwer, pp. 245–52 (1988).

Dung, N.L. and Mahrenholtz, O. Numerical method for plastic working processes and die design. *Proc. 1st Int. Conf. on Technology of Plasticity*, ed. H. Kudo, JSTP/JSPE, pp. 574–9 (1984).

Dung, N.L. and Mahrenholtz, O. A criterion for ductile fracture in cold forging. *Proc. 2nd Int. Conf. on Technology of Plasticity*, ed. K. Lange, Springer, pp. 1013–20 (1987).

Gelin, J.C. Application of implicit methods for the analysis of damage with temperature effects in large strain plasticity problems. *Numerical Methods for Non-Linear Problems*, Vol.3, ed. C. Taylor *et al.*, Pineridge Press, pp. 494–507 (1986).

Gelin, J.C. Numerical analyses of strain rate and temperature effects on localization of plastic flow and ductile fracture – applications to metal forming processes. *Proc. Int. Conf. on Computational Methods for Predicting Material Processing Defects*, ed. M. Predeleanu, Elsevier, pp. 123–32 (1987).

Gelin, J.C., Oudin, J. and Ravalard, Y. An improved finite element method for the analysis of damage and ductile fracture in cold forming processes. *Annals of CIRP* **34**, 209–13 (1985).

Hartley, P., Clift, S.E., Salimi-Namin, J., Sturgess, C.E.N. and Pillinger, I. The prediction of ductile fracture initiation in metalforming using a finite-element model and various fracture criteria. *Res Mechanica* **28**, 269–93 (1989).

Kawka, M. Strain localization problems in the plane strain tension test. *Proc. 2nd Int. Conf. on Numerical Methods in Industrial Forming Processes*, ed. K. Mattiasson, A. Samuelsson, R.D. Wood and O.C. Zienkiewicz, Balkema Press, pp. 125–30 (1986).

Kobayashi, S. and Lee, C.H. Deformation mechanics and workability in upsetting solid circular cylinders. *Proc. 1st Nth. American Metalworking Res. Conf.*, SME, pp. 185–204 (1973).

Mathur, K.K. and Dawson, P.R. Damage evolution modeling in bulk forming processes. *Proc. Int. Conf. on Computational Methods for Predicting Material Processing Defects*, ed. M. Predeleanu, Elsevier, pp. 251–62 (1987).

Mathur, K.K. and Dawson, P.R. On modeling damage evolution during the drawing of metals. *Mech. Mater.* **6**, 179–96 (1987).

Needleman, A. and Tvergaard, V. Finite element analysis of localization in plasticity. *Finite elements – Special Problems in Solid Mechanics 5*, ed. J.T. Oden and G.F. Carey, Prentice-Hall, pp. 95–157 (1984).

Oh, S.I. and Kobayashi, S. Workability of aluminium alloy 7075-T6 in upsetting and rolling. *J. Eng. Ind.* **98**, 800–6 (1976).

Osakada, K. and Mori, K. Prediction of ductile fracture in cold forging. *Annals of CIRP* **27**, 135–9 (1978).

Predeleanu, M. Finite strain plasticity analysis of damage effects in metal forming processes. *Proc. Int. Conf. on Computational Methods for Predicting Material Processing Defects*, ed. M. Predeleanu, Elsevier, pp. 295–308 (1987).

Predeleanu, M., Cordebois, J.P. and Belkhiri, L. Failure analysis of cold upsetting by computer and experimental simulation. *Proc. 2nd Int. Conf. on Numerical Methods in Industrial Forming Processes*, ed. K. Mattiasson, A. Samuelsson, R.D. Wood and O.C. Zienkiewicz, Balkema Press, pp. 277–282 (1986).

Forming of new materials (composites, superalloys)

Abouaf, M., Chenot, J.L., Raisson, G. and Bauduin, P. Finite element simulation of hot isostatic pressing of metal powders. *Int. J. Num. Meth. Eng.* **25**,

Special Issue on Numerical Methods in Industrial Forming Processes, 191–212 (1988).

Argyris, J.H. and Doltsinis, J.St., A primer on superplasticity in natural formulation. *Comp. Meth. Appl. Mech. Eng.* **46**, 83–131 (1984).

Fukui, Y., Yonezawa, J., Yamaguchi, Y., Mizuta, A. and Tsuda, O. Analysis of forging effect and closing of internal cavities in free forging by rigid-plastic finite element method. *J. Jap. Soc. for Techn. of Plasticity* **21**, 975 (1980).

Hartley, C.S., Dehghani, M., Iyer, N., Male, A.T. and Lovic, W.R. Defects at interfaces in coextruded metals. *Proc. Int. Conf. on Computational Methods for Predicting Material Processing Defects*, ed. M. Predeleanu, Elsevier, pp. 357–66 (1987).

Im, Y.T. and Kobayashi, S. Finite element analysis of plastic deformation of porous materials. *Metal Forming and Impact Mechanics*, ed. S.R. Reid, Pergamon, pp. 103–22 (1985).

Im, Y.T. and Kobayashi, S. Coupled thermo-viscoplastic finite element analysis of plane-strain compression of porous materials. *Advanced Manuf. Processes* **1**, 269 (1986).

Im, Y.T. and Kobayashi, S. Analysis of axisymmetric forging of porous materials by the finite element method. *Advanced Manuf. Processes* **1**, 473–99 (1986).

Im, Y.T. and Kobayashi, S. Finite element applications in forming billet and P/M preforms. *Modelling of Metal Forming Processes, Proc. Euromech 233 Colloquium*, ed. J.L. Chenot and E. Onate, Kluwer, pp. 217–26 (1988).

Kumar, A., Jain, P.C. and Mehta, M.L. Modified formulation for extrusion of porous powder preforms using finite element method. *Proc. 1st Int. Conf. on Numerical Methods in Industrial Forming Processes*, ed. J.F.T. Pittman, R.D. Wood, J.M. Alexander and O.C. Zienkiewicz, Pineridge Press, pp. 257–66 (1982).

Kumar, A., Jha, S., Jain, P.C. and Mehta, M.L. An investigation into product defects in the hot extrusion of aluminium powder preforms. *J. Mech. Wkg. Techn.* **11**, 275–90 (1985).

Levaillant, C., Querbes, J.L., Laugee, C. and Amouroux, P. Modelling of aluminium alloy powder hot forging process in view of computer aided design (CAD) of preform. *Proc. 2nd Int. Conf. on Technology of Plasticity*, ed. K. Lange, Springer, pp. 953–60 (1987).

Mori, K., Shima, S. and Osakada, K. Analysis of free forging by rigid-plastic finite-element method based on the plasticity equation for porous materials. *Bull. JSME* **23**, 523 (1980).

Mori, K., Shima, S. and Osakada, K. Finite element method for the analysis of plastic deformation of porous metals. *Bull. JSME* **23**, no. 178, 516–22 (1980).

Mori, K., Oda, T., Shima, S. and Osakada, K. Analysis of supressing of internal holes in rolling by the rigid-plastic finite element method. *Proc. 32nd Japanese Joint Conf. for Technology of Plasticity*, Osaka, p. 49 (1981).

Oh, S.I. and Gegel, H.L. Modeling of P/M forming by the finite element method. *Proc. 14th Nth. American Manufacturing Res. Conf.*, SME, (1986).

Oh, S.I., Wu, W.T. and Park, J.J. Application of the finite element method to P/M forming processes. *Proc. 2nd Int. Conf. on Technology of Plasticity*, ed. K. Lange, Springer, pp. 961–8 (1987).

Pacheco, L.A. and Alexander, J.M. On the hydrostatic extrusion of copper-covered aluminium rods. *Proc. 1st Int. Conf. on Numerical Methods in Industrial Forming Processes*, ed. J.F.T. Pittman, R.D. Wood, J.M. Alexander and O.C. Zienkiewicz, Pineridge Press, pp. 205–16 (1982).

Rowe, G.W. Problems involved in the numerical analysis of plastic working processes for new materials. *Proc. 1st Int. Conf. on Technology of Plasticity*, ed. H. Kudo, JSTP/JSPE, pp. 27–38 (1984).

Shima, S., Inamoto, J., Osakada, K. and Narutaki, R. The finite element analysis of elastic–plastic deformation of porous metals. *J. Jap. Soc. for Techn. of Plasticity* **16**, 660–7 (1975).

Tanaka, M., Ono, S., Tsuneno, M. and Iwadate, T. An analysis of void crushing during flat die free forging. *Proc. 2nd Int. Conf. Technology of Plasticity*, ed. K. Lange, Springer, pp. 1035–42 (1987).

Tayal, A.K. and Natarajan, R. A finite element analysis of axisymmetric extrusion of composite rods. *Int. J. Machine Tool Des. Res.* **21**, no 3/4, 227–35 (1981).

Thermo-mechanical analyses

Alexander, J.M. and Price, J.W.H. Finite element analysis of hot metal forming. *Proc. 18th Int. Machine Tool Des. Res. Conf.*, ed. J.M. Alexander, Macmillan, pp. 267–74 (1977).

Altan, T. Process simulation of hot die forging processes. *Proc. 2nd Int. Conf. on Technology of Plasticity*, ed. K. Lange, Springer, pp. 1021–34 (1987).

Argyris, J.H. and Doltsinis, J.St. On the natural formulation and analysis of large deformation coupled thermomechanical problems. *Comp. Meth. Appl. Mech. Eng.* **25**, 195–253 (1981).

Argyris, J.H., Doltsinis, J.St., Pimenta, P.M. and Wustenburg, H. Thermo-mechanical response of solids at high strains – natural approach. *Comp. Meth. Appl. Mech. Eng.* **32**, 3–57 (1982).

Beynon, J.H., Ponter, A.R.S. and Sellars, C.M. Metallographic verification of computer modelling of hot rolling. *Modelling of Metal Forming Processes, Proc. Euromech 233 Colloquium*, ed. J.L. Chenot and E. Onate, Kluwer, pp. 321–8 (1988).

Cescotto, S. and Charlier, R. Numerical simulation of elasto-visco-plastic large strains of metals at high temperature. *Proc. 8th Conf. on Structural Mechanics in Reactor Technology, Brussels, (1985)*.

Cescutti, J.P., Soyris, N., Surdon, G. and Chenot, J.L. Thermo-mechanical finite element calculation of three dimensional hot forging with re-meshing. *Proc. 2nd Int. Conf. on Technology of Plasticity*, ed. K. Lange, Springer, pp. 1051–8 (1987).

Chandra, A. and Mukherjee, S. A finite-element analysis of metal forming problems with thermomechanical coupling. *Int. J. Mech. Sci.* **21**, 661–76 (1984).

Cornfield, G.C. and Johnson, R.H. Theoretical predictions of plastic flow in hot rolling including the effect of various temperature distributions. *J. Iron and Steel Inst.* **211**, 567–73 (1973).

Dawson, P.R. and Thompson, E.G. Steady state thermomechanical finite element analysis of elasto-viscoplasticity metal forming processes. *Numerical Modelling of Manufacturing Processes*, ed. R.F. Jones Jr., H. Armen and J.T. Fong, ASME, pp. 167–82 (1977).

Doltsinis, J.St. Physical and numerical aspects in large deformation coupled thermomechanical problems. *18th Israel Conf. on Mechanical Engineering*, ed. P. Greenberg, Weizmann Science Press, (1984).

Dung, N.L. and Mahrenholtz, O. Numerical method for plastic working processes and die design. *Proc. 1st Int. Conf. on Technology of Plasticity*, ed. H. Kudo, JSTP/JSPE, pp. 574–79 (1984).

Hartley, P., Sturgess, C.E.N. and Rowe, G.W. Finite-element predictions of the influence of strain-rate and temperature variations on the properties of forged products. *Proc. 8th Nth. American Metalworking Res. Conf.*, SME, 121–8 (1980).

Huetink, J., Van Der Lugt, J. and Vreede, P.T. A mixed Eulerian–Lagrangian FEM for simulation of thermo-mechanical forming processes. *Modelling of Metal Forming Processes, Proc. Euromech 233 Colloquium*, ed. J.L. Chenot and E. Onate, Kluwer, pp. 57–64 (1988).

Kato, T., Masashi, A. and Tozawa, Y. Thermal analysis of cold upsetting. *J. Jap. Soc. for Techn. of Plasticity* **28**, 791–8 (1987).

Kato, T., Tanaka, T. and Tozawa, Y. Thermal analysis of backward cold extrusion. *J. Jap. Soc. for Techn. of Plasticity* **28**, 1239–44 (1987).

Kato, T., Tozawa, J. and Shinagawa, K. Thermal analysis for backward cold extrusion of steel taking blue brittleness into consideration. *Proc. 2nd Int. Conf. on Technology of Plasticity*, ed K. Lange, Springer, pp. 559–64 (1987).

Kobayashi, S. Thermo-viscoplastic analysis of metal forming problems by the finite element method. *Numerical Analysis of Forming Processes*, ed. J.F.T. Pittman, O.C. Zienkiewicz, R.D. Wood and J.M. Alexander, Wiley, pp. 45–69 (1984).

Mahrenholtz, O., Westerling, C. and Dung, N.L. Thermomechanical analysis of metalforming processes through the combined approach FEM/FDM. *Proc. 1st Int. Workshop on Simulation of Metal Forming Processes by the Finite Element Method (SIMOP–I)*, ed. K. Lange, Springer, pp. 19–49 (1986).

Mahrenholtz, O., Westerling, C., Klie, W. and Dung, N.L. Finite element approach to large plastic deformation at elevated temperatures. *Constitutive Equations: Macro and Computational Aspects, Winter Annual Meeting of ASME*, ed. K.J. William, ASME, pp. 165–78 (1984).

Marten, J., Westerling, C., Dung, N.L. and Mahrenholtz, O. A coupled analysis of plastic deformation and heat transfer. *Proc. 25th Int. Machine Tool Des. Res. Conf.*, ed. S.A. Tobias, Macmillan, pp. 397–403 (1985).

Oden, J.T., Bhanari, D.R., Yagewa, G. and Chung, T.J. A new approach to the finite element formulation and solution of a class of problems in coupled thermo-elastoviscoplasticity of solids. *Nuc. Eng. Des* **24**, 420 (1973).

Oh, S.I., Park, J.J., Kobayashi, S. and Altan, T. Application of FEM modelling to simulate metal flow in forging a titanium alloy engine disc. *J. Eng. Ind.* **105**, 251–8 (1983).

Pillinger, I., Hartley, P., Sturgess, C.E.N. and Rowe, G.W. Thermo-mechanical finite-element analysis of metalforming. *Proc. 4th Int. Conf. on Numerical Methods in Thermal Problems*, ed. R.W. Lewis and K. Morgan, Pineridge Press, pp. 1176–88 (1985).

Pillinger, I., Hartley, P., Sturgess, C.E.N. and Rowe, G.W. Thermo-mechanical modelling of metalforming using the finite-element method. *Proc. Int. Conf. on Computational Plasticity*, ed. D.R.J. Owen, E. Hinton and E. Onate, Pineridge Press, pp. 1073–86 (1987).

Pillinger, I., Hartley, P., Sturgess, C.E.N. and Rowe, G.W. Finite-element modelling of metal flow in three-dimensional forming. *Int. J. Num. Meth. Eng.* **25**, *Special Issue on Numerical Methods in Industrial Forming Processes*, 87–98 (1988).

Price, J.W.H. and Alexander, J.M. A study of the isothermal forming of a titanium alloy. *Proc. 4th Nth. American Metalworking Res. Conf.*, SME, pp. 46–53 (1976).

Price, J.W.H. and Alexander, J.M. The finite-element analysis of two high-temperature metal deformation processes. *Proc. 2nd Int. Symp. on Finite Elements in Fluid Problems*, Santa Margherita Ligure, Italy, pp. 715–20 (1976).

Raghavan, K.S. and Wagoner, R.H. Analysis of nonisothermal tensile tests using measured temperature distributions. *Int. J. Plasticity* **3**, 33–49 (1987).

Rebelo, N. and Kobayashi, S. A coupled analysis of viscoplastic deformation and heat transfer – part I. *Int. J. Mech Sci.* **22**, 699–705 (1980).

Rebelo, N. and Kobayashi, S. A coupled analysis of viscoplastic deformation and heat transfer – part II. *Int. J. Mech. Sci.* **22**, 707–18 (1980).

Theodosiu, C., Soos, E. and Rosu, I. A finite element model of the hot working of axially symmetric products. 2 Determination of the velocity and temperature fields during hot extrusion. *Rev. Roum. Sci. Techn. Mec. Appl.* **29**, (1984).

Van Der Lugt, J. and Huetink, J. Thermal-mechanically coupled finite element analysis in metal forming processes. *Comp. Meth. Appl. Mech. Eng.* **54**, 145–60 (1986).

Wertheimer, T.B. Thermal mechanically coupled analysis in metal forming processes. *Proc. 1st Int. Conf. on Numerical Methods in Industrial Forming Processes*, ed. J.F.T. Pittman, R.D. Wood, J.M. Alexander and O.C. Zienkiewicz, Pineridge Press, pp. 425–34 (1982).

Wu, W.T. and Oh, S.I. ALPIDT: A general purpose FEM code for simulation of nonisothermal forming processes. *Proc. 13th Nth. American Manufacturing Res. Conf.*, SME, (1985).

Zienkiewicz, O.C., Onate, E. and Heinrich, J.C. A general formulation for coupled thermal flow of metals using finite elements. *Int. J. Num. Meth. Eng.* **17**, 1497–514 (1981).

Miscellaneous Reports

Andresen, K. Three dimensional forging of rectangular blocks. *Arch Eisenhuettenwes* **45**, 297–300 (1974).

Argyris, J.H. Elasto-plastic matrix displacement analysis of three-dimensional continua. *J. Roy. Aero. Soc.* **69**, 633–6 (1965).

Argyris, J.H., Balmer, H., Doltsinis, J.St., Dunne, P.C., Haase, M., Kleiber, M., Malejannakis, G.A., Mlejnek, H-P., Muller, M. and Scharpf, D.W. Finite element method – the natural approach. *Comp. Meth. Appl. Mech. Eng.* **17/18**, 1–106 (1979).

Argyris, J.H., Doltsinis, J.St. and Wustenberg, H. Analysis of thermoplastic forming processes – natural formulation. *Numerical Analysis of Forming Processes*, ed. J.F.T. Pittman, O.C. Zienkiewicz, R.D. Wood and J.M. Alexander, Wiley, pp. 89–115 (1984).

Argyris, J.H., Dunne, P.C. and Muller, M. Isochoric constant strain finite elements. *Comp. Meth. Appl. Mech. Eng.* **13**, 245–78 (1978).

Argyris, J.H., Dunne, P.C. and Muller, M. Note on large strain applications of modified constant strain finite elements. *Comp. Meth. Appl. Mech. Eng.* **15**, 389–405 (1978).

Arnaudeau, F. and Zarka, J. Numerical simulations of dynamical metal forming. *Proc. Int. Conf. on Computational Methods for Predicting Material Processing Defects*, ed. M. Predeleanu, Elsevier, pp. 1–8 (1987).

Balmer, H.A. and Doltsinis, J.St. Extensions to the elastoplastic analysis with the ASKA program system. *Comp. Meth. Appl. Mech. Eng.* **13**, 363–401 (1978).

Bathe, K.J. An assessment of current finite element analysis of nonlinear problems in solid mechanics. *Numerical Solution of Partial Differential Equations* III, ed. B. Hubbard, Academic Press, pp. 117–64 (1976).

Bathe, K.J. and Ozdemir, H. Elastic–plastic large deformation static and dynamic analysis. *Comput. Struct.* **6**, 81–92 (1976).

Bathe, K.J., Ramm, E. and Wilson, E.L. Finite element formulations for large deformation dynamic analysis. *Int. J. Num. Meth. Eng.* **9**, 353–86 (1975).

Belytschko, T. and Bazant, Z.P. Strain softening materials and finite element solutions. *Constitutive Equations: Macro and Computational Aspects, Winter Annual Meeting of ASME*, ed. K.J. William, ASME, pp. 253–72 (1984).

Bourdouxhe, M., Charlier, R. and Cescotto, S. A finite element for thermo mechanical problems. *Proc. 2nd Int. Conf. on Numerical Methods in Industrial Forming Processes*, ed. K. Mattiasson, A. Samuelsson, R.D. Wood and O.C. Zienkiewicz, Balkema Press, pp. 97–102 (1986).

Bruhns, O. and Lehmann, TH. Optimum deformation rate in large inelastic deformations. *Metal Forming Plasticity*, ed. H. Lippmann, Springer, pp. 120–138 (1979).

Casey, J. and Naghdi, P.M. A remark on the use of the decomposition $F = F_e F_p$ in plasticity. *J. Appl. Mech.* **47**, 672–5 (1980).

Chabrand, P., Pinto, Y. and Raous, M. Numerical modelling of friction for metal forming processes. *Modelling of Metal Forming Processes, Proc. Euromech 233 Colloquium*, ed. J.L. Chenot and E. Onate, Kluwer, pp. 93–9 (1988).

Dackweiler, G. and El-Magd, E. Simulation of impact tension deformation of metals by FEM. *Modelling of Metal Forming Processes, Proc. Euromech 233 Colloquium*, ed. J.L. Chenot and E. Onate, Kluwer, pp. 19–26 (1988).

Dafalias, Y.F. Corotational rates for kinematic hardening at large plastic deformations. *J. Appl. Mech.* **50**, 561–5 (1983).

Dashner, P.A. Invariance considerations in large strain elasto-plasticity. *J. Appl. Mech.* **53**, 55–60 (1986).

Emery, A.F., Sugihara, K. and Jones, A.T. A comparison of some of the thermal characteristics of finite-element and finite-difference calculations of transient problems. *Num. Heat Transfer* **2**, 97–113 (1979).

Feijoo, R.A. and Zouain, N. Variational formulations for rates and increments in plasticity. *Proc. Int. Conf. on Computational Plasticity*, ed. D.R.J. Owen, E. Hinton and E. Onate, Pineridge Press, pp. 33–58 (1987).

Fengels, W. The superimposement of the finite element solutions with respect to Lagrange's and Euler's reference system. *Mech. Res. Comm.* **5**, pp. 341–6 (1978).

Gadala, M.S. and Dokainish, M.A. Formulation methods of geometric and material nonlinearity problems. *Int. J. Num. Meth. Eng.* **20**, 887–914 (1984).

Gadala, M.S., Oravas, G.AE. and Dokainish, M.A. Continuum bases and consistent numerical formulations of nonlinear continuum mechanics problems. *Solid Mech. Arch.* **9**, 1–52 (1984).

Gray, W.H., and Schurr, N.W. A comparison of the finite-element and finite-difference methods for the analysis of steady two-dimensional heat conduction problems. *Comp. Meth. Appl. Mech. Eng.* **6**, 243–5 (1975).

Gunasekera, J.S., Gegel, H.L., Malas, J.C., Doraivelu, S.M. and Alexander, J.M. Materials modelling and intrinsic workability for simulation of bulk deformation. *Proc. 2nd Int. Conf. on Technology of Plasticity*, ed. K. Lange, Springer, pp. 1243–52 (1987).

Gunther, H. Large elastic–plastic deformations in compressible metal-plasticity. *Proc. Int. Conf. on Computational Plasticity*, ed. D.R.J. Owen, E. Hinton and E. Onate, Pineridge Press, pp. 329–38 (1987).

Hayes, D.J. and Marcal, P.V. Determination of upper bounds for problems in plane stress using finite element techniques. *Int. J. Mech. Sci.* **9**, 245–51 (1967).

Hrycaj, P., Lochegnies, D., Oudin, J., Gelin, Y. and Ravalard, Y. Finite element analysis of two-roll hot piercing. *Modelling of Metal Forming Processes, Proc.Euromech 233 Colloquium*, ed. J.L. Chenot and E. Onate, Kluwer, pp. 329–36 (1988).

Hughes, T.J.R. and Winget, J. Finite rotation effects in numerical integration of rate-constitutive equations arising in large-deformation analysis. *Int. J. Num. Meth. Eng.* **15**, 1862–7 (1980).

Jimma, T., Tomita, Y. and Shimamura, S. Benchmark test on a plastic deformation problem application of numerical methods of analysis to the uniaxial tension of a block or cylindrical bar with both ends fixed. *Proc. 2nd Int. Conf. on Technology of Plasticity*, ed. K. Lange, Springer, pp. 73–80 (1987).

Keifer, B.V. and Hilton, P.D. Combined viscous and plastic deformations in two-dimensional large strain finite element analysis. *J. Eng. Mater. Techn.* **107**, 13–18 (1985).

Kershaw, D.S. The incomplete Cholesky-conjugate gradient method for the iterative solution of systems of linear equations. *J. Comput. Phys.* **26**, 43–65 (1978).

Key, S.W. A finite element procedure for the large deformation dynamic response of axisymmetric solid. *Comp. Meth. Appl. Mech. Eng.* **4**, 195–218 (1974).

Kher, S.N. and Amateau, M.F. Deformation model of a thermomechanical gear finishing process. *J. Mater. Shap. Techn.* **6**, 149–57 (1989).

Kobayashi, S. The role of the finite element method in metal forming technology. *Proc. 1st Int. Conf. on Technology of Plasticity*, ed. H. Kudo, JSTP/JSPE, pp. 1035–40 (1984).

Kropp, P.K. and Lahoti, G.D. Application of a non-linear finite element method to the forming of bearing races and rollers. *Proc. Int. Conf. on Computers in Engineering*, Chicago, pp. 109–14 (1986).

Kunar, R.R. and Minowa, N. A comparison between finite element and finite difference formulations for triangular and quadrilateral plane-strain elements. *Int. J. Num. Anal. Methods Geomech.* **5**, 217–24 (1981).

Lahoti, G.D. and Altan, T. Prediction of temperature distributions in axisymmetric compression and torsion. *J. Eng. Mater. Techn.* **97**, 113–20 (1975).

Larsen, P.K. and Popov, E.P. A note on incremental equilibrium equations and approximate constitutive relations in large deformations. *Acta Mechanica* **19**, pp. 1–14 (1979).

Lubliner, J. Normality rules in large-deformation plasticity. *Mech. of Mater.* **5**, 29–34 (1986).

Mamalis, A.G. and Johnson, W. Defects in the processing of metals and composites. *Proc. Int. Conf. on Computational Methods for Predicting Material Processing Defects*, ed. M. Predeleanu, Elsevier, pp. 231–50 (1987).

Marcal, P.V. and King, I.P. Elastic plastic analysis of two dimensional stress systems by the finite element method. *Int. J. Mech. Sci.* **9**, 143–55 (1967).

Marti, J., Kalsi, G. and Atkins, A.G. A numerical and experimental study of deep elastoplastic indentation. *Proc. 1st Int. Conf. on Numerical Methods in Industrial Forming Processes*, ed. J.F.T. Pittman, R.D. Wood, J.M. Alexander and O.C. Zienkiewicz, Pineridge Press, pp. 279–87 (1982).

Nemat-Nasser, S. On finite plastic flow of crystalline solids and geomaterials. *J. Appl. Mech.* **50**, 1114–26 (1983).

Oden, J.T. Finite element formulation of problems of finite deformation and irreversible thermodynamics of nonlinear continua. *Recent Advances in Matrix Methods of Structural Analysis and Design*, ed. R.H. Gallagher, Y. Yamada and J.T. Oden, pp. 383–414 (1970).

Oden, J.T. and Martins, J.A.C. Models and computational methods for dynamic friction phenomena. *Comp. Meth. Appl. Mech. Eng.* **52**, 527–634 (1985).

Oden, J.T., Bhanari, D.R. Yagewa, G. and Chung, T.J. A new approach to the finite element formulation and solution of a class of problems in coupled thermo-elastoviscoplasticity of solids. *Nuc. Eng. Des.* **24**, 420 (1973).

Osakada, K. and Mori, K. The use of micro-supercomputers for simulation of metal forming processes. *Annals of the CIRP* **34**, 241 (1985).

Osman, F.H. and Bramley, A.N. An incremental analytical technique for forging and extrusion of metals *Proc. 1st Int. Conf. on Numerical Methods in Industrial Forming Processes*, ed. J.F.T. Pittman, R.D. Wood, J.M. Alexander and O.C. Zienkiewicz, Pineridge Press, pp. 333–42 (1982).

Perzyna, P. Fundamental problems in viscoplasticity. *Adv. in Appl. Mech.* **9**, 243 (1966).

Pillinger, I., Hartley, P., Sturgess, C.E.N. and Rowe, G.W. An elastic–plastic finite element analysis of the radial expansion of a thick-walled cylinder and analytical validation. *Proc. 1st Int. Conf. on Numerical Methods in Industrial Forming Processes*, ed. J.F.T. Pittman, R.D. Wood, J.M. Alexander and O.C. Zienkiewicz, Pineridge Press, pp. 123–33 (1982).

Pinsky, P.M., Ortiz, M. and Pister, K.S. Numerical integration of rate constitutive equations in finite deformation analysis. *Comp. Meth. Appl. Mech. Eng.* **40**, 137–58 (1983).

Prager, W. An elementary discussion of definitions of stress rate. *Quart. Appl. Math.* **18**, 403–7 (1961).

Rice, J.R. and Tracey, D.M. Computational fracture mechanics. *Numerical and Computer Methods in Fracture Mechanics*, ed. S.J. Fenves, N. Perrone, A.R. Robinson and W.C. Schnobrich, Academic Press, pp. 585–623 (1973).

Richmond, O., Devenpeck, M.L. and Appleby, E.J. Large tensile deformations in sheets and rods: a comparison of observations and predictions. *Formability: Analysis, Modeling and Experimentation, AIME/ASM Fall Meeting*, Chicago (1977).

Rubinestein, R. and Atluri, S.N. Objectivity of incremental constitutive relations over finite time steps in computational finite deformation analysis. *Comp. Meth. Appl. Mech. Eng.* **36**, 277–90 (1983).

Runesson, K, Saran, M. and Mattiasson, K. Generalized midpoint integration schemes in finite deformation plasticity using convected co-ordinates. *Proc. Int. Conf. on Computational Plasticity*, ed. D.R.J. Owen, E. Hinton and E. Onate, Pineridge Press, pp. 281–96 (1987).

Runesson, K., Samuelsson, A. and Bernspang, L. Numerical technique in plasticity including solution advancement control. *Int. J. Num. Meth. Eng.* **22**, 769–88 (1986).

Samuelsson, A. and Froier, M. Finite elements in plasticity – a variational inequality approach. *The Mathematics of Finite Elements and Applications*, ed. J.R. Whiteman, Academic Press, 105–15 (1979).

Santiago, J.M. and Wisniewski, H.L. Implementing the Besseling–White plasticity model in the ADINA finite-element program. *Proc. 1st Int. Conf. on Numerical Methods in Industrial Forming Processes*, ed. J.F.T. Pittman, R.D. Wood, J.M. Alexander and O.C. Zienkiewicz, Pineridge Press, pp. 435–47 (1982).

Shimazeki, Y. and Thompson, E.G. Elasto-viscoplastic flow with special attention to boundary conditions. *Int. J. Num. Meth. Eng.* **17**, 97–112 (1981).

Simo, J.C. and Ortiz, M. A unified approach to finite deformation elasto-plastic analysis based on the use of hyperelastic constitutive equations. *Comp. Meth. Appl. Mech. Eng.* **49**, 221–45 (1985).

Simo, J.C. and Pister, K.S. Remarks on rate constitutive equations for finite deformation problems: computational implications. *Comp. Meth. Appl. Mech. Eng.* **46**, 201–15 (1984).

Tekkaya, A.E. and Roll, K. Analysis of metal forming processes by different finite element methods. *Proc. 2nd Int. Conf. on Numerical Methods for Nonlinear Problems*, Barcelona, pp. 450–61 (1984).

Tomita, Y. Displacement constraints in large strain elastic–plastic finite element analysis. *J. Jap. Soc. for Techn. of Plasticity* **22**, 410–18 (1981).

Tracey, D.M. Finite element solutions for crack tip behaviour in small-scale yielding. *J. Eng. Mater. Techn.* **98**, 146–51 (1976).

Tseng, A.A. Finite element simulation and material characterization for forming miniature metal parts. *Friction and Material Characterizations, Winter Annual Meeting of ASME*, MD 10, ed. I. Haque, J.E. Jackson Jr, A.A. Tseng and J.L. Rose, ASME, pp. 63–72 (1988).

Wesner, P.J. and Weinstein, A.S. Computerized relaxation applied to the plain strain indenter. *ASME pub.* 69-MET-M, ASME (1969).

Wifi, A.S. Finite element correction matrices in metalforming analysis (with application to hydrostatic bulging of a circular sheet). *Int. J. Mech. Sci.* **24**, 393–406 (1982).

Wilkins, M.L. Calculation of elastic–plastic flow. *Methods in Computational Physics*, ed. B. Alder *et al.*, Academic Press, pp. 211–63 (1964).

Yamada, Y. Finite element analysis of nonlinear problems. *J. Jap. Soc. for Techn. of Plasticity* **14**, 758–65 (1973).

Yamada, Y., Hirakawa, T. and Wifi, A.S. Analysis of large deformation and bifurcation in plasticity problems by the finite element method. *Finite Elements in Non-linear Mechanics*, Tapir, pp. 393–412 (1978).

Yamaguchi, K., Takakura, N. and Fukuda, M. FEM simulation of surface roughening and its effects on forming limit in stretching of aluminium sheet. *Proc. 2nd Int. Conf. on Technology of Plasticity*, ed. K. Lange, Springer, pp. 1267–74 (1987).

Zhang, L. and Owen, D.R.J. A modified secant Newton method for non-linear problems. *Comput. Struct.* **15**, 543–7 (1982).

Zienkiewicz, O.C. and Godbole, P.N. A penalty function approach to problems of plastic flow of metals with large surface deformation. *J. Strain Analysis* **10**, 180–5 (1975).

Zienkiewicz, O.C., Onate, E. and Heinrich, J.C. Plastic flow in metal forming: I. Coupled thermal, II. Thin sheet forming. *Application of Numerical Methods to Forming Processes, Winter Annual Meeting of ASME*, AMD 28, ed. H. Armen and R.F. Jones, ASME, pp. 107–20 (1978).

Processes

Drawing

Strip drawing

Gordon, J.L. and Weinstein, A.S. A finite element analysis of the plane strain drawing problem. *Proc. 2nd Nth. American Metalworking Res. Conf.*, SME, (1974).

Lu, S.C-Y., Appleby, E.J., Rao, R.S., Devenpeck, M.L., Wright, P.K. and Richmond, O. A numerical solution of strip drawing employing measured boundary conditions obtained with transparent sapphire dies. *Proc. 1st Int. Conf. on Numerical Methods in Industrial Forming Processes*, ed. J.F.T. Pittman, R.D. Wood, J.M. Alexander and O.C. Zienkiewicz, Pineridge Press, pp. 735–46 (1982).

Lu, S.C-Y. and Wright, P.K. Finite element modelling of plane-strain strip drawing. *J. Eng. Ind.* **110**, 101–10 (1988).

Mathur, K.K. and Dawson, P.R. Damage evolution modeling in bulk forming processes. *Proc. Int. Conf. on Computational Methods for Predicting Material Processing Defects*, ed. M. Predeleanu, Elsevier, pp. 251–62 (1987).

Wire drawing

Chen, C.C. and Kobayashi, S. Deformation analysis of multi-pass bar drawing and extrusion. *Annals of CIRP* **27**, 151–5 (1978).

Gelin, J.C. Numerical analyses of strain rate and temperature effects on localization of plastic flow and ductile fracture – applications to metal forming processes. *Proc. Int. Conf. on Computational Methods for Predicting Material Processing Defects*, ed. M. Predeleanu, Elsevier, pp. 123–32 (1987).

Gerhardt, J. and Tekkaya, A.E. Simulation of drawing-processes by the finite element method. *Proc. 2nd Int. Conf. on Technology of Plasticity*, ed. K.Lange, Springer, pp. 841–8 (1987).

Huetink, J. Analysis of metal forming processes on a combined Eulerian–Lagrangian finite element formulation. *Proc. 1st Int. Conf. on Numerical Methods in Industrial Forming Processes*, ed. J.F.T. Pittman, R.D. Wood, J.M. Alexander and O.C. Zienkiewicz, Pineridge Press, pp. 501–9 (1982).

Klie, W., Lung, M. and Mahrenholtz, O. Axisymmetric plastic deformation using finite element method. *Mech. Res. Commun.* **1**, 315–20 (1974).

Pietrzyk, M., Luksza, J. and Sadok, L. Finite element analysis of the shear strain in the axisymmetric drawing. *Proc. 2nd Int. Conf. on Technology of Plasticity*, ed. K. Lange, Springer, pp. 835–40 (1987).

Rigaut, J.M., Lochegnies, D., Oudin, J., Gelin, J.C. and Ravalard, Y. Numerical analysis of cold drawing of tubes. *Modelling of Metal Forming Processes, Proc. Euromech 233 Colloquium*, ed. J.L. Chenot and E. Onate, Kluwer, pp. 261–8 (1988).

Extrusion

Backward extrusion

Braudel, H.J., Abouaf, M. and Chenot, J.L. An implicit and incrementally objective formulation for solving elastoplastic problems at finite strain by the FEM application to cold forging. *Proc. 2nd Int. Conf. on Numerical Methods in Industrial Forming Processes*, ed. K. Mattiasson, A.

Samuelsson, R.D. Wood and O.C. Zienkiewicz, Balkema Press, pp. 255–61 (1986).

Dung, N.L., Klie, W. and Mahrenholtz, O. Analysis of plastic flow with a simplified finite element method. *Mech. Res. Commun.* **7**, 33–8 (1980).

Hartley, P., Sturgess, C.E.N. and Rowe, G.W. A finite element analysis of extrusion-forging. *Proc. 6th Nth. American Metalworking Res. Conf.*, SME, pp. 212–19 (1978).

Hartley, P., Sturgess, C.E.N. and Rowe, G.W. Prediction of deformation and homogeneity in rim-disc forging. *J. Mech. Wkg. Techn.* **4**, 145–54 (1980).

Kato, K., Okada, T., Murota, T. and Itoh, H. Finite element analysis of non-symmetric extrusion using some experimental results. *Proc. 2nd Int. Conf. on Technology of Plasticity*, ed. K. Lange, Springer, pp. 523–30 (1987).

Kato, T., Tozawa, J. and Shinagawa, K. Thermal analysis for backward cold extrusion of steel taking blue brittleness into consideration. *Proc. 2nd Int. Conf. on Technology of Plasticity*, ed. K. Lange, Springer, pp. 559–64 (1987).

Mori, K., Osakada, K. and Fukuda, M. Simulation of severe plastic deformation by finite element method with spatially fixed elements. *Int. J. Mech. Sci.* **26**, 515–25 (1984).

Mori, K., Osakada, K., Ieda, T. and Fukuda, M. Finite element simulation of metal flow around round corner of tool. *Proc. 2nd Int. Conf. on Technology of Plasticity*, ed. K. Lange, Springer, pp. 1105–10. (1987).

Oh, S.I., Lahoti, G.D. and Altan, T. Analysis of backward extrusion process by the finite element method. *Proc. 10th Nth. American Metalworking Res. Conf.*, SME, p. 143 (1982).

Price, J.W.H. and Alexander, J.M. Specimen geometries predicted by computer model of high deformation forging. *Int. J. Mech. Sci.* **21**, 417–30 (1979).

Sato, K., Tanaka, S. and Uchida, F. Numerical simulation system for plastic forming process based on a FEM introducing slide between elements. *Proc. 1st Int. Conf. on Technology of Plasticity*, ed. H. Kudo, JSTP/JSPE, pp. 1021–6 (1984).

Zienkiewicz, O.C. and Godbole, P.N. Flow of plastic and visco-plastic solids with special reference to extrusion and forming processes. *Int. J. Num. Meth. Eng.* **8**, 3–16 (1974).

Combined extrusion

Osen, W. Possibilities and limitations of cold extrusion processes combined with radial extrusion. *Proc. 2nd Int. Conf. on Technology of Plasticity*, ed. K. Lange, Springer, pp. 575–82 (1987).

Tayal, A.K. and Natarajan, R. Analysis of combined backward-forward extrusion by the finite element method. *Proc. 1st Int. Conf. on Numerical Methods in Industrial Forming Processes*, ed. J.F.T. Pittman, R.D. Wood, J.M. Alexander and O.C. Zienkiewicz, Pineridge Press, pp. 247–55 (1982).

Forward extrusion

Abo-Elkhier, M., Oravas, G. AE. and Dokainish, M.A. A consistent Eulerian formulation for large deformation with reference to metal-extrusion process. *Int. J. Non-linear Mech.* **23**, 37–52 (1988).

Braudel, H.J., Abouaf, M. and Chenot, J.L. An implicit and incrementally objective formulation for solving elastoplastic problems at finite strain by the FEM application to cold forging. *Proc. 2nd Int. Conf. on Numerical Methods in Industrial Forming Processes*, ed. K. Mattiasson, A.

Samuelsson, R.D. Wood and O.C. Zienkiewicz, Balkema Press, pp. 255–61 (1986).

Chanda, S.K., Hartley, P., Sturgess, C.E.N., Rowe, G.W. and Simkin, J. Deformation in low-reduction plane-strain containerless extrusion. *Proc. 12th Nth. American Manufacturing Res. Conf.*, SME, pp. 120–7 (1984).

Chandra, A. and Mukherjee, S. A finite-element analysis of metal forming problems with thermomechanical coupling. *Int. J. Mech. Sci.* **21**, 661–76 (1984).

Chevalier, L. and Le Nevez, P. Simulation of cold extrusion process: a numerical-experimental comparison. *CAD/CAM in Metal Working*, ed. S.K. Ghosh and A. Niku-Lari, Pergamon, pp. 25–34 (1988).

Derbalian, K.A., Lee, E.H., Mallett, R.L. and McMeeking, R.M. Finite element metal forming analysis with spacially fixed mesh. *Application of Numerical Methods to Forming Processes, Winter Annual Meeting of ASME*, AMD 28, ed. H. Armen and R.F. Jones Jr., ASME, pp. 39–47 (1978).

Gadala, M.S. and El-Madany, M.M. Finite element formulation and application to metal flow problems. *Proc. Int. Conf. on Computational Plasticity*, ed. D.R.J. Owen, E. Hinton and E. Onate, Pineridge Press, pp. 989–1004 (1987).

Gelin, J.C., Lochegnies, D., Oudin, J., Picart, P. and Ravalard, Y. Finite element simulation of metal flow in axisymmetric extrusion process. *CAD/CAM in Metal Working*, ed. S.K. Ghosh and A. Niku-Lari, Pergamon, pp. 47–60 (1988).

Gerhardt, J. and Tekkaya, A.E. Applications of the finite element method on the determination of residual stresses in drawing and extrusion. *Proc. Int. Conf. on Computational Plasticity*, ed. D.R.J. Owen, E. Hinton and E. Onate, Pineridge Press, pp. 1037–50 (1987).

Hartley, C.S. Residual stresses in axisymmetrically formed products. *Proc. 2nd Int. Conf. on Technology of Plasticity*, ed. K. Lange, Springer, pp. 605–14 (1987).

Hartley, C.S., Dehghani, M., Iver, N., Male, A.T. and Lovic, W.R. Defects at interfaces in coextruded metals. *Proc. Int. Conf. on Computational Methods for Predicting Material Processing Defects*, ed. M. Predeleanu, Elsevier, pp. 357–366 (1987).

Hata, K., Ishikawa, H. and Yamamoto, K. An analysis of extrusion by the finite element method. *J. Jap. Soc. for Techn. of Plasticity* **15**, 1003–10 (1974).

Huang, G.C., Liu, Y.C. and Zienkiewicz, O.C. Error control, mesh updating schemes and automatic adaptive remeshing for finite element analysis of unsteady extrusion processes. *Modelling of Metal Forming Processes, Proc. Euromech 233 Colloquium*, ed. J.L. Chenot and E. Onate, Kluwer, pp. 75–84 (1988).

Iwata, K., Osakada, K. and Fujino, S. Analysis of hydrostatic extrusion by the finite element method. *J. Eng. Ind.* **94**, 697–703 (1972).

Kanetake, N and Lange, K. Metal flow in the rod extrusion of rate sensitive materials. *Proc. 2nd Int. Conf. on Technology of Plasticity*, ed. K. Lange, Springer, pp. 493–8 (1987).

Kumar, A., Jain, P.C. and Mehta, M.L. Modified formulation for extrusion of porous powder preforms using finite element method. *Proc. 1st Int. Conf. on Numerical Methods in Industrial Forming Processes*, ed. J.F.T. Pittman, R.D. Wood, J.M. Alexander and O.C. Zienkiewicz, Pineridge Press, pp. 257–66 (1982).

Kumar, A., Jha, S., Jain, P.C. and Mehta, M.L. An investigation into product defects in the hot extrusion of aluminium powder preforms. *J. Mech. Wkg. Techn.* **11**, 275–90 (1985).

Lee, E.H., Mallett, R.L. and McMeeking, R.M. Stress and deformation analysis of metal forming processes. *Numerical Modelling of Manufacturing Processes*, ed. R.F. Jones Jr., H. Armen and J.T. Fong, ASME, p. 19 (1977).

Lee, E.H., Mallett, R.L. and McMeeking, R.M. Stress and deformation analysis of metal-forming processes. *Metal Forming Plasticity*, ed. H. Lippmann, Springer, pp. 177–89 (1979).

Lee, E.H., Mallett, R.L. and Yang, W.H. Stress and deformation analysis of the metal extrusion process. *Comp. Meth. App. Mech. Eng.* **10**, 339–53 (1977).

Mallett, R.L. *Finite element selection for finite deformation elastic–plastic analysis.* SUDAM rep. no. 80-4, Stanford University (1980).

Pacheco, L.A. and Alexander, J.M. On the hydrostatic extrusion of copper-covered aluminium rods. *Proc. 1st Int. Conf. on Numerical Methods in Industrial Forming Processes*, ed. J.F.T. Pittman, R.D. Wood, J.M. Alexander and O.C. Zienkiewicz, Pineridge Press, pp. 205–16 (1982).

Roll, K. Possibilities for the use of the finite element method for the analysis of bulk metalforming processes. *Annals of CIRP* **31**, 145–50 (1982).

Roll, K. and Neitzert, Th. On the application of different numerical methods to calculate cold forming processes. *Proc. 1st Int. Conf. on Numerical Methods in Industrial Forming Processes*, ed. J.F.T. Pittman, R.D. Wood, J.M. Alexander and O.C. Zienkiewicz, Pineridge Press, pp. 97–107 (1982).

Saga, J. and Nojima, H. Some applications of the finite element method in cold extrusion. *J. Jap. Soc. for Techn. of Plasticity* **14**, p. 838 (1973).

Said, M.E., Wifi, A.S. and El-Fadaly, M.S. Experimental and finite element analysis of the hydrostatic extrusion process. *CAD/CAM in Metal Working*, ed. S.K. Ghosh and A. Niku-Lari, Pergamon, pp. 71–85 (1988).

Sato, K., Tanaka, S. and Uchida, F. Numerical simulation system for plastic forming process based on a FEM introducing slide between elements. *Proc. 1st Int. Conf. Technology of Plasticity*, ed. H. Kudo, JSTP/JSPE, pp. 1021–6 (1984).

Sebastian, M.A., Rodrigues, P. and Sanchez, A.M. A method of discretisation and an approach to three-dimensional deformation analysis of extrusion by the finite element method. *Proc. 1st Int. Conf. on Numerical Methods in Industrial Forming Processes*, ed. J.F.T. Pittman, R.D. Wood, J.M. Alexander and O.C. Zienkiewicz, Pineridge Press, pp. 227–36 (1982).

Sharman, F.W. *Analysis of metal deformation-extrusion.* ECRC rep. M537, Electricity Council Research Centre, UK (1972).

Sharman, F.W. *Hot bar extrusion and the effect of radial temperature profiles in the billet.* ECRC rep. R851, Electricity Council Research Centre, UK (1975).

Smelser, R.E., Richmond, O. and Thompson, E.G. A numerical study of the effects of die profile on extrusion. *Proc. 2nd Int. Conf. on Numerical Methods in Industrial Forming Processes*, ed. K. Mattiasson, A. Samuelsson, R.D. Wood and O.C. Zienkiewicz, Balkema Press, pp. 305–12 (1986).

Tayal, A.K. and Natarajan, R. Plane strain extrusion studies by the finite element method. *J. Inst. India* **63**, 31–3 (1982).

Tekkaya, A.E. and Gerhardt, J. Residual stresses in cold-formed workpieces. *Annals of CIRP* **34**, 225–30 (1985).

Tekkaya, A.E. and Lange, K. Determination of residual stresses in industrial extrusion. *Proc. 12th Nth. American Manufacturing Res. Conf.*, SME, pp. 103–10 (1984).

Wertheimer, T.B. Thermal mechanically coupled analysis in metal forming processes. *Proc. 1st Int. Conf. on Numerical Methods in Industrial Forming Processes*, ed. J.F.T. Pittman, R.D. Wood, J.M. Alexander and O.C. Zienkiewicz, Pineridge Press, pp. 425–34 (1982).

Yamada, Y. and Hirakawa, H. Large deformation and instability analysis in metal forming process. *Applications of Numerical Methods to Forming Processes, Winter Annual Meeting of ASME*, AMD 28, ed. H. Armen and R.F. Jones Jr. ASME, (1978).

Zienkiewicz, O.C. and Godbole, P.N. Flow of plastic and viscoplastic solids with special reference to extrusion and forming processes. *Int. J. Num. Meth. in Eng.* **8**, 3–16 (1974).

Zienkiewicz, O.C., Jain, P.C. and Onate, E. Flow of solids during forming and extrusion: some aspects of numerical solutions. *Int. J. Solids Struct.* **14**, 15–38 (1978).

Zienkiewicz, O.C., Nakazawa, S. and Vilotte, J.P. Finite elements in forming processes. *Proc. 1st Int. Conf. on Technology of Plasticity*, ed. H. Kudo, JSTP/JSPE, pp. 1041–50 (1984).

Zienkiewicz, O.C., Onate, E. and Heinrich, J.C. Plastic flow in metal forming: I. Coupled thermal, II. Thin sheet forming. *Application of Numerical Methods to Forming Processes, Winter Annual Meeting of ASME*, AMD 28, ed. H. Armen and R.F. Jones Jr., ASME, pp. 107–20 (1978).

Forging

Axi-symmetric Forging

Al-Sened, A.A.K., Hartley, P., Sturgess, C.E.N. and Rowe, G.W. Forming sequences in axi-symmetric cold forging. *Proc. 12th Nth. American Manufacturing Res. Conf.*, SME, pp. 151–8 (1984).

Alberti, N., Cannizzaro, L. and Riccobono, R. A new numerical method for axisymmetrical forming processes. *Annals of CIRP* **36**, 131–3 (1987).

Altan, T. Process simulation of hot die forging processes. *Proc. 2nd Int. Conf. on Technology of Plasticity*, ed. K. Lange, Springer, pp. 1021–34 (1987).

Duggirala, R. and Badawy, A. Finite element approach to forging process design. *J. Mater. Shap. Techn.* **6**, 81–9 (1988).

Ficke, J.A., Oh, S.I. and Altan, T. Use of electronic geometry transfer and FEM simulation in the design of hot forming dies for a gear blank. *Annals of CIRP* **33**, 123–7 (1984).

Germain, Y., Mosser, P.E. and Chenot, J.L. Finite element analysis of shaped lead-tin disk forgings. *Proc. 2nd Int. Conf. on Numerical Methods in Industrial Forming Processes*, ed. K. Mattiasson, A. Samuelsson, R.D. Wood and O.C. Zienkiewicz, Balkema Press, pp. 271–6 (1986).

Groche, P. and Weiss, U. Numerical identification of forging parameters. *Modelling of Metal Forming Processes, Proc. Euromech 233 Colloquium*, ed. J.L. Chenot and E. Onate, Kluwer, pp. 237–45 (1988).

Gunasekera, J.S., Gegel, H.L., Malas, J.C., Doraivelu, S.M. and Morgan, J.T. Computer-aided process modelling of hot forging and extrusion of aluminium alloys. *Annals of CIRP* **31**, 131–5 (1982).

Hartley, P., Sturgess, C.E.N. and Rowe, G.W. Prediction of deformation and homogeneity in rim-disc forging. *J. Mech. Wkg. Techn.* **4**, 145–54 (1980).

Hartley, P., Sturgess, C.E.N. and Rowe, G.W. Computer simulation of metal-forming processes. *Advances in Engineering Software Journal* **4**, 20–5 (1982).

Hwang, S.M. and Kobayashi, S. Preform design in disk forging. *Int. J. Machine Tool Des. Res.* **26**, no. 3, 231–43 (1986).

Hwang, S.M. and Kobayashi, S. Preform design in shell nosing at elevated temperatures. *Int. J. Machine Tools Manufact.* **27**, no. 1, 1–14 (1987).

Jackson Jr., J.E., Haque, I., Gangjee, T. and Ramesh, M. Some numerical aspects of frictional modeling in material forming processes. *Friction and Material Characterizations, Winter Annual Meeting of ASME*, MD 10, ed. I. Haque, J.E. Jackson Jr, A.A. Tseng and J.L. Rose, ASME, pp. 39–46 (1988).

Kobayashi, S. Rigid-plastic finite element analysis of axisymmetric metal forming processes. *Numerical Modelling of Manufacturing Processes*, ed. R.F. Jones Jr., H. Armen and J.T. Fong, ASME, pp. 49–65 (1977).

Kobayashi, S. Process design in metal forming by the finite element method. *Proc. 2nd Int. Conf. on Technology of Plasticity*, ed. K. Lange, Springer, pp. 1213–20 (1987).

Kopp, R., Cho, M.L. and de Souza, M. Multi-level simulation of metal forming process. *Proc. 2nd Int. Conf. on Technology of Plasticity*, ed. K. Lange, Springer, pp. 1229–34 (1987).

Lung, M. and Mahrenholtz, O. A finite element procedure for analysis of metal forming processes. *Trans. CSME* **2**, 31–6 (1973).

Nagtegaal, J.C. and Rebelo, N. On the development of a general purpose finite element program for analysis of forming processes. *Proc. 2nd Int. Conf. on Numerical Methods in Industrial Forming Processes*, ed. K. Mattiasson, A. Samuelsson, R.D. Wood and O.C. Zienkiewicz, Balkema Press, pp. 41–50 (1986).

Nagtegaal, J.C. and Rebelo, N. On the development of a general purpose finite element program for analysis of forming processes. *Int. J. Num. Meth. Eng.* **25**, *Special Issue on Numerical Methods in Industrial Forming Processes*, 113–31 (1988).

Oh, S.I., Lahoti, G.D. and Altan, T. ALPID – a general purpose FEM program. *Proc. 9th Nth. American Metalworking Res. Conf.*, SME, pp. 83–8 (1981).

Oh, S.I., Lahoti, G.D. and Altan, T. Application of FEM to industrial metal forming processes. *Proc. 1st Int. Conf. on Numerical Methods in Industrial Forming Processes*, ed. J.F.T. Pittman, R.D. Wood, J.M. Alexander and O.C. Zienkiewicz, Pineridge Press, pp. 145–53 (1982).

Oh, S.I., Park, J.J., Kobayashi, S. and Altan, T. Application of FEM modeling to simulate metal flow in forging a titanium alloy engine disc. *J. Eng. Ind.* **105**, 251–8 (1983).

Park, J.J., Rebelo, N. and Kobayashi, S. A new approach to preform design in metal forming with the finite element method. *Int. J. Machine Tool Des. Res.* **23**, no. 1, 71–9 (1983).

Price, J.W.H. and Alexander, J.M. Specimen geometries predicted by computer model of high deformation forging. *Int. J. Mech. Sci.* **21**, 417–30 (1979).

Roll, K. Comparison of different numerical methods for the calculation of metal forming processes. *Annals of CIRP* **28**, 141–5 (1979).

Rowe, G.W. Problems involved in the numerical analysis of plastic working processes for new materials. *Proc. 1st Int. Conf. on Technology of Plasticity*, ed. H. Kudo, JSTP/JSPE, pp. 27–38 (1984).

Rydstad, H. and Boer, C.R. Forging of an axisymmetrical shape – modelling of material flow, die loads and die stresses using FEM. *Proc. 2nd Int. Conf. on Numerical Methods in Industrial Forming Processes*, ed. K. Mattiasson, A. Samuelsson, R.D. Wood and O.C. Zienkiewicz, Balkema Press, pp. 283–6 (1986).

Schroder, G., Rydstad, H. and Gessinger, G.H. Improved properties of forged superalloys – results of a better modelling and controlling of the forging process. *Proc. 2nd Int. Conf. on Technology of Plasticity*, ed. K. Lange, Springer, pp. 1073–80 (1987).

Surdon, G. and Baroux, M. Comparison of simulation techniques and industrial data in forging applications. *Modelling of Metal Forming Processes, Proc. Euromech 233 Colloquium*, ed. J.L. Chenot and E. Onate, Kluwer, pp. 199–206 (1988).

Wu, W.T., Oh, S.I. and Altan, T. Investigation of defect formation in rib-web type forging by ALPID. *Proc. 12th Nth. American Manufacturing Res. Conf.*, SME, pp. 159–65 (1984).

Flat tool forging

Paukert, R. Investigations into metal flow in radial-forging. *Annals of CIRP* **32**, 211–14 (1983).

Rodic, T., Stok, B., Gologranc, F. and Owen, D. Finite element modelling of a radial forging process. *Proc. 2nd Int. Conf. on Technology of Plasticity*, ed. K. Lange, Springer, pp. 1065–72 (1987).

Sun, J-X. Analysis of special forging processes for heavy ingots by finite element method. *Int. J. Machine Tools Manufact.* **28**, no. 2, 173–9 (1988).

Sun, J-X., Li, G.J. and Kobayashi, S. Analysis of spread in flat-tool forging by the finite element method. *Proc. 11th Nth. American Metalworking Res. Conf.*, SME, pp. 224–31 (1983).

Plane-strain forging

Dung, N.L. and Mahrenholtz, O. Progress in the analysis of unsteady metal forming processes using the finite element method. *Proc. 1st Int. Conf. on Numerical Methods in Industrial Forming Processes*, ed. J.F.T. Pittman, R.D. Wood, J.M. Alexander and O.C. Zienkiewicz, Pineridge Press, pp. 187–96 (1982).

Jain, V.K., Matson, L.E., Gegel, H.L. and Srinivasan, R. Physical modelling of metalworking processes – II: Comparison of visioplastic modelling and computer simulation. *J. Mater. Shap. Techn.* **5**, 249–57 (1988).

Kato, K., Rowe, G.W., Sturgess, C.E.N., Hartley, P. and Pillinger, I. Fundamental deformation modes in open die forging – Finite element analysis of open die forging I. *J. Jap. Soc. for Techn. of Plasticity* **27**, No. 311, 1383–9 (in Japanese) (1986).

Kato, K., Rowe, G.W., Sturgess, C.E.N., Hartley, P. and Pillinger, I. Classification of deformation modes and deformation property diagrams in open die forging – Finite element analysis of open die forging II. *J. Jap. Soc. for Techn. of Plasticity* **28**, No. 312, 67–74 (in Japanese) (1987).

Rebelo, N. and Boer, C.R. A process modelling study of the influence of friction during rib forging. *Proc. 12th Nth. American Manufacturing Res. Conf.*, SME, pp. 146–50 (1984).

Rebelo, N., Rdystad, H. and Schroder, G. Simulation of material flow in closed-die forging by model techniques and rigid-plastic FEM. *Proc. 1st Int. Conf. on Numerical Methods in Industrial Forming Processes*, ed. J.F.T. Pittman,

R.D. Wood, J.M. Alexander and O.C. Zienkiewicz, Pineridge Press, pp. 237–46 (1982).

Tanaka, S. and Sato, K. Extended two-dimensional analysis applied to forging process. *J. Jap. Soc. for Techn. of Plasticity* **28**, 257–63 (1987).

Three-dimensional forging

Argyris, J.H. and Doltsinis, J.St. Computer simulation of metal forming processes. Supplement to *Proc. 2nd Int. Conf. on Numerical Methods in Industrial Forming Processes*, ed. K. Mattiasson, A. Samuelsson, R.D. Wood and O.C. Zienkiewicz, Balkema Press, (1986).

Argyris, J.H., Doltsinis, J.St. and Luginsland, J. Three-dimensional thermomechanical analysis of metal forming processes. *Proc. 1st Int. Workshop on Simulation of Metal Forming Processes by the Finite Element Method (SIMOP–I)*, ed. K. Lange, Springer, pp. 125–60 (1986).

Desai, C.S. and Phan, H.V. Three-dimensional finite element analysis including material and geometrical nonlinearities. *Computational Methods in Nonlinear Mechanics. Proc. 2nd TICOM Conf.*, ed. J.T. Oden, North-Holland, pp. 205–24 (1979).

Mori, K. and Osakada, K. Approximate analysis of three dimensional deformation in edge rolling by the rigid-plastic finite element method. *J. Jap. Soc. for Techn. of Plasticity* **23**, 897 (1982).

Mori, K., Osakada, K., Nakadoi, K. and Fukuda, M. Simulation of three-dimensional deformation in metal forming by the rigid-plastic finite element method. *Proc. 1st Int. Conf. on Technology of Plasticity*, ed. H. Kudo, JSTP/JSPE, pp. 1009–14 (1984).

Nagamatsu, A. Analysis of contact pressure and deformation of square blocks in elastic–plastic compression by the finite element method. *J. Jap. Soc. for Techn. of Plasticity* **14**, no. 146, 220–9 (1973).

Park, J.J. and Kobayashi, S. Three-dimensional finite element analysis of block compression. *Int. J. Mech. Sci.* **26**, 165–76 (1984).

Park, J.J. and Oh, S.I. Application of three-dimensional finite element analysis to metal forming processes. *Proc. 15th Nth. American Manufacturing Res. Conf.*, SME, pp. 296–303 (1987).

Pillinger, I., Hartley, P., Sturgess, C.E.N. and Rowe, G.W. Thermo-mechanical finite-element analysis of metalforming. *Proc. 4th Int. Conf. on Numerical Methods in Thermal Problems*, ed. R.W. Lewis and K. Morgan, Pineridge Press, pp. 1176–88 (1985).

Pillinger, I., Hartley, P., Sturgess, C.E.N. and Rowe, G.W. Finite-element analyses of three-dimensional forgings. *Applications of Non-Linear Stress Analysis*, Institute of Physics, London (1985).

Pillinger, I., Hartley, P., Sturgess, C.E.N. and Rowe, G.W. A three-dimensional finite-element analysis of the cold forging of a model connecting rod. *Proc. Instn. Mech. Engrs.* **199**, no.C4, 319–24 (1985).

Pillinger, I., Hartley, P., Sturgess, C.E.N. and Rowe, G.W. An elastic–plastic three-dimensional finite-element analysis of the upsetting of rectangular blocks and experimental comparison. *Int. J. Machine Tool Des. Res.* **25**, no.3, 229–43 (1985).

Pillinger, I., Hartley, P., Sturgess, C.E.N. and Rowe, G.W. Elastic–plastic three-dimensional finite-element analysis of bulk metalforming processes. *Proc. 1st Int. Workshop on the Simulation of Metal Forming Processes by the*

Finite-Element Method (SIMOP–I), ed. K. Lange, Springer, pp. 91–124 (1986).

Pillinger, I., Hartley, P., Sturgess, C.E.N. and Rowe, G.W. Finite-element modelling of metal flow in three-dimensional and temperature-dependent forming. *Proc. 2nd Int. Conf. on Numerical Methods in Industrial Forming Processes*, ed. K. Mattiasson, A. Samuelsson, R.D. Wood and O.C. Zienkiewicz, Balkema Press, pp. 151–6 (1986).

Pillinger, I., Hartley, P., Sturgess, C.E.N. and Rowe, G.W. Finite-element modelling of metal flow in three-dimensional forming. *Int. J. Num. Meth. Eng.* **25**, *Special Issue on Numerical Methods in Industrial Forming Processes*, 87–98 (1988).

Shiau, Y.C. and Kobayashi, S. Three-dimensional finite element analysis of open-die forging. *Int. J. Num. Meth. Eng.* **25**, *Special Issue on Numerical Methods in Industrial Forming Processes*, 67–86 (1988).

Soyris, N., Cescutti, J.P., Coupez, T., Bracchotte, G. and Chenot, J.L. Three dimensional finite element calculation of the forging of a connecting rod. *Modelling of Metal Forming Processes, Proc. Euromech 233 Colloquium*, ed. J.L. Chenot and E. Onate, Kluwer, pp. 227–36 (1988).

Sun, J-X. Analysis of special forging processes for heavy ingots by finite element method. *Int. J. Machine Tools Manufact.* **28**, no. 2, 173–9 (1988).

Sun, J.X. and Kobayashi, S. Analysis of block compression with simplified three-dimensional elements. *Proc. 1st Int. Conf. on Technology of Plasticity*, ed. H. Kudo, JSTP/JSPE, pp. 1027–34 (1984).

Surdon, G. and Chenot, J.L. Finite element calculation of three-dimensional forging. *Proc. 2nd Int. Conf. on Numerical Methods in Industrial Forming Processes*, ed. K. Mattiasson, A. Samuelsson, R.D. Wood and O.C. Zienkiewicz, Balkema Press, pp. 287–92 (1986).

Surdon, G. and Chenot, J.L. Finite element calculation of three-dimensional forging. *Int. J. Num. Meth. Eng.* **24**, 2107–17 (1987).

Tanaka, M., Ono, S-I. and Tsuneno, M. A numerical analysis on void crushing during side compression of round bar by flat dies. *J. Jap. Soc. for Techn. of Plasticity* **28**, 238–44 (1987).

Webster, W. and Davis, R. Finite element analysis of round to square extrusion processes. *Proc. 6th Nth. American Metalworking Res. Conf.*, SME, pp. 166–70 (1978).

Upsetting

Al-Khattat, I.M. Plane strain upsetting: a new approach to the general contact problem in mechanics. *Proc. 1st Int. Conf. on Numerical Methods in Industrial Forming Processes*, ed. J.F.T. Pittman, R.D. Wood, J.M. Alexander and O.C. Zienkiewicz, Pineridge Press, pp. 109–22 (1982).

Belkhiri, L., Charlier, R. and Detraux, J.M. On the large deformation of axisymmetric bodies: experimental and numerical studies. *Proc. Int. Conf. on Numerical Methods in Engineering, Theory and Applications (NUMETA)*, Swansea (1985).

Boer, C.R., Gudmundson, P. and Rebelo, N. Comparison of elasto-plastic FEM, rigid-plastic FEM and experiments for cylinder upsetting. *Proc. 1st Int. Conf. on Numerical Methods in Industrial Forming Processes*, ed. J.F.T. Pittman, R.D. Wood, J.M. Alexander and O.C. Zienkiewicz, Pineridge Press, pp. 217–26 (1982).

Brouha, M., de Jong, J.E. and Van der Weide, K.J.A. Experimental verification of finite element analysis on axisymmetric deformation processes. *Proc. 7th Nth. American Metalworking Res. Conf.* SME, (1979).

Creus, G.J. and Groehs, A.G. Finite elements analysis of large plastic deformation in metals. *Modelling of Metal Forming Processes, Proc. Euromech 233 Colloquium*, ed. J.L. Chenot and E. Onate, Kluwer, pp. 27–36 (1988).

Doltsinis, J.St., Luginsland, J. and Nolting, S. Some developments in the numerical simulation of metal forming processes. *Proc. Int. Conf. on Computational Plasticity*, ed. D.R.J. Owen, E. Hinton and E. Onate, Pineridge Press, pp. 875–900 (1987).

Dung, N.L. Finite element modelling of upsetting tests. *VDI-Forsch. Ing.-Wes.* **50**, 55–62 (1984).

Evans, R.W. The plane strain compression test for the determination of hot working properties. *Proc. 1st Int. Conf. on Numerical Methods in Industrial Forming Processes*, ed. J.F.T. Pittman, R.D. Wood, J.M. Alexander and O.C. Zienkiewicz, Pineridge Press, pp. 481–90 (1982).

Fadzil, M., Chuah, K.C., Hartley, P., Sturgess, C.E.N. and Rowe, G.W. Metalforming analysis on desk-top microcomputers using non-linear elastic-plastic finite element techniques. *Proc. 22nd Int. Machine Tool Des. Res. Conf.*, ed. B.J. Davies, Macmillan, pp. 533–9 (1981).

Hartley, P., Sturgess, C.E.N. and Rowe, G.W. Influence of friction on the prediction of forces, pressure distributions and properties in upset forging. *Int. J. Mech. Sci.* **22**, 743–53 (1980).

Hartley, P., Sturgess, C.E.N. and Rowe, G.W. Finite-element predictions of the influence of strain-rate and temperature variations on the properties of forged products. *Proc. 8th Nth. American Metalworking Res. Conf.*, SME, pp. 121–8 (1980).

Kobayashi, S. and Lee, C.H. Deformation mechanics and workability in upsetting solid circular cylinders. *Proc. 1st Nth. American Metalworking Res. Conf.*, SME, pp. 185–204 (1973).

Kobayashi, S., Lee, C.H. and Shah, S.N. Analysis of rigid-plastic deformation problems by the matrix method. *J. Jap. Soc. for Techn. of Plasticity* **15**, 770–8 (1973).

Kudo, H. The use of finite element method in the analysis of plastic deformation of some metal forming processes. *Annals of CIRP* **23**, 219–25 (1974).

Kudo, H. and Matsubara, S. Joint examination project of validity of various numerical methods for the analysis of metal forming processes. *Metal Forming Plasticity*, ed. H. Lippmann, Springer, pp. 378–401 (1979).

Lee, C.H. and Kobayashi, S. Analyses of axisymmetric upsetting and plane strain side-pressing of solid cylinders by the finite element method. *J. Eng. Ind.* **93**, 445–54 (1971).

Lee, C.H. and Kobayashi, S. New solutions to rigid-plastic deformation problems using a matrix method. *J. Eng. Ind.* **95**, 865–73 (1973).

Lee, C.H., Shah, S.N. and Kobayashi, S. Analysis of rigid-plastic deformation problems by the matrix method. *J. Jap. Soc. for Techn. of Plasticity* **14**, 770 (1973).

Lush, A. and Anand, L. Implicit time-integration procedures for a set of internal variable constitutive equations for hot-working. *Proc. 2nd Int. Conf. on Numerical Methods in Industrial Forming Processes*, ed. K. Mattiasson, A. Samuelsson, R.D. Wood and O.C. Zienkiewicz, Balkema Press, pp. 131–8 (1986).

Madnaik, S.D., Maiti, S.K. and Chaturvedi, R.C. A finite element solution to the problem of cold plane strain compression with large reductions. *Proc. 1st Int. Conf. on Numerical Methods in Industrial Forming Processes*, ed. J.F.T. Pittman, R.D. Wood, J.M. Alexander and O.C. Zienkiewicz, Pineridge Press, pp. 155–63 (1982).

Mori, K., Shima, S. and Osakada, K. Finite element method for the analysis of plastic deformation of porous metals. *Bull. JSME* **23**, no. 178, 516–22 (1980).

Mori, K., Shima, S. and Osakada, K. Analysis of free forging by rigid-plastic finite element method based on the plasticity equation for porous metals. *Bull. JSME* **23**, no.178, 523–9 (1980).

Murakawa, M., Koga, N., Makinouchi, A. and Ike, H. Experimental examination of finite element deformation analysis in the flattening of surface asperities by flat rigid dies. *Proc. 15th Nth. American Manufacturing Res. Conf.*, SME, pp. 380–5 (1987).

Nagamatsu, A., Murota, T. and Jimma, T. On the non-uniform deformation of a block in plane-strain compression caused by friction (part 3). *Bull. JSME* **14**, no.70, 314–21 (1971).

Nagamatsu, A., Murota, T. and Jimma, T. On the non-uniform deformation of a block in plane-strain compression caused by friction (part 4). *Bull. JSME* **14**, no.70, 322–30 (1971).

Nagamatsu, A., Murota, T. and Jimma, T. On the non-uniform deformation of material in axially symmetric compression caused by friction (part 2) *Bull. JSME* **14**, no.70, 339–47 (1971).

Oh, S.I. and Kobayashi, S. Calculation of frictional stress distribution at the die-workpiece interface in simple upsetting. *Proc. 3rd Nth. American Metalworking Res. Conf.*, SME, pp. 159–67 (1975).

Price, J.W.H. and Alexander, J.M. A study of isothermal forming or creep forming of a titanium alloy. *Proc. 4th Nth. American Metalworking Res. Conf.*, SME, pp. 46–53 (1976).

Price, J.W.H. and Alexander, J.M. Specimen geometries predicted by computer model of high deformation forging. *Int. J. Mech. Sci.* **21**, 417–30 (1979).

Sakuta, H., Suzuki, T. and Barber, B. Estimation method for metal forging. *Proc. Int. Conf. on Computational Plasticity*, ed. D.R.J. Owen, E. Hinton and E. Onate, Pineridge Press, pp. 1793–802 (1987).

Shah, S.N., Lee, C.H. and Kobayashi, S. Compression of tall, circular, solid cylinders between parallel flat dies. *Proc. Int. Conf. Prod. Eng.*, Tokyo, pp. 295–300 (1974).

Shima, S., Mori, K. and Osakada, K. Analysis of metal forming by the rigid-plastic finite element method based on plasticity theory for porous metals. *Metal Forming Plasticity* ed. H. Lippmann, Springer, pp. 305–17 (1979).

Takahashi, H. and Kobayashi, S. Some aspects of finite-element analysis of plastic compression. *Proc. 5th Nth. American Metalworking Res. Conf.*, SME, p. 87 (1977).

Tekkaya, A.E., Roll, K., Gerhardt, J., Herrmann, M. and Du, G. Finite-element-simulation of metal forming processes using two different material-laws. *Proc. 1st Int. Workshop on Simulation of Metal Forming Processes by the Finite Element Method (SIMOP–I)*, ed. K. Lange, Springer, pp. 50–85 (1986).

Vardan, O.C., Bagchi, A. and Altan, T. Investigations of die wear in upsetting using the FEM code ALPID. *Proc. 15th Nth. American Manufacturing Res. Conf.*, SME, pp. 386–9 (1987).

Wifi, A.S. Large strain elasto-plastic analysis of some hydrostatic pressure-aided deformation processes. *Proc. 1st Int. Conf. on Numerical Methods in Industrial Forming Processes*, ed. J.F.T. Pittman, R.D. Wood, J.M. Alexander and O.C. Zienkiewicz, Pineridge Press, pp. 135–44 (1982).

Yamada, Y., Wifi, A.S. and Hirakawa, T. Analysis of large deformation and stress in metal forming processes by the finite element method. *Metal Forming Plasticity* ed. H. Lippmann, Springer, pp. 158–76 (1979).

Heading

Al-Sened, A.A.K., Hartley, P., Sturgess, C.E.N. and Rowe, G.W. Finite-element analysis of a five-stage cold heading process. *J. Mech. Wkg, Techn.* **14**, 225–34 (1987).

Argyris, J.H. and Doltsinis, J. St. On the large strain inelastic analysis in natural formulation. *Comp. Meth. Appl. Mech. Eng.* **20**, 213–51 (1979).

Hussin, A.A.M., Hartley, P., Sturgess, C.E.N. and Rowe, G.W. Simulation of industrial cold forming processes. *J. Comm. Appl. Num. Meth.* **3**, 415–26 (1987).

Kher, S.N. and Amateau, M.F. Finite element modelling of a thermomechanical surface forming process. *Proc. Int. Conf. on Computational Plasticity*, ed. D.R.J. Owen, E. Hinton and E. Onate, Pineridge Press, pp. 975–88 (1987).

Reinhall, Per G., Ramulu, M. and Ghassaei, S. Numerical and experimental analysis of electromagnetic riveting process. *Proc. 12th All India Mach. Tool Des. Res. Conf.*, ed. U.R.K. Rao, P.N. Rao and N.K. Tiwari, Tata McGraw-Hill, pp. 401–5 (1986).

Roll, K. Possibilities for the use of the finite element method for the analysis of bulk metalforming processes. *Annals of CIRP* **31**, 145–50 (1982).

Schwab, W. Effect of process parameters on metalforming on fatigue behaviour. *Annals of CIRP* **34**, 215–19 (1985).

Schwab, W. and Hager, B. Influence of metal forming parameters on the fatigue behaviour of extruded shafts. *Proc. 2nd Int. Conf. on Technology of Plasticity*, ed. K. Lange, Springer, pp. 597–604 (1987).

Sekhon, G.S., Arora, K.L. and Shishodia, K.S. Application of finite-difference and finite element methods to the axisymmetric cold heading process. *Proc. Int. Conf. on Computational Plasticity*, ed. D.R.J. Owen, E. Hinton and E. Onate, Pineridge Press, pp. 1061–72 (1987).

Shah, S.N. and Kobayashi, S. Rigid-plastic analysis of cold heading by the matrix method. *Proc. 15th Int. Machine Tool Des. Res. Conf.*, ed. S.A. Tobias and F. Koenigsberger, Macmillan, pp. 603–10 (1974).

Indentation

Lee, C.H. and Kobayashi, S. Elastoplastic analysis of plane strain and axisymmetric flat punch indentation by the finite element method. *Int. J. Mech. Sci.* **12**, 349–70 (1970).

Lee, C.H., Masaki, S. and Kobayashi, S. Analysis of ball indentation. *Int. J. Mech. Sci.* **14**, 417–26 (1972).

Madnaik, S.D. and Chaturvedi, R.C. A finite-element solution to the problem of plane-strain compression of pre-shaped material with inclined tools. *Proc. 2nd Int. Conf. on Numerical Methods in Industrial Forming Processes*, ed. K. Mattiasson, A. Samuelsson, R.D. Wood and O.C. Zienkiewicz,

Balkema Press, pp. 139–44 (1986).

Prevost, J.H. and Hughes, T.J.R. Finite element solution of elastic–plastic boundary-value problems. *J. Appl. Mech.* **48**, 69–74 (1981).

Sun, J. and Guo, H. Analysis of the closing and consolidation of internal cavities in heavy rotor forgings by finite element method. *Proc. 2nd Int. Conf. on Technology of Plasticity*, ed. K. Lange, Springer, pp. 1059–64 (1987).

Wesner, P.J. and Weinstein, A.S. Computerized relaxation applied to the plane strain indenter. *ASME pub.* '69-MET-M (1969).

Rolling

Plane-strain rolling

Beynon, J.H., Brown, P.R., Mizban, S.I., Ponter, A.R.S. and Sellars, C.M. An Eulerian finite element method for the thermal and visco plastic deformation of metals in industrial hot rolling. *Proc. Int. Conf. on Computational Methods for Predicting Material Processing Defects*, ed. M. Predeleanu, Elsevier, pp. 19–28 (1987).

Beynon, J.H., Ponter, A.R.S. and Sellars, C.M. Metallographic verification of computer modelling of hot rolling. *Modelling of Metal Forming Processes, Proc. Euromech 233 Colloquium*, ed. J.L. Chenot and E. Onate, Kluwer, 321–8 (1988).

Cornfield, G.C. and Johnson, R.H. Theoretical prediction of plastic flow in hot rolling including the effect of various temperature distribution. *J. Iron and Steel Inst.* **211**, 567–73 (1973).

Dawson, P.R. Viscoplastic finite-element analysis of steady state forming processes including strain history and stress flux dependence. *Application of Numerical Methods to Forming Processes, Winter Annual Meeting of ASME*, AMD 28, ed. H. Armen and R.F. Jones Jr., ASME pp. 55–66 (1978).

Dawson, P.R. On modeling material property changes during flat rolling of aluminium. *Int. J. Solids Struct.* **23**, 947–68 (1987).

Dawson, P.R. and Thompson, E.G. Finite element analysis of steady-state elastic-visco-plastic flow by the initial stress-rate method. *Int. J. Num. Meth. Eng.* **12**, 47–57 (1978).

Grober, H. Finite element simulation of hot flat rolling of steel. *Proc. 2nd Int. Conf. on Numerical Methods in Industrial Forming Processes*, ed. K. Mattiasson, A. Samuelsson, R.D. Wood and O.C. Zienkiewicz, Balkema Press, pp. 225–30 (1986).

Grober, H., Cescotto, R., Charlier, M., Bourdouxhe, M. and Habraken, A.M. Numerical simulation of metal forming processes. *Proc. Int. Conf. on Nonlinear Mechanics*, Shanghai (1985).

Hamauzu, S., Yamada, K., Kawanami, T. and Mori, K. Rigid-plastic finite element analysis of asymmetric rolling. *Proc. Int. Conf. on Computational Plasticity*, ed. D.R.J. Owen, E. Hinton and E. Onate, Pineridge Press, pp. 1087–96 (1987).

Hirakawa, T., Fujita, F., Kamata, M. and Yamada, Y. Analysis of strip rolling by the finite element method. *Proc. 1st Int. Conf. on Technology of Plasticity*, ed. H. Kudo, JSTP/JSPE, pp. 1132–7 (1984).

Hwang, S.M. and Kobayashi, S. Preform design in plane-strain rolling by the finite-element method. *Int. J. Machine Tool Des. Res.* **24**, no. 4, 253–66 (1984).

Hwu, Y-J and Lenard, J.G. A finite element study of flat rolling. *J. Eng. Mater. Techn.* **110**, 22–7 (1988).

Key, S.W., Krieg, R.D. and Bathe, K-J. On the application of the finite element method to metal forming processes. *Comp. Meth. Appl. Mech. Eng.* **17/18**, 597–608 (1979).

Li, G.J. and Kobayashi, S. Rigid-plastic finite-element analysis of plane strain rolling. *J. Eng. Ind.* **104**, 55–64 (1982).

Liu, C., Hartley, P., Sturgess, C.E.N. and Rowe, G.W. Elastic–plastic finite-element modelling of cold rolling of strip. *Int. J. Mech. Sci.* **27**, 531–41 (1985).

Liu, C., Hartley, P., Sturgess, C.E.N. and Rowe, G.W. Simulation of the cold rolling of strip using an elastic–plastic finite element technique. *Int. J. Mech. Sci.* **27**, 829–39 (1985).

Mori, K. and Osakada, K. Finite element analysis of non steady state deformation in flat rolling. *J. Iron and Steel Institute of Japan* **67**, 2337 (1981).

Mori, K., Osakada, K. and Oda, T. Simulation of plane-strain rolling by the rigid-plastic finite element method. *Int. J. Mech. Sci.* **24**, 519–27 (1982).

Mori, K., Osakada, K., Nakadoi, K. and Fukuda, M. Coupled analysis of steady state forming process with elastic tools. *Proc. 2nd Int. Conf. on Numerical Methods in Industrial Forming Processes*, ed. K. Mattiasson, A. Samuelsson, R.D. Wood and O.C. Zienkiewicz, Balkema Press, pp. 237–42 (1986).

Pietrzyk, M. Rigid-plastic finite element simulation of plane-strain rolling with significantly non-uniform flow of metal. *Proc. 2nd Int. Conf. on Numerical Methods in Industrial Forming Processes*, ed. K. Mattiasson, A. Samuelsson, R.D. Wood and O.C. Zienkiewicz, Balkema Press, pp. 243–8 (1986).

Pietrzyk, M. and Lenard, J.G. Thermal-mechanical modelling for hot rolling: experimental substantiation. *Modelling of Metal Forming Processes, Proc. Euromech 233 Colloquium*, ed. J.L. Chenot and E. Onate, Kluwer, pp. 281–8 (1988).

Rao, S.S. and Kumar, A. Finite element analysis of cold strip rolling. *Int. J. Mech. Sci.* **17**, 159–68 (1977).

Shima, S., Mori, K., Oda, T. and Osakada, K. Rigid-plastic finite element analysis of strip rolling. *Proc. 4th Int. Conf. on Production Engineering*, Tokyo, pp. 82–7 (1980).

Tamano, T. Finite element analysis of steady flow in metal processing. *J. Jap. Soc. for Techn. of Plasticity* **14**, 766 (1973).

Tamano, T. and Yanagimoto, S. Finite element analysis of steady metal flow problems. *Trans. JSME.* **41**, 1130 (1975).

Thompson, E.G. Inclusion of elastic strain-rate in the analysis of viscoplastic flow during rolling. *Int. J. Mech. Sci.* **24**, 655–9 (1982).

Thompson, E.G. and Berman, H.M. Steady state analysis of elasto-viscoplastic flow during rolling. *Proc. 1st Int. Conf. on Numerical Methods in Industrial Forming Processes*, ed. J.F.T. Pittman, R.D. Wood, J.M. Alexander and O.C. Zienkiewicz, Pineridge Press, pp. 29–37 (1982).

Thompson, E.G. and Berman, H.M. Steady state analysis of elasto-viscoplastic flow during rolling. *Numerical Analysis of Forming Processes*, ed. J.F.T. Pittman, O.C. Zienkiewicz, R.D. Wood and J.M. Alexander, Wiley, pp. 269–84 (1984).

Yarita, I., Mallett, R.L. and Lee, E.H. Stress and deformation analysis of metal rolling processes. *Proc. 1st Int. Conf. on Technology of Plasticity*, ed. H. Kudo, JSTP/JSPE, pp. 1126–31 (1984).

Zienkiewicz, O.C., Jain, P.C. and Onate, E. Flow of solids during forming and extrusion: some aspects of numerical solutions. *Int. J. Solids Struct.* **14**, 15–38 (1978).

Shape rolling

Bertrand, C., David, C., Chenot, J.L. and Buessler, P. Stresses calculation in finite element analysis of three-dimensional hot shape rolling. *Proc. 2nd Int. Conf. on Numerical Methods in Industrial Forming Processes*, ed. K. Mattiasson, A. Samuelsson, R.D. Wood and O.C. Zienkiewicz, Balkema Press, pp. 207–12 (1986).

Bertrand-Corsini, C., David, C., Bern, A., Montmitonnet, P., Chenot, J.L., Buessler, P. and Fau, F. A 3D thermomechanical analysis of steady flows in hot forming processes. Application to hot flat rolling and hot shape rolling. *Modelling of Metal Forming Processes, Proc. Euromech 233 Colloquium*, ed. J.L. Chenot and E. Onate, Kluwer, pp. 271–80 (1988).

Maekawa, Y., Hirai, T., Katayama, T. and Hawkyard, J.B. Modelling and numerical analysis of cross rolling and profile ring rolling processes. *Proc. 1st Int. Conf. on Technology of Plasticity*, ed. H. Kudo, JSTP/JSPE, pp. 930–5 (1984).

Three-dimensional rolling

Bertrand, C., David, C., Chenot, J.L. and Buessler, P. A transient three dimensional finite element analysis of hot rolling of thick slabs. *Proc. 2nd Int. Conf. on Numerical Methods in Industrial Forming Processes*, ed. K. Mattiasson, A. Samuelsson, R.D. Wood and O.C. Zienkiewicz, Balkema Press, pp. 219–24 (1986).

Branswyck, O., David, C., Levaillant, C., Chenot, J.L., Billard, J.P., Weber, D. and Guerlet, J.P. Surface defects in hot rolling of copper based heavy ingots: three dimensional simulation including failure criterion. *Proc. Int. Conf. on Computational Methods for Predicting Material Processing Defects*, ed. M. Predeleanu, Elsevier, pp. 29–38 (1987).

Hartley, P., Sturgess, C.E.N., Liu, C. and Rowe, G.W. Mechanics of metal flow in cold rolling. *CAD/CAM in Metal Working*, ed. S.K. Ghosh and A. Niku-Lari, Pergamon, pp. 35–46 (1988).

Hartley, P., Sturgess, C.E.N., Liu, C. and Rowe, G.W. Experimental and theoretical studies of workpiece deformation, stress and strain, during flat rolling. *Int. Mater. Rev.* **34**, 19–34 (1989).

Huisman, H.J. and Huetink, J. A combined Eulerian–Langrangian three-dimensional finite-element analysis of edge rolling. *J. Mech. Wkg. Techn.* **11**, 333–53 (1985).

Kanazawa, K. and Marcal, P.V. Finite element analysis of the steel rolling process. *Application of Numerical Methods to Forming Processes, Winter Annual Meeting of ASME*, AMD 28, ed. H. Armen and R.F. Jones Jr., ASME, p. 81 (1978).

Kiefer, B.V. Three-dimensional finite element prediction of material flow and strain distributions in rolled rectangular billets. *Proc. 1st Int. Conf. on Technology of Plasticity*, ed. H. Kudo, JSTP/JSPE, pp. 1116–25 (1984).

Li, G-J. and Kobayashi, S. Spread analysis in rolling by the rigid-plastic finite element method. *Proc. 1st Int. Conf. on Numerical Methods in Industrial*

Forming Processes, ed. J.F.T. Pittman, R.D. Wood, J.M. Alexander and O.C. Zienkiewicz, Pineridge Press, pp. 777–86 (1982).

Li, G.J. and Kobayashi, S. Analysis of spread in rolling by the rigid-plastic, finite-element method. *Numerical Analysis of Forming Processes*, ed. F.T. Pittman, O.C. Zienkiewicz, R.D. Wood and J.M. Alexander, Wiley, pp. 71–88 (1984).

Liu, C., Hartley, P., Sturgess, C.E.N. and Rowe, G.W. Analysis of stress and strain distributions in slab rolling using an elastic–plastic finite-element method. *Proc. 2nd Int. Conf. on Numerical Methods in Industrial Forming Processes*, ed. K. Mattiasson, A. Samuelsson, R.D. Wood and O.C. Zienkiewicz, Balkema Press, pp. 231–6 (1986).

Liu, C., Hartley, P., Sturgess, C.E.N. and Rowe, G.W. Finite-element modelling of deformation and spread in slab rolling. *Int. J. Mech. Sci.* **29**, 271–83 (1987).

Liu, C., Hartley, P., Sturgess, C.E.N. and Rowe, G.W. Analysis of stress and strain distributions in slab rolling using an elastic–plastic finite-element method. *Int. J. Num. Meth. Eng.* **25**, *Special Issue of Numerical Methods in Industrial Forming Processes*, 55–66 (1988).

Mori, K. and Osakada, K. Simulation of three-dimensional rolling by the rigid-plastic finite element method. *Proc. 1st Int. Conf. on Numerical Methods in Industrial Forming Processes*, ed. J.F.T. Pittman, R.D. Wood, J.M. Alexander and O.C. Zienkiewicz, Pineridge Press, pp. 747–56 (1982).

Mori, K. and Osakada, K. Simulation of three-dimensional deformation in rolling by the finite-element method. *Int. J. Mech. Sci.* **26**, 515–25 (1984).

Mori, K., Osakada, K., Nikaido, H., Naoi, T. and Aburtani, Y. Finite element simulation of non-steady state deformation in rolling of slab and plate. *Proc. 4th Int. Steel Rolling Conf.*, Deauville (1987).

Nikaido, H., Naoi, T., Shibata, K., Osakada, K. and Mori, K. FEM simulation of non-steady deformation in edge rolling. *J. Jap. Soc. for Techn. of Plasticity* **24**, p. 486 (1983).

Takuda, H., Mori, K., Hatta, N. and Kokado, J. Experiment and finite element analysis of hot rolling of slab in width direction with flat roll. *J. Jap. Soc. for Techn. of Plasticity* **23**, 1103–8 (1982).

Yukawa, N., Ishikawa, T. and Tozawa, Y. Numerical analysis of the shape of rolled strip. *Proc. 2nd Int. Conf. on Numerical Methods in Industrial Forming Processes*, ed. K. Mattiasson, A. Samuelsson, R.D. Wood and O.C. Zienkiewicz, Balkema Press, pp. 249–54 (1986).

Review papers

Cheng, J-H. and Kikuchi, N. An analysis of metal forming processes using large deformation elastic-plastic formulations. *Comp. Meth. Appl. Mech. Eng.* **49**, 71–108 (1985).

Haque, I., Jackson Jr., J.E., Gangjee, T. and Raikar, T. Empirical and finite element approaches to forging die design: a state-of-the-art survey. *J. Mater. Shaping Techn.* **5**, 23–33 (1987).

Jimma, T., Tomita, Y. and Shimamura, S. Benchmark test on a plastic deformation problem, application of numerical methods of analysis to the uniaxial tension of a block or cylindrical bar with both ends fixed. *Proc. 2nd Int. Conf. on Technology of Plasticity*, ed. K. Lange, Springer, pp. 73–80 (1987).

Kobayashi, S. A review on the finite-element method and metal forming process modelling. *J. Appl. Metalworking* **2**, no.3, 163–9 (1982).

Kobayashi, S. Metal forming and the finite element method – past and future. *Proc. 25th Int. Machine Tool Des. Res. Conf.*, ed. S.A. Tobias, Macmillan, pp. 17–32 (1985).

Kobayashi, S. Advances in forging technology by the finite element method. *Proc. 2nd Int. Conf. on Numerical Methods in Industrial Forming Processes*, ed. K. Mattiasson, A. Samuelsson, R.D. Wood and O.C. Zienkiewicz, Balkema Press, pp. 19–28 (1986).

Kudo, H. The use of finite element method in the analysis of plastic deformation of some metal forming processes. *Annals of CIRP* **23**, 219–25 (1974).

Mahrenholtz, O. and Dung, N.L. Mathematical modelling of metal forming processes by numerical methods. *Proc. 2nd Int. Conf. on Technology of Plasticity*, ed. K. Lange, Springer, pp. 3–10 (1987).

Sturgess, C.E.N., Hartley, P. and Rowe, G.W. Finite-element modelling of forging. *Proc. 7th International Cold Forging Congress*, Institution of Sheet Metal Engineers, pp. 39–47 (1985).

Theoretical background

Finite-element theory

Elastic–plastic finite-element theory

Abo-Elkhier, M., Dokainish, M.A. and Oravas, G.AE. Analysis of large elasto-plastic deformation. *Material Non-linearity in Vibration Problems*, AMD 71, ed. M. Sathyamoorthy, ASME, pp. 113–24 (1985).

Agah-Tehrani, A., Lee, E.H., Mallett, R.L. and Onat, E.T. The theory of elastic–plastic deformation at finite strain with induced anisotropy modeled as combined isotropic–kinematic hardening. *J. Mech. Phys. Solids* **35**, 519–39 (1987).

Argyris, J.H. and Doltsinis, J. St. On the large strain inelastic analysis in natural formulation – part I Quasistatic problems. *Comp. Meth. Appl. Mech. Eng.* **20**, 213–51 (1979).

Argyris, J.H. and Doltsinis, J.St. On the large strain inelastic analysis in natural formulation – part II. Dynamic problems. *Comp. Meth. Appl. Mech. Eng.* **21**, 91–128 (1979).

Argyris, J.H. and Doltsinis, J. St. The natural concept of material description. *Constitutive Equations: Macro and Computational Aspects, Winter Annual Meeting of ASME*, ed. K.J. Willam, ASME, pp. 1–24 (1984).

Argyris, J.H. and Kleiber, M. Incremental formulation in nonlinear mechanics and large strain elasto-plasticity – natural approach, part I. *Comp. Meth. Appl. Mech. Eng.* **11**, 215–47 (1977).

Argyris, J.H. and Kleiber, M. Incremental formulation in nonlinear mechanics and large strain elasto-plasticity – natural approach, part II. *Comp. Meth. Appl. Mech. Eng.* **14**, 259–94 (1978).

Atluri, S.N. On constitutive relations at finite strain hyper-elasticity and elasto-plasticity with isotropic or kinematic hardening. *Comp. Meth. Appl. Mech. Eng.* **43**, 137–71 (1984).

Bathe, K.J. and Ozdemir, H. Elastic and plastic large deformation static and dynamic analysis. *Comput. Struct.* **6**, 81–92 (1976).

Casey, J. and Naghdi, P.M. Constitutive results for finitely deforming elastic–plastic materials. *Constitutive Equations: Macro and Computational Aspects, Winter Annual Meeting of ASME*, ed. K.J. Willam, ASME, pp. 53–72 (1984).

Dafalias, Y.F. A missing link in the formulation and numerical implementation of finite transformation elastoplasticity. *Constitutive Equations: Macro and Computational Aspects, Winter Annual Meeting of ASME*, ed. K.J. Willam, ASME, pp. 25–40 (1984).

Dienes, J.K. The analysis of rotation and stress rate in deforming bodies. *Acta Mech.* **32**, 217–32 (1979).

Dogui, A. and Sidoroff, F. Large strain formulation of anisotropic elasto-plasticity for metal forming. *Proc. Int. Conf. on Computational Methods for Predicting Material Processing Defects*, ed. M. Predeleanu, Elsevier, 81–92 (1987).

Fardshisheh, F. and Onat, E.T. Representation of elastoplastic behavior by means of state variables. *Int. Symp. on Foundations of Plasticity, Problems of Plasticity*, ed. A. Sawczuk, pp. 89–115 (1972).

Green, A.E. and Naghdi, P.M. Some remarks on elastic–plastic deformation at finite strain. *Int. J. Eng. Sci.* **9**, 1219–29 (1971).

Hibbitt, H.D., Marcal, P.V. and Rice, J.R. A finite element formulation for problems of large strain and large displacement. *Int. J. Solids Struct.* **6**, 1069–86 (1970).

Hofmeister, L.D., Greenbaum, G.A. and Evensen, D.A. Large strain elastic–plastic finite element analysis. *AIAA J.* **9**, 1248–54 (1971).

Kitagawa, H. and Tomita, Y. Application of finite element method to large strain elastic–plastic problems and its experimental verification. *J. Jap. Soc. for Techn. of Plasticity* **14**, p. 788 (1973).

Kleiber, M. Kinematics of deformation processes in materials subjected to finite elastic–plastic strains. *Int. J. Eng. Sci.* **13**, 513–25 (1975).

Krieg, R.D. and Key, S.W. On the accurate representation of large strain non-proportional plastic flow in ductile materials. *Constitutive Equations: Macro and Computational Aspects, Winter Annual Meeting of ASME*, ed. K.J. Willam, ASME, pp. 41–52 (1984).

Krieg, R.D. and Krieg, D.B. Accuracies of numerical solution methods for the elastic–perfectly plastic model. *J. Press. Vessel Techn.* **99**, 510–15 (1977).

Lee, E.H. Elastic–plastic deformations at finite strains. *J. Appl. Mech.* **36**, 1–6 (1969).

Lee, E.H. *The basis of an elastic–plastic code*. SUDAM rep. no. 76-1, Stanford University (1976).

Lee, E.H. *Some comments on elastic–plastic analysis*. SUDAM rep. no. 80-5, Stanford University (1980).

Lee, E.H. Some comments on elastic–plastic analysis. *Int. J. Solids Structures* **17**, 859–72 (1981).

Lee, E.H. Finite deformation effects in plasticity analysis. *Proc. 1st Int. Conf. on Numerical Methods in Industrial Forming Processes*, ed. J.F.T. Pittman, R.D. Wood, J.M. Alexander and O.C. Zienkiewicz, Pineridge Press, pp. 39–50 (1982).

Lee, E.H. The structure of elastic–plastic constitutive relations for finite deformation. *Constitutive Equations: Macro and Computational Aspects, Winter Annual Meeting of ASME*, ed. K.J. Willam, ASME, pp. 103–10 (1984).

Lee, E.H. Finite deformation effects in plasticity analysis. *Numerical Analysis of Forming Processes*, ed. J.F.T. Pittman, O.C. Zienkiewicz, R.D. Wood and J.M. Alexander, Wiley, pp. 373–86 (1984).

Lee, E.H. Finite element modelling of elastic–plastic large deformation for associated and non-associated modified von Mises criteria. *Proc. Int. Conf. on Computational Plasticity*, ed. D.R.J. Owen, E. Hinton and E. Onate, Pineridge Press, pp. 315–28 (1987).

Lee, E.H. and Agah-Tehrani, A. The structure of constitutive equations for finite deformation of elastic–plastic materials with strain-induced anisotropy. *Proc. 2nd Int. Conf. on Numerical Methods in Industrial Forming Processes*, ed. K. Mattiasson, A. Samuelsson, R.D. Wood and O.C. Zienkiewicz, Balkema Press, pp. 29–36 (1986).

Lee, E.H. and Agah-Tehrani, A. The structure of constitutive equations for finite deformation of elastic–plastic materials involving strain-induced anisotropy with applications. *Int. J. Num. Meth. Eng.* **25**, *Special Issue on Numerical Methods in Industrial Forming Processes*, 133–46 (1988).

Lee, E.H. and Germain, P. Elastic–plastic theory at finite strain. *Symp. on Problems of Plasticity*, Warsaw, pp. 117–33 (1972).

Lee, E.H. and Liu, D.T. Finite strain elastic–plastic theory. *Proc. IUTAM Symp.*, Springer, pp. 213–22 (1968).

Lee, E.H. and McMeeking, R.M. Concerning elastic and plastic components of deformation. *Int. J. Solids Struct.* **16**, 715–21 (1980).

Lee, E.H., Mallett, R.L. and McMeeking, R.M. Stress and deformation analysis of metal forming processes. *Numerical Modelling of Manufacturing Processes*, ed. R.F. Jones Jr., H. Armen and J.T. Fong, ASME, pp. 19–33 (1977).

Lee, E.H., Mallett, R.L. and Wertheimer, T.B. *Stress analysis for kinematic hardening in finite-deformation plasticity*. SUDAM rep. no. 81-11, Stanford University (1981).

Lee, E.H., Mallett, R.L. and Wertheimer, T.B. Stress analysis for anisotropic hardening in finite-deformation plasticity. *J. Appl. Mech.* **50**, 554–60 (1983).

Lubarda, V.A. Elastic–plastic deformation with plastic anisotropy. *Theor. and Appl. Mech.* **8**, 95 (1982).

Lubarda, V.A. On the strain-hardening elasto-plastic constitutive equation. *Proc. Int. Conf. on Computational Plasticity*, ed. D.R.J. Owen, E. Hinton and E. Onate, Pineridge Press, pp. 93–104 (1987).

Lubarda, V.A. and Lee, E.H. *A correct definition of elastic and plastic deformation and its computational significance*. SUDAM rep. no. 80-1, Stanford University (1980).

Lubarda, V.A. and Lee, E.H. A correct definition of elastic and plastic deformation and its computational significance. *J. Appl. Mech.* **48**, 35–40 (1981).

Mallett, R.L. *Finite-element selection for finite deformation elastic–plastic analysis*. SUDAM rep. no. 80-4, Stanford University (1980).

Mallett, R.L. Finite increment formulation of the Prandtl–Reuss constitutive equations. *Trans. 2nd Army Conf. on applied Mathematics and Computing*, pp. 285–301 (1985).

Marcal, P.V. A comparative study of numerical methods of elastic–plastic analysis. *AIAA J.* **6**, 157–8 (1966).

Marcal, P.V. and King, I.P. Elastic–plastic analysis of two-dimensional stress systems by the finite element method. *Int. J. Mech. Sci.* **9**, 143–55 (1967).

McMeeking, R.M. and Rice, J.R. Finite-element formulations for problems of large elastic–plastic deformation. *Int. J. Solids Struct.* **11**, 601–16 (1975).

Nagtegaal, J.C. Some recent developments in combined geometric and nonlinear finite element analysis. *Recent Advances in Non-Linear Computational*

Mechanics, ed. E. Hinton, D.R.J. Owen and C. Taylor, Pineridge Press, pp. 87–117 (1982).

Nagtegaal, J.C. On the implementation of inelastic constitutive equations with special reference to large deformation problems. *Comp. Meth. Appl. Mech. Eng.* **33**, 469–84 (1982).

Nagtegaal, J.C. and de Jong, J.E. Some computational aspects of elastic–plastic large strain analysis. *Proc. 2nd TICOM Conf., Computational Methods in Nonlinear Mechanics*, ed. J.T. Oden, North-Holland, pp. 303–34 (1980).

Nagtegaal, J.C. and de Jong, J.E. Some computational aspects of elastic–plastic large strain analysis. *Int. J. Num. Meth. Eng.* **17**, 15–41 (1981).

Nagtegaal, J.C. and Veldpaus, F.E. On the implementation of finite strain plasticity equations in a numerical model. *Numerical Analysis of Forming Processes*, ed. J.F.T. Pittman, O.C. Zienkiewicz, R.D. Wood and J.M. Alexander, Wiley, pp. 351–71 (1984).

Nagtegaal, J.C. and Wertheimer, T.B. Constitutive equations for anisotropic large strain plasticity. *Constitutive Equations: Macro and Computational Aspects, Winter Annual Meeting of ASME*, ed. K.J. Willam, ASME, pp. 73–86 (1984).

Nagtegaal, J.C., Parks, D.M. and Rice, J.R. On numerically accurate finite element solutions in the fully plastic range. *Comp. Meth. Appl. Mech. Eng.* **4**, 153–77 (1974).

Nayak, G.C. and Zienkiewicz, O.C. Elastic plastic stress analysis: generalization of various constitutive relations including strain softening. *Int. J. Num. Meth. Eng.* **5**, 113–35 (1972).

Needleman, A. On finite element formulations for large elastic–plastic deformations. *Comput. Struct.* **20**, 247–57 (1985).

Nemat-Nasser, S. Decomposition of strain measures and their rates in finite deformation elastoplasticity. *Int. J. Solids Struct.* **15**, 155–66 (1979).

Nemat-Nasser, S. On finite deformation elasto-plasticity. *Int. J. Solids Struct.* **18**, 857–72 (1982).

Onat, E.T., Creus, G.J. and Groehs, A.G. Representation of elastic–plastic behaviour in the presence of finite deformations and anisotropy. *Proc. Conf. on Structural Analysis and Design of Nuclear Power Plants*, Porto Alegre, Brazil (1984).

Ortiz, M. Some computational aspects of finite deformation plasticity. *Proc. Int. Conf. on Computational Plasticity*, ed. D.R.J. Owen, E. Hinton and E. Onate, Pineridge Press, pp. 1717–56 (1987).

Ortiz, M. and Simo, J.C. An analysis of a new class of integration algorithms for elastoplastic constitutive relations. *Int. J. Num. Meth. Eng.* **23**, pp. 353–66 (1986).

Osias, J.R. and Swedlow, J.L. Finite elasto-plastic deformation I – theory and numerical examples. *Int. J. Solids Struct.* **10**, 321–39 (1974).

Pecherski, R.B. Constitutive description and numerical approach in modelling for metal forming processes. *Modelling of Metal Forming Processes, Proc. Euromech 233 Colloquium*, ed. J.L. Chenot and E. Onate, Kluwer, pp. 11–18 (1988).

Pillinger, I., Hartley, P., Sturgess, C.E.N. and Rowe, G.W. Use of a mean-normal technique for efficient and numerically-stable large-strain elastic-plastic finite-element solutions. *Int. J. Mech. Sci.* **28**, 23–9 (1986).

Pillinger, I., Hartley, P., Sturgess, C.E.N. and Rowe, G.W. A new linearized definition of strain increment for the finite-element analysis of deformations involving finite rotation. *Int. J. Mech. Sci.* **28**, 253–62 (1986).

Pillinger, I., Hartley, P., Sturgess, C.E.N., Van Bael, A., Van Houtte, P. and Aernoudt, E. Finite element analysis of anisotropic material deformation. *Supplement to Proc. 3rd Int. Conf. on Numerical Methods in Industrial Forming Processes*, ed. E.G. Thompson, R.D. Wood, O.C. Zienkiewicz and A. Samuelsson, Balkema Press, (1989).

Reed, K.W. and Atluri, S.N. Constitutive modeling and stress analysis for finite deformation inelasticity. *Constitutive Equations: Macro and Computational Aspects, Winter Annual Meeting of ASME*, ed. K.J. Willam, ASME, pp. 111–30 (1984).

Rice, J.R. *A note on the 'small strain' formulation for elastic–plastic problems.* Tech. rep. no. N00014-67-A-0191-00318, Brown University (1970).

Rolph, III, W.D. and Bathe, K.J. On a large strain finite element formulation for elastic–plastic analysis. *Constitutive Equations: Macro and Computational Aspects, Winter Annual Meeting of ASME*, ed. K.J. Willam, ASME, pp. 131–48 (1984).

Sidoroff, F. Incremental constitutive equations for large strain elastoplasticity. *Int. J. Eng. Sci.* **20**, 19–26 (1982).

Simo, J.C. Maximum plastic dissipation and the multiplicative decomposition in finite strain plasticity. *Proc. Int. Conf. on Computational Plasticity*, ed. D.R.J. Owen, E. Hinton and E. Onate, Pineridge Press, pp. 1821–36 (1987).

Simo, J.C. A framework for finite strain elastoplasticity based on maximum plastic dissipation and the multiplicative decomposition. Part II: Computational aspects. *Comp. Meth. Appl. Mech. Eng.* **68**, 1–31 (1988).

Simo, J.C. and Ortiz, M. A unified approach to finite deformation elastoplasticity based on the use of hyperelastic constitutive equations. *Comp. Meth. Appl. Mech. Eng.* **49**, 221–45 (1985).

Simo, J.C., Ortiz, M., Pister, K.S. and Taylor, R.L. Rate constitutive equations for finite deformation plasticity: Are they necessary? *Constitutive Equations: Macro and Computational Aspects, Winter Annual Meeting of ASME*, ed. K.J. Willam, ASME, pp. 87–101 (1984).

Simo, J.C. and Taylor, R.L. Consistent tangent operators for rate independent elastoplasticity. *Comp. Meth. Appl. Mech. Eng.* **48**, 101–18 (1984).

Simo, J.C. and Taylor, R.L. A return mapping algorithm for plane strain elastoplasticity. *Int. J. Num. Meth. Eng.* **22**, 649–70 (1986).

Simo, J.C., Taylor, R.L. and Pister, K.S. Variational and projection methods for the volume constraint in finite deformation elastoplasticity. *Comp. Meth. Appl. Mech. Eng.* **51**, 177–208 (1985).

Tekkaya, A.E. and Roll, K. Analysis of metal forming processes by different finite element methods. *Proc. 2nd Int. Conf. on Numerical Methods for Nonlinear Problems*, Barcelona, pp. 450–61 (1984).

Van Houtte, P., Van Bael, A., Aernoudt, E., Pillinger, I., Hartley, P. and Sturgess, C.E.N. The introduction of anisotropic yield loci derived from texture measurements in elasto-plastic finite element calculations. *Proc. 2nd Int. Symp. on Plasticity*, ed. A.S. Khan and M. Tokuda, Pergamon, pp. 111–14 (1989).

Willis, J.R. Elastic–plastic deformations at finite strains. *J. Mech. Phys. Solids* **17**, 359–69 (1969).

Yamada, Y. Constitutive modelling of inelastic behavior and numerical solution of nonlinear problems by the finite element method. *Comput. Struct.* **8**, 533–43 (1978).

Yamada, Y. and Hirakawa, H. Large deformation and instability analysis of the metal extrusion process. *Applications of Numerical Methods to Forming*

Processes, Winter Annual Meeting of ASME, AMD 28, ed. H. Armen and R.F. Jones Jr., ASME, pp. 27–38 (1978).

Yamada, Y., Hirakawa, T. and Wifi, A.S. Analysis of large deformation and bifurcation in plasticity problems by the finite element method. *Finite Elements in Nonlinear Mechanics*, Tapir, pp. 393–412 (1978).

Yamada, Y., Wifi, A.S. and Hirakawa, T. Analysis of large deformation and stress in metal forming processes by the finite element method. *Metal Forming Plasticity* ed. H. Lippmann, Springer, pp. 158–76 (1979).

Yamada, Y., Yoshimura, N. and Sakurai, T. Plastic stress–strain matrix and its application for the solution of elastic–plastic problems by the finite element method. *Int. J. Mech. Sci.* **10**, 343–54 (1968).

Zienkiewicz, O.C., Valliappan, S. and King, I.P. Elasto-plastic solutions of engineering problems 'initial stress', finite element approach. *Int. J. Num. Meth. Eng.* **1**, 75–100 (1969).

Rigid-plastic finite-element theory

Alexander, J.M. and Price, J.W.H. Finite element analysis of hot metal forming. *Proc. 18th Int. Machine Tool Des. Res. Conf.*, ed. J.M. Alexander, Macmillan, pp. 267–74 (1977).

Jackson Jr., J.E., Haque, I. and Gangjee, T. The CFORM finite element code for material forming analysis. SECTAM Conf., Biloxi, MS., *Developments in Theoretical and Applied Mechanics*, Vol. XIV, pp. 555–66 (1988).

Klie, W. Plastic deformation with free boundaries – a finite element approach. *Metal Forming Plasticity*, ed. H. Lippmann, Springer, pp. 260–72 (1979).

Kobayashi, S. Thermoviscoplastic analysis of metal forming problems by the finite element method. *Proc. 1st Int. Conf. on Numerical Methods in Industrial Forming Processes*, ed. J.F.T. Pittman, R.D. Wood, J.M. Alexander and O.C. Zienkiewicz, Pineridge Press, pp. 17–25 (1982).

Lee, C.H. and Kobayashi, S. New solutions to rigid-plastic deformation problems using a matrix method. *J. Eng. Ind.* **95**, 865–873 (1973).

Matsumoto, H., Oh, S.I. and Kobayashi, S. A note on the matrix method for rigid-plastic analysis of ring compression. *Proc. 18th Int. Machine Tool Des. Res. Conf.*, ed. J.M. Alexander, Macmillan, pp. 3–9 (1977).

Mori, K., Shima, S. and Osakada, K. Finite element method for the analysis of plastic deformation of porous metals. *Bull. JSME* **23**, no. 178, 516–22 (1980).

Mori, K., Shima, S. and Osakada, K. Some improvements of the rigid-plastic finite element method. *J. Jap. Soc. for Techn. of Plasticity* **21**, 593 (1980).

Osakada, K., Nakano, J. and Mori, K. Finite element method for rigid-plastic analysis of metal forming – formulation for finite deformation. *Int. J. Mech. Sci.* **24**, 459–68 (1982).

Visco-plastic finite-element theory

Argyris, J.H., Doltsinis, J.St., Knudson, W.C., Szimmat, J., Willam, K.J. and Wustenburg, H. Eulerian and Lagrangian techniques for elastic and inelastic deformation processes. *Computational Methods in Nonlinear Mechanics (TICOM 1979)*, ed. J.T. Oden, North-Holland, pp. 13–66 (1980).

Bohatier, C. and Chenot, J.L. Finite element formulation for non-steady state viscoplastic deformation. *Int. J. Num. Meth. Eng.* **21**, 1697–708 (1985).

Chandra, A. and Mukherjee, S. A finite element analysis of metal forming problems with an elastic–viscoplastic material model. *Int. J. Num. Meth. Eng.* **20**, 1613–28 (1984).

Cormeau, I. Numerical stability in quasi static elasto-visco-plasticity. *Int. J. Num. Meth. Eng.* **9**, 109–28 (1975).

Dawson, P.R. and Thompson, E.G. Finite element analysis of steady-state elasto-visco-plastic flow by the initial stress-rate method. *Int. J. Num. Meth. Eng.* **12**, 45–57 (1978).

Dewhurst, T.B. and Dawson, P.R. Analysis of large plastic deformations at elevated temperatures using state variable models for viscoplastic flow. *Constitutive Equations: Macro and Computational Aspects, Winter Annual Meeting of ASME*, ed. K.J. Willam, ASME, pp. 149–64 (1984).

Eggert, G.M. and Dawson, P.R. A viscoplastic formulation with elasticity for transient metal forming. *Comp. Meth. Appl. Mech. Eng.* **70**, 165–90 (1988).

Owen, D.R.J. and Damjanic, F. Viscoplastic analysis of solids. *Recent Advances in Non-linear Computational Mechanics*, ed. E. Hinton, D.R.J. Owen and C. Taylor, Pineridge Press, pp. 225–53 (1982).

Perzyna, P. and Wojno, W. On the constitutive equations of elastic-viscoplastic materials at finite strain. *Arch. Mech. Stos.* **18**, 85–99 (1966).

Shimazeki, Y. and Thompson, E.G. Elasto-viscoplastic flow with special attention to boundary conditions. *Int. J. Num. Meth. Eng.* **17**, 97–112 (1981).

Zienkiewicz, O.C. Flow formulation for numerical solution of forming processes. *Numerical Analysis of Forming Processes*, ed. J.F.T. Pittman, O.C. Zienkiewicz, R.D. Wood and J.M. Alexander, Wiley, pp. 1–44 (1984).

Zienkiewicz, O.C. Viscoplasticity, plasticity and creep in elastic solids, a unified numerical solution approach. *Int. J. Num. Meth. Eng.* **8**, 821–45 (1987).

Zienkiewicz, O.C. and Cormeau, I.C. Viscoplasticity solution by finite element process. *Arch. Mech.* **24**, 873–88 (1972).

Zienkiewicz, O.C. and Cormeau, I.C. Viscoplasticity, and creep in elastic solids, a unified numerical solution approach. *Int. J. Num. Meth. Eng.* **8**, 821–45 (1974).

Zienkiewicz, O.C. and Godbole, P.N. Flow of plastic and viscoplastic solids with special reference to extrusion and forming processes. *Int. J. Num. Meth. Eng.* **8**, 3–16 (1974).

Zienkiewicz, O.C. and Godbole, P.N. Viscous, incompressible flow with special reference to non-Newtonian (plastic) fluids. *Finite Elements in Fluids*, pp. 25–55 (1975).

Zienkiewicz, O.C., Jain, P.C. and Onate, E. Flow of solids during forming and extrusion: some aspects of numerical solutions. *Int. J. Solids Struct.* **14**, 15–38 (1978).

Zienkiewicz, O.C., Toyoshima, S., Liu, Y.C. and Zhu, J.Z. Flow formulation for numerical solution of forming processes II – some new directions. *Proc. 2nd Int. Conf. on Numerical Methods in Industrial Forming Processes*, ed. K. Mattiasson, A. Samuelsson, R.D. Wood and O.C. Zienkiewicz, Balkema Press, pp. 3–10 (1986).

Plasticity theory

Argyris, J.H. and Scharpf, D.W.N. Methods of elastoplastic analysis. *J. Appl. Math. Phys.* **23**, 517–51 (1972).

Chu, E. and Sowerby, R. Some aspects of finite deformation. *Proc. 1st Int. Conf. on Technology of Plasticity*, ed. H. Kudo, JSTP/JSPE, 1065–9 (1984).

Dafalias, Y.F. A missing link in the macroscopic constitutive formulation of large plastic deformations. *Plasticity Today*, ed. A. Sawczuk and G. Bianchi, Elsevier, pp. 483–502 (1985). (From the Int. Symposium on Current Trends and Results in Plasticity, Udine, Italy, 1983).

Dafalias, Y.F. The plastic spin. *J. Appl. Mech.* **107**, 865–71 (1985).

Drucker, D.C. A more fundamental approach to plastic stress–strain relations. *Proc. 1st National Congress on Applied Mechanics*, ASME, pp. 487–91 (1951).

Green, A.E. and Naghdi, P.M. A general theory of an elastic–plastic continuum. *Arch. Rat. Mech. Ann.* **18**, 251–81 (1965).

Hill, R. A variational principle of maximum plastic work in classical plasticity. *Quart. J. Appl. Math.* **1**, 18–28 (1948).

Hill, R. New horizons in the mechanics of solids. *J. Mech. Phys. Solids* **5**, 66 (1956).

Hill, R. Some basic principles in the mechanics of solids without a natural time. *J. Mech. Phys. Solids* **7**, 209–25 (1959).

Hughes, T.R.J. and Winget, J. Finite rotation effects in numerical integration of rate constitutive equations arising in large deformation analysis. *Int. J. Num. Meth. Eng.* **15**, 1862–7 (1980).

Nagtegaal, J.C. On the implementation of inelastic constitutive equations with special reference to large deformation problems. *Comp. Meth. Appl. Mech. Eng.* **33**, 469–84 (1982).

Paulun, J.E. and Percherski, R.B. Study of corotational rates for kinematic hardening in finite deformation plasticity. *Arch. Mech.* **37**, 661 (1985).

Pecherski, R.B. The disturbed plastic spin concept and its consequences in plastic instability. *Proc. 2nd Int. Conf. on Numerical Methods in Industrial Forming Processes*, ed. K. Mattiasson, A. Samuelsson, R.D. Wood and O.C. Zienkiewicz, Balkema Press, pp. 145–50 (1986).

Prager, W. The theory of plasticity: a survey of recent achievements. *Proc. Instn. Mech. Engrs.* **169**, 41–57 (1955).

Prager, W. An elementary discussion of definitions of stress rates. *Quart. Appl. Math.* **18**, 403–7 (1961).

Simo, J.C. On the computational significance of the intermediate configuration and hyperelastic relations in finite deformation elastoplasticity. *Mech. of Mater.* **4**, 439–51 (1986).

Taylor, L.M. and Becker, E.B. Some computationl aspects of large deformation, rate-dependent plasticity problems. *Comp. Meth. Appl. Mech. Eng.* **41**, 251–77 (1983).

Thomas, T.Y. On the structure of the stress-strain relations. *Proc. Natl. Acad. Sci. USA* **41**, 716–20 (1955).

Thomas, T.Y. Combined elastic and Prandtl–Reuss stress-strain relations. *Proc. Natl. Acad. Sci. USA* **41**, 720–6 (1955).

Thomas, T.Y. Combined elastic and von Mises' stress-strain relations. *Proc. Natl. Acad. Sci. USA* **41**, 908–10 (1955).

Unpublished theses

Abo-Elkhier, M. On the numerical solution of nonlinear problems in continuum mechanics. Ph.D. thesis, McMaster University (1985).

Al-Khattat, I.M. A continuous model for friction with applications in metal forming plasticity. Ph.D. thesis, Stanford University (1981).

Al-Sened, A.A.K. Development of simulation techniques for cold forging sequences. Ph.D. thesis, University of Birmingham, UK (1984).

Cescotto, S. Etude par éléments finis des grand déplacement et grandes déformation. Application aux problèmes spécifiques des matériaux quasi incompressibles. Thèse de doctorat, Université de Liège (1978).

Chanda, S.K. FE analysis of orthogonal deformation processes. Ph.D. thesis, University of Birmingham, UK (1989).

Chandra, A. Finite element and boundary element analyses of large strain displacement problems in elastoplasticity. Ph.D. thesis, Cornell University (1983).

Chandrasekaran, N. Finite-element analysis of superplastic metal forming processes. Ph.D. thesis, Texas A&M University (1986).

Chen, C-C. Finite-element analysis of plastic deformation in metal forming processes. Ph.D. thesis, University of California, Berkeley (1978).

Cheng, J-H. Finite-element simulations of metal forming processes by large deformation thermo-elastic–plastic formulations with grid adaptive and mesh rezoning methods. Ph.D. thesis, University of Michigan, Ann Arbor (1985).

Chiou, J-H. Finite element analysis of large strain elastic–plastic solids. Ph.D. thesis, University of Minnesota (1987).

Clift, S.E. Identification of defect locations in forged products using the finite-element method. Ph.D. thesis, University of Birmingham, UK (1986).

Cormeau, I.C. Viscoplasticity and plasticity in the finite element method. Ph.D. thesis, University of Wales, Swansea (1974).

Dehghani, M. Simulation of hydrostatic co-extrusion of aluminium and copper. Ph.D. thesis, Louisiana State University (1987).

Dewhurst, T.B. The simulation of three-dimensional metalforming processes using state variable constitutive models for large strain, viscoplastic flow at elevated temperature. Ph.D. thesis, Cornell University (1985).

Eames, A.J. A computer system for forging die design and flow simulation. Ph.D. thesis, University of Birmingham, UK (1986).

Eggert, G.M. A large strain thermomechanical forming model for elastic–visco-plastic behavior in metals. Ph.D. thesis, Cornell University (1987).

Fu, C-C. Efficient finite element methods for large displacement elasto-plastic problems. Ph.D. thesis, University of Missouri–Rolla (1987).

Gangjee, T. Friction and lubrication in finite element analysis of forging processes. Ph.D. thesis, Clemson University (1987).

Gelin, J.C. Modèles numériques et expérimentaux en grandes déformations plastiques et endommagement de rupture ductile. Thèse de doctorat d'Etat, Université de Paris (1985).

Germain, Y. Finite element analysis of viscoplastic flow with friction boundary conditions – application to metalforming. Dr. Eng. thesis, Ecole des Mines de Paris (1985).

Gordon, J.L. A finite element analysis of the plane strain sheet drawing problem. Ph.D. thesis, Carnegie–Mellon University (1971).

Gortemaker, P.C.M. Applications of an elastic plastic finite element program for large deformations. Dissertation, Twente University of Technology, The Netherlands (1980).

Groehs, A.G. ESFINGE, a computer system for the analysis of finite elastic–plastic deformations. Ph.D. thesis, COPPE, UFRJ, Rio de Janeiro (1983).

Guha, R.M. Finite element large deformation analysis of hydrostatic extrusion. Ph.D. thesis, Imperial College, London (1977).

Habraken, A-M. Contribution a la modèlisation du formage des métaux par la methode des éléments finis. Doctor of Science thesis, Liège University (1989).

Hartley, P. Metal flow and homogeneity in extrusion-forging. Ph.D. thesis, University of Birmingham, UK (1979).

Herbertz, R. Zur Braucharbeit eines starr-viscoplastischen Materialgesetzes der Lösung umformtechnischer mit Hilfe der finite elemente Methode. Ph.D. thesis, Technische Hochschule Aachen (1982).

Huetink, J. On the simulation of thermo-mechanical forming processes. Dissertation, University of Twente, Netherlands (1986).

Hussin, A.A.M. Transference of mainframe finite-element elastic-plastic analysis to microcomputers, and its application to forging and extrusion. Ph.D. thesis, University of Birmingham, UK (1986).

Hwang, S.M. Preform design and friction evaluation in metal forming by the finite element method. Ph.D. thesis, University of California, Berkeley (1985).

Im, Y-T. Finite element modelling of plastic deformation of porous materials. Ph.D. thesis, University of California, Berkeley (1985).

Kher, S.N. Finite element modelling of large plastic strains in a rolling contact metal forming process. Ph.D. thesis, The Pennsylvania State University (1987).

Kim, J-H. Analysis of sheet metal forming by the finite-element method. Ph.D. thesis, University of California, Berkeley (1977).

Lammering, R. Beiträge zur Theorie und Numerik grosser plastischer und kleiner elastischer Deformation mit Schadigungseinfluss. Ph.D. thesis, University of Hannover (1987).

Lin, K-N. Nonlinear finite element analysis for plastic deformation by local random yielding. Ph.D. thesis, The University of Nebraska–Lincoln (1987).

Liu, C. Simulation of strip and slab rolling using an elastic–plastic finite-element method. Ph.D. thesis, University of Birmingham, UK (1985).

Lu, S.C-Y. Finite element modelling of large-strain elasto-plastic deformation processes with tool interface friction. Ph.D. thesis, Carnegie–Mellon University (1984).

Lubarda, F.A. Elastic–plastic deformation at finite-strain. Ph.D. thesis, Stanford University (1980).

Lung, M. Ein Verfahren zur Berechnung des Geschwindigkeits – und Spannungsfeldes bei stationären starr – plastischen Formänderungen mit finiten Elementen. Dissertation, Technische Universität Hannover (1971).

Madnaik, S.D. Finite element analysis of the processes of plane strain compression, indentation by inclined tools and rolling for a linearly and exponentially workhardening materials. Ph.D. thesis, Indian Institute of Technology, Bombay (1982).

Mattiasson, K. On the co-rotational finite-element formulation for large deformation problems. Doctor thesis, Chalmers University of Technology, Gothenburg (1983).

Mori, K-I. Analysis of metal forming processes by finite element method for compressible rigid-plastic materials. Doctor of Engineering thesis, Kyoto University (1982).

Mosavi-Mashadi, M. A study of the plastic flow around a hardness indentation using the finite element method. Ph.D. thesis, University of Arkansas (1984).

Nakano, J. A study of rigid-plastic finite element method for analysis of metal forming processes. Master thesis, Kobe University, (1981).

Onate, E. Plastic flow in metals. Ph.D. thesis, University of Wales, Swansea (1978).

Pacheco, L.A. Rigid-plastic finite element analysis of some metal forming processes. Ph.D. thesis, Imperial College, London (1980).

Park, J.J. Applications of the finite-element method to metal forming problems. Ph.D. thesis, University of California, Berkeley (1982).

Parks, D.M. Some problems in elastic–plastic finite element analysis of cracks. Ph.D. thesis, Brown University (1975).

Phu, N.D. A finite element method for three-dimensional analysis of metal forming. Ph.D. thesis, University of Missouri–Rolla (1974).

Pillinger, I. The prediction of metal flow and properties in three-dimensional forgings using the finite-element method. Ph.D. thesis, University of Birmingham, UK (1984).

Price, J.W.H. A study of dieless drawing and isothermal forming. Ph.D. thesis, Imperial College, London (1977).

Rajiyah, H. An analysis of large deformation axisymmetric inelastic problems by finite element and boundary element methods. Ph.D. thesis, Cornell University (1987).

Rebelo, N. Finite-element modeling of metalworking processes for thermo-viscoplastic analysis. Ph.D. thesis, University of California, Berkeley (1980).

Schreurs, P.G.J. Numerical simulation of forming processes. The use of the arbitrary Eulerian–Lagrangian method. Ph.D. thesis, Eindhoven University of Technology, Netherlands (1983).

Shah, M.K. A nonlinear finite element analysis of plane strain and axisymmetric forging. Ph.D. thesis, North Carolina State University at Raleigh (1986).

Shivpuri, R. An explicit time integration finite-element method for metalrolling problems. Ph.D. thesis, Drexel University (1986).

Taylor, L.M. A finite element analysis for large deformation metal forming problems involving contact and friction. Ph.D. thesis, University of Texas at Austin (1981).

Toh, C.H. Process modeling of sheet metal forming of general shapes by the finite element method based on large strain formulation. Ph.D. thesis, University of California, Berkeley (1983).

Toyoshima, S. Iterative mixed methods and their application to analysis of metal forming processes. Ph.D. thesis, University of Wales, Swansea (1985).

Van der Lugt, J. A finite element for simulation of thermo-mechanical contact problems in forming processes. Dissertation, University of Twente, Netherlands (1988).

Webster, W.D. A three-dimensional analysis of extrusion and metal forming by the finite-element method. Ph.D. thesis, University of Missouri–Rolla (1978).

Weiss, U. Numerische simulation von Prazisionsschmiedeprozessen mit der finite-elemente Methode. Dissertation, University of Hannover (1987).

Wertheimer, T.B. Problems in large deformation elasto-plastic analysis using the finite element method. Ph.D. thesis, Stanford University (1982).

Wifi, A.S. Studies on large strain elasto-plasticity and finite element analysis of deformation processes. Ph.D. thesis, University of Tokyo (1978).

Wu, L. A dynamic relaxation finite element method for metal forming processes. Ph.D. thesis, Drexel University (1986).

Index